T0289652

Horse Nutrition and Feeding Management

Horse Nutrition and Feeding Management

Edited by Jenna Matthews

SYRAWOOD
PUBLISHING HOUSE

New York

Published by Syrawood Publishing House,
750 Third Avenue, 9ᵗʰ Floor,
New York, NY 10017, USA
www.syrawoodpublishinghouse.com

Horse Nutrition and Feeding Management
Edited by Jenna Matthews

International Standard Book Number: 978-1-64740-367-6 (Hardback)

Trademark Notice: Registered trademark of products or corporate names are used only for explanation and identification without intent to infringe.

Cataloging-in-publication Data

Horse nutrition and feeding management / edited by Jenna Matthews.
    p. cm.
Includes bibliographical references and index.
ISBN 978-1-64740-367-6
1. Horses--Feeding and feeds. 2. Horses--Nutrition. 3. Horses--Nutrition--Requirements.
4. Horses. 5. Feeds. 6. Nutrition--Requirements. I. Matthews, Jenna.
SF285.5 .H67 2023
636.108 5--dc23

TABLE OF CONTENTS

PREFACE

The horse is a type of domesticated and hoofed mammal with odd-toed feet. It falls under the taxonomic family Equidae and belongs to the extant subspecies of Equus ferus. Horses are used for a variety of purposes such as agriculture, therapy, police work and entertainment. Appropriate nutrition and feeding management is required to ensure their well-being. This has an impact on a horse's performance capacity, reproduction, growth, and overall health. Horse feeding is a complex task since poor nutrition can lead to horses becoming overweight and suffering from diseases, such as nutritional secondary hyperparathyroidism, laminitis and developmental orthopedic disease. Various other aspects of horse management include environmental and stable conditions, and feeding methods. This book unravels the recent studies on horse nutrition and feeding management. It is appropriate for students seeking detailed information on this topic as well as for experts.

This book unites the global concepts and researches in an organized manner for a comprehensive understanding of the subject. It is a ripe text for all researchers, students, scientists or anyone else who is interested in acquiring a better knowledge of this dynamic field.

I extend my sincere thanks to the contributors for such eloquent research chapters. Finally, I thank my family for being a source of support and help.

Editor

Health and Body Conditions of Riding School Horses Housed in Groups or Kept in Conventional Tie-Stall/Box Housing

Jenny Yngvesson [1],*[iD], Juan Carlos Rey Torres [1], Jasmine Lindholm [1], Annika Pättiniemi [1], Petra Andersson [2] and Hanna Sassner [1]

[1] Department of Animal Environment & Health, Swedish University of Agricultural Sciences, P.O.B. 234, SE-53223 Skara, Sweden; jcreytorres@gmail.com (J.C.R.T.); jasmine.lindholm@hotmail.com (J.L.); annika.pattiniemi@live.se (A.P.); hanna.sassner@slu.se (H.S.)

[2] Department of Philosophy, Linguistics and Theory of Science, University of Gothenburg, P.O.B. 100, SE-40530 Gothenburg, Sweden; petra.andersson@filosofi.gu.se

* Correspondence: jenny.yngvesson@slu.se.

Simple Summary: We compared welfare indicators of riding school horses in group housing and tie-stalls/boxes. Of a total of 207 health conditions in 158 horses, tie-stall/box horses tended to have more small skin lesions at the saddling and girth sites, and in commissures of the lips. Tie-stall/box horses had had more respiratory problems and colic, possibly because of not having similar access to outdoor movement and water as group-housed horses. Many horses in both housing systems were above optimal weight. We conclude that group-housed riding school horses have better health and that all riding school horses would benefit from independent feed advice to maintain a healthy weight.

Abstract: We compared welfare measures of horses among Swedish riding schools (RS) during winter where horses were kept either in group housing ($n = 8$) or in tie-stalls/boxes ($n = 8$), Health data for six previous months were obtained for all horses at each RS from their records. Ten horses per RS were examined, with the exception of one where only 8 horses were examined. Health conditions and body condition score (BCS) using the Henneke scale were recorded and management factors were quantified (health check routines, feeding, housing-related risk factors, time outside). RS-recorded health data (for 327 horses in total) revealed that lameness was the most common issue in both systems. Respiratory problems and colic were significantly more common in tie-stall/box horses. The percentage of horses with respiratory problems (mean ± SEM) was 5.8 ± 1.4 in tie-stall/box systems and 1.1 ± 0.8 in group housing ($F = 8.65$, $p = 0.01$). The percentage with colic was 2.38 ± 0.62 in tie-stall/box systems and 0.38 ± 0.26 in group housing ($F = 8.62$, $p = 0.01$). Clinical examination of 158 horses revealed 207 conditions in these horses, the most common being minor skin injuries in areas affected by tack (i.e., saddle and bridle, including bit). Such injuries tended to be more prevalent in horses housed in tie-stalls/boxes (1.8 ± 0.6) than in group housing (0.5 ± 0.3) ($F=3.14$, $p = 0.01$). BCS was similar between systems (tie-stall/box 6.2 ± 0.1, group 6.3 ± 0.1), but the average BCS exceeded the level that is considered optimal (BCS 4–6). In conclusion, we found that Swedish RS horses are generally in good health, particularly when group-housed. However, 25%–32% were overweight. Riding schools would thus benefit from having an independent feeding expert performing regular body condition scoring of all horses and advising on feeding regimens.

Keywords: equine; health; welfare; riding school; colic; feeding; body condition

1. Introduction

One useful definition of animal welfare, and hence horse welfare, is to include health, behavioral physiology, and production/reproduction when assessing the welfare of an individual [1]. This paper focuses mainly on the health aspects of horse welfare. The housing requirements and general welfare of ridden horses have been debated [2,3] and researchers are approaching consensus on the basic needs of horses, such as social interaction with conspecifics, access to roughage, and free movement [4]. Hence, many horse owners are re-evaluating conventional individual housing systems and, at least in Sweden, increasing numbers of riding school horses are being kept in loose, group housing enclosures [5].

Group housing designs for horses vary, but the system is generally characterized by a large or small paddock, mostly outdoors, preferably with a drained surface, a shelter, some sort of roughage (often straw combined with haylage), and ad libitum access to water. Shelter is a legal requirement for horses in Sweden during winter [6]. Group housing systems may be more or less complex and mechanized, e.g., some have automatic feeding stations and horses are managed somewhat similarly to dairy cows in modern systems. A common feature of group housing systems is that they aim to enable the horses to move about more freely and interact more naturally with conspecifics, thus improving horse welfare.

In Sweden, around 18 000 horses are kept at approximately 500 riding schools [5]. The vast majority of these riding schools house their horses in individual boxes or tie-stalls at night and in some form of paddock during daytime. However, an increasing number of riding schools are now choosing to house their horses continually in groups.

In the public debate in Sweden, there are a number of potential welfare concerns about housing horses, particularly riding school horses, in groups [7]. To our knowledge, only a few studies (e.g., reference [8]) have compared feeding regimens, general health, and other indicators of welfare of riding school horses and even fewer have considered group-housed horses. Hence, there is a knowledge gap about the major health problems in riding school horses, how these horses are fed, or how this affects their welfare. One fear expressed in the popular media and discussion threads online is that horses kept in group housing become overweight and suffer from undiscovered health issues, based on a belief that group-housed horses receive less individual attention than tie-stall/box horses.

The aim of this study, which is part of a larger project examining horse behavior and human working conditions in riding schools with different housing systems (the results of which will be published elsewhere), was to describe and compare the health and body condition of riding school horses kept either in group housing or in conventional tie-stall/box housing.

2. Materials and Methods

Riding schools (RS) were selected through the Swedish Equestrian Federation, advertisements in a horse magazine, personal contacts, and browsing the internet. First, group housing RS were identified and enrolled. Inclusion criteria were that their group housing system had been in use at least six months and that they housed 10 or more horses. These RS were then matched with tie-stall/box RS with the same target group, the same type of horses, of similar size, in a similar geographical area and, when possible, with a manager with the same educational background.

The RS were visited during the winter season (November–March) in 2016–2017 and 2017–2018 (except one that was visited in April), with eight each winter (Supplementary Table S2). The team collecting the data comprised an agronomist specializing in horse feeding, an agronomist specializing in housing-related injuries, an ethologist, and an equine veterinarian. The team also had special training from the Swedish Trotting Association regarding risks of injury to horses in housing systems.

Group housing systems and tie-stall/box systems were visited alternately. Visits lasted from 08.00–10.00 h to 18.00–20.00 h.

2.1. Data Obtained from Riding School Managers

RS managers were interviewed, using an open questionnaire, about general horse health over the previous six months. Additionally, any journals or notes on horse health were evaluated, management routines were quantified, and feeding routines and feed quality (both hygienic and nutritional) were recorded. Water supply was quantified in terms of placement of water sources, water flow in automatic water cups, and number of hours the horses were without access to water (Supplementary Table S3 and S5).

2.2. Selection of Horses for Clinical Examination

Each RS selected 10 animals (five horses, five ponies), currently working, for clinical examination by the experienced equine veterinarian (Supplementary Table S1 and S2). It included rectal temperature recording and heart and lung auscultation to determine resting heart and respiratory rates. The cough reflex was checked and the eyes, mucus membranes, and lymph nodes were examined. Mouth health was superficially checked. The skin and coat and general appearance were examined. Any wounds or swellings on the body, as well as cleanliness of the horse (Appendix A) were noted. The muscular skeletal system was examined in walk and trot on a straight line. The 10 selected horses per RS were also scored for body condition according to the 1–9 Hennecke scale [9], where 1 denotes emaciated and 9 denotes obese.

2.3. Housing-Related Risks of Injury to the Horses

A thorough facility inspection was performed at each RS, both indoors and outdoors (including stables, lying halls, corridors for moving horses indoors and outdoors, ventilation, doors, aisles, flooring, outdoor ground surfaces, hay nets, feed racks, hooks, and any other fittings) for risks of injuries in the housing system. The method used was developed in collaboration with the Swedish Trotting Association and the protocol can be found in Appendix B and data is found in Supplementary Table S4.

Data were compiled using Excel and statistical analysis was performed in Minitab®Statistical Software 2016 (PA, USA) The data were checked for normality using the Anderson–Darling procedure. When large graphical differences between the housing systems were found, the data were analyzed for differences using one-way analysis of variance (ANOVA), a method that is robust to varying distribution of the data.

3. Results

We visited a total of 16 RS, eight with group housing and eight with the horses in tie-stalls/boxes. Of the population of RS in Sweden ($n = $ ~500), our sample included over 50% of those with group housing and $\geq$10 horses, and only 1.6% of those with tie-stalls/boxes. In total, 158 horses were clinically examined (10 at all RS except one, which only had eight school horses at the time of the visit).

3.1. Management and Feeding Routines

3.1.1. Management Routines

Management routines were quantified through both observations and interviews. At the RS with the horses in tie-stalls/boxes, all horses were fed at approximately 07.00 h and then let out into the paddock/paddocks at around 08.00 h. Most RS gave the horses roughage in the paddock. Horses were then generally brought back indoors again at around 14.00 h, fed, and prepared for lessons. In four of the eight RS with tie-stalls/boxes, the horses ran freely back to the stable, where they were fed upon arrival. Some horses were ridden by the instructors. Lessons generally started at 17.00 h. After lessons (at 20.00–22.00 h), the horses were fed.

In the RS with group housing, the horses were given a group-level health check when staff arrived at the RS, the paddock and lying halls were cleaned and bedded. Some horses were ridden by the instructors. Approximately 1–2 hours before lessons, the horses were brought from the group housing into a stable by the staff. Some horses were given extra concentrate in the stable. In two cases, the students collected their horses from the group housing themselves (one of these cases was a high school with a riding specialization). Lessons generally started at 17.00 h. The horses were led out into the paddock/system after lessons, in some cases by the students.

3.1.2. Feed Rations and Hygienic Quality of the Feed

All RS used haylage for the majority of horses. A few individual horses received hay, as they had special requirements. It was impossible to determine how much nutrients all horses in this study obtained from roughage and from concentrate, as this information was unavailable in most RS. In seven RS, all using group housing, the horses had free access to some type of roughage (mainly straw). When hay or haylage was available all the time for the group-housed horses, different types of nets were used to reduce the rate of intake. In the other cases, access to roughage for group-housed horses was limited in time or by feeding them in an automated feeding station, where each horse was identified through an individual tag.

Forage quality analysis data were available at 11 RS (four group housing and seven tie-stall/box). However, only five RS (two group housing and three tie-stall/box) used these data to calculate the feed ration for some or all horses. Ten RS fed concentrate for most or all of their horses, and eight of these RS had forage analysis data (Table 1).

Table 1. Riding school (RS) feeding strategy. All RS that did not use concentrate indicated that they would use it if needed to fulfil the horses' needs. Four RS (#6, 7, 8, and 9) used a scientifically supported strategy with roughage nutrient and hygienic quality analysis as a basis for calculating the needs of individual horses.

RS	Housing System	Type of Roughage	Analysis	Use Analysis to Calculate Ration	Concentrate	Individual Ration
1	Group	Haylage	No	Not applicable	No	No
2	Tie-stall/box	Haylage	Yes	No	Yes	Yes
3	Group	Haylage	No	Not applicable	Yes	Yes
4	Tie-stall/box	Haylage	Yes	For some	No	Yes
5	Group	Haylage	Yes	No	No	No
6	Tie-tall/box	Haylage	Yes	Yes	Yes	Yes
7	Group	Haylage & hay	Yes	Yes	Yes	Yes
8	Tie-stall/box	Haylage & hay	Yes	Yes	No	Yes
9	Group	Haylage	Yes	Yes	Yes	Yes
10	Tie-stall/box	Haylage	Yes	No	Yes	Yes
11	Group	Hay	No	Not applicable	No	No
12	Tie-stall/box	Haylage & hay	Yes	No	Yes	Yes
13	Group	Haylage	No	Not applicable	Yes	No
14	Tie-stall/box	Haylage	No	Not applicable	No	No
15	Group	Haylage	Yes	No	Yes	No
16	Tie-stall/box	Haylage	Yes	No	Yes	No

RS: Riding schools.

Hygienic quality in the haylage was found to be good in all RS. However, in three cases (all group housing), we found straw of insufficient quality (mold and dust).

3.1.3. Body Condition Scoring

Body condition score (BCS) was found to be similar between housing systems, but exceeded the level considered optimal from a health perspective. For the group-housed horses, BCS (mean ± SEM) was 6.4 ±1 (range 4.5–8.5) and for the tie-stall/box horses it was 6.0 ±1 (range 4.0–9.0). For 37%

of group-housed horses and 25% of tie-stall/box horses BCS was greater than 6, which is the ideal maximum.

3.2. Horse Health—Clinical Examinations

The sample of 158 horses evaluated according to a 27-point protocol (Appendix A) represented 56% of all horses on RS with group housing and 46% of all horses on those RS with tie stall/box housing.

Overall, we found 207 abnormal health conditions in the 158 horses, or on average 1.3 ± 0.23 per horse, with the difference between the two housing systems not being significant (Figure 1).

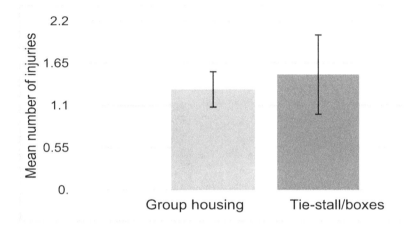

Figure 1. Average number of abnormal health condition findings per horse and riding school (bars indicate standard error of the mean).

The most common finding was minor skin lesions located in areas in contact with the tack, i.e., saddle and bridle (including bit). Such lesions tended to be more common in horses housed in tie-stalls/boxes (mean ± SEM 1.8 ± 0.6) than in group housing (0.5 ± 0.3) (ANOVA GLM F = 3.14, $p = 0.09$).

Mild lameness was found in four horses, two in each type of housing system. No severe lameness was found.

3.3. Horse Health—Retrospective Health Data Obtained from RS Managers

The data obtained from RS managers included all horses housed at the RS for six months before the visits. This included a total of 150 horses in the RS with group housing and 177 horses in the RS with tie-stall/box housing (327 in all). The most common RS-reported health issue in each type of RS housing was lameness, while other health issues differed in prevalence (Table 2).

Table 2. The health issues most commonly reported by riding school (RS) managers within the six months preceding our visit. The RS with group housing had a total of 150 horses and those with tie-stall/box housing had a total of 177 horses.

Health Issues Found	Group Housing (n = 150 horses)	Tie-Stall/Box Housing (n = 177 horses)
Most common health issue	Lameness 8%	Lameness 9.6%
Second most common health issue	Skin lesions 7.3%	Hoof injuries 7.3%
Third most common health issue	Wounds, cause unknown 6%	Skin lesions, respiratory problems, and wounds cause unknown, all 6%

Over the previous six months, the number of cases of colic was significantly greater for the RS with tie-stalls/boxes (19 cases in 177 horses; mean per RS ± SEM 2.38 ± 0.62) than for the RS with group housing (3 cases in 150 horses; 0.38 ± 0.26) (F = 8.62, $p = 0.01$).

The tie-stall/box horses spent more hours/day confined (17 ± 0.4 h) than the group-housed horses (3.8 ± 0.9 h). Furthermore, none of the tie-stall/box horses had access to water in the field during winter, whereas all group-housed horses did. Of the eight RSs with tie-stall/box housing, four had an open water surface and four had water cups where the horses started the water flow by muzzle manipulation. All eight RS with group housing had an open water supply.

Respiratory airway problems (mainly coughing) recorded by RS managers during the previous six months were significantly greater for the RS with tie-stalls/boxes (Figure 2). The percentage of horses in the RS with tie-stall/boxes (11 cases in 177 horses) with airway problems was 5.8 ± 1.4 (mean $\pm$ SEM), compared to 1.1 ± 0.8 for the RS with group housing (two reported cases in 150 horses, $F = 8.65$, $p = 0.01$).

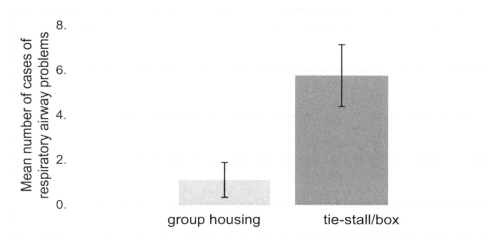

Figure 2. Mean number of horses with airway problems recorded by riding school managers during the previous six months.

For numbers of euthanized horses, housing-related injuries, bite and kick injuries, lameness and skin disorders recorded by RS managers over the previous six months, differences between the housing systems were not significant.

We asked the RS managers to give us an estimate of how many overweight or underweight horses they had. Only eight of the 16 RS (five with group housing and three with tie-stall/box housing) provided BCS estimates that were similar to those that we obtained on measuring BCS in 10 horses.

The horses were ridden for, on average, 8.6 ± 1.5 hours per week in the RS with group housing and on average 12.3 ± 1.6 hours per week in the RS with tie-stall/box housing. This difference was not statistically significant. The mean number of weeks on summer pasture was 3.3 ± 1.1 weeks for the group housing RS and 4.9 ± 0.9 weeks for the tie-stall/box RS.

3.4. Housing-Related Risk Factors in the Different Environments

Injury risks found in the housing indoors, paddocks, lying halls, and lanes or corridors for moving the horses are listed in Table 3.

Table 3. Safety risks observed and photographed on visits to the riding schools. Statistical comparisons between housing methods were not made. There were slippery surfaces both outdoors and indoors. 'Low ceiling' in this case includes low parts of the box doors.

Risks of Injuries Found	Group Housing	Tie-Stall/Box Housing
Most common risks of injury	Hay nets, rugs in the paddock, low ceiling, slippery surfaces (6/8)	Low ceiling (8/8)
Second most common risks of injury	Sharp edges, halters in the paddock, weak bars (e.g., on windows), poor fencing (4/8)	Weak bars, slippery surfaces, rugs in the paddock (7/8)

Halters may be a risk factor when the horses are in the paddocks or group housing. However, halters were worn by very few horses in our sample.

Four of the RS using tie-stall/box housing allowed their horses run freely into the stable before the lessons started. Horses were fed some concentrate in the tie-stalls/boxes and therefore were very willingly returned to the stable. However, on at least three occasions we saw horses falling due to slippery footing. This is not something that we can analyze statistically, however, it is an important risk factor to take into consideration when the ground is slippery, e.g., in the winter.

4. Discussion

Horse welfare is attracting increased public and research attention. It has been reported that people involved with horses identify health and management as two major areas of concern, in particular, e.g., body condition and stabling [3]. Both these factors were found to have a significant effect on the welfare of the RS horses in our study. This study was not designed to find specific causal factors, but to investigate welfare in horses housed in different housing systems. We used previous research to discuss potential causes of the health differences but stress that the causes of the differences need to be investigated in further detail.

As pointed out in a previous study [8], an important factor to take into consideration is the prevalence of injuries and lesions either caused by the tack or present in areas where riding tack is positioned. The tie-stall/box horses in our study tended to have more small injuries apparently caused by the riding tack. The reason for this difference remains to be investigated, but one possible explanation is that the generally better health observed in group-housed horses results in more rapid wound healing.

It is known that locomotor problems are an important cause of death in Swedish RS horses [10]. In the working RS horses that we examined, lameness was the most common health issue recorded by managers, and it was similar between systems. Hence the welfare issue of lameness in RS horses needs to be addressed, regardless of the housing system.

Body condition scores did not differ between the housing systems studied. However, 28.5% of the scored horses had BCS ≥ 7, indicating that most horses consumed more energy than needed for maintenance and the work they perform. This indicates that overweight condition may be a welfare problem in Swedish RS horses. From an international perspective too, overweight and obesity seems to be an increasing problem in horses [11,12]. There are potential explanations for this finding on different meta-levels. Technical competence on how to feed horses may be lacking or RS managers may have this competence but, because of customer demands, do not apply it. At several of the RS surveyed, the staff said that they needed to have the horses overweight, because otherwise the students and parents would complain about the horses being too thin. We did not collect any data on students' and parents' views, so it may be important in future studies to investigate whether and how public views affect horse management staff through consumer pressure. A recent study found that obesity was less common in professional equine establishments such as RS and studs [13], possibly since they are more likely to employ staff with great experience of horses and their needs. However, it could also be the case that private horse owners are under less pressure from the public, students, and parents, and can therefore apply existing knowledge on correct feeding management of horses.

However, the picture is more complex than just social pressure. For example, it has been shown that horse owners in general are unskilled at estimating horse BCS [11]. We found that around one-third of the group-housed horses and one-quarter of the tie-stall/box horses included in the present study had BCS ≥ 7, which is overweight to obese according the standardized BCS thresholds [9]. At the same time, only half of RS staff estimated their horses' BCS correctly (relative to our measured values). In this regard, it is important to note that we only measured BCS on roughly half the horses kept at each RS, so conclusions cannot be drawn about the actual knowledge of RS staff about BCS. However, as very few horses were found to be underweight, although some were clearly not very muscular, overweight is a pronounced problem in the RS horses that we investigated.

As this study was carried out in Sweden in the winter, the horses received very little, if any, of their nutrient intake from grazing. All the food was provided by RS staff. Forage analysis data were available in most RS. When not available it was of course impossible to calculate the balance between roughage and concentrates regarding energy and digestible crude protein. A surprising finding was that even in the 11 RS where roughage analysis data were available, less than half of these RS actually used the data. There is plenty of evidence of the importance of using forage analysis data to ensure that the nutritional demands of horses are met [14,15]. Although not statistically confirmed, at the four RS where roughage analysis data were used to calculate the individual ration, only a small proportion of the horses were overweight.

Some health problems, e.g., colic, were more common in tie-stall/box horses. Colic is a non-specific term for abdominal pain in horses [16,17]. It may be mild and treatable by just walking and resting the horse, or it may require advanced veterinary care. Nutritional risk factors for colic include a rapid change in diet, type of hay [18], or hay batch, and poor hygienic quality [19]. Amount of grain and/or concentrate and rapid changes in the amount have also been shown to increase the risk of colic [19–21]. Horses not fed grain at all did not acquire colic in one study [19].

Increased time on pasture per day and increased paddock area are associated with significantly reduced risk of colic [19,20]. Moreover, there is an important effect of water intake, whereby reduced intake and not providing horses with water in the paddock greatly increase the incidence of colic [21,22]. We found significant differences between our RS horses in terms of time spent outdoors and access to water, with the tie-stall/box horses spending more than twice as much time indoors compared with the group-housed horses. When outdoors at temperatures below freezing, none of the tie-stall/box horses had access to water. However, all tie-stall/box horses had ad libitum access to water indoors. We did not find significant differences in the amount of concentrate fed or sudden changes in the feeding regimen. This study was not designed to study causal factors, but the finding that colic was more common in the tie-stall/box horses indicates that each individual RS could benefit from analyzing their routines to prevent colic. One action could be to provide water outdoors in the paddocks for tie-stall/box horses, as this can be done cheaply and easily with modern equipment.

Colic can also be associated with infestation with intestinal parasites [23,24]. To significantly reduce the parasitic pressure, removing droppings from pastures is recommended to be done twice weekly as a preventive measure [25,26]. Only half the RS with group housing in our study reported removing droppings from their paddocks and none of the RS with tie-stall/box housing did so. However, we studied the RS during only one season and during the winter when parasites are less active and hence parasite infestation or paddock management may or may not be relevant for the difference found in colic. It is not a disadvantage, however, to remove the manure and there are technical solutions available today that make removing manure from paddocks both easy and work-efficient when housing many horses in small areas.

To conclude, several of the well-known risk-factors for colic were present to a higher extent in the RS with tie-stall/box housing than in the RS with group housing. As colic was also more common in the RS with tie-stalls/boxes, it is important to point out that some risk factors can be easily alleviated without changing the housing system. The risk factor of immobility/time spent in the stable might require individual solutions designed for each RS.

We also found respiratory airway problems to be more common in horses housed in tie-stalls/boxes. This could potentially be connected to the fact that these horses were found to spend more than twice as long indoors compared with group-housed horses. It has previously been found that air quality in the stable can be an important risk factor for respiratory health problems in both horses and humans [26]. Hence, this is a One Welfare [27] issue, affecting the performance of staff and horses, thus directly affecting the quality of the riding at the RS and indirectly affecting RS economics. This is also a problem that has more simple solutions than building a group housing system. One such solution is to provide adequate ventilation. As in the case of risk factors for colic discussed above, this would improve the welfare of RS and also offer RS clients more healthy horses. Respiratory airway problems could also have other causes.

As RS in Sweden are often part of the activities made available for young people by local councils [28], the responsibility for investments could be shared between the RS and the council, which could work together to improve both horse and human welfare.

Methods Used

As we covered a large proportion of the RS in Sweden with group housing, but only a small proportion of the RS with tie-stall/box housing, the results should be interpreted with caution.

Furthermore, data collection was done on one occasion only. If the RS had been visited several times, we would have gained information that could have indicated what the causal factors for health problems in the horses were. At present, we can only speculate and relate to earlier research on the causal factors.

Another limitation is that the RS themselves chose the horses to be clinically examined and this may have introduced bias, as they could either have chosen the best-looking horses available or horses with health issues when they had the chance to receive cost-free advice on health and body condition. However, our impression was that all RS managers were genuinely interested and talked to us about their horses as individuals (some of which were not working due to health issues) and wanted to discuss preventive health measures. We definitely did not get the impression that they were hiding any horses or any information from us. An earlier study [29] comparing two different methods of welfare assessment found discrepancies in results when looking at some horses compared with looking at all horses, with more welfare issues being found when all horses were examined. We attempted to deal with these issues by using the same method for both systems studied. As we wanted to examine horses that were currently working, we did not look at horses that were ill, injured, or under treatment for any condition, and hence not working at the time of the visit. However, such horses were included in data recorded by RS managers.

We used the Hennecke scale to estimate BCS. This is a well validated scale for measuring the amount of adipose tissue in live horses [9]. However, we found that a method for estimating muscle build-up would be very beneficial for working horse welfare assessment. To our knowledge, there is no easily used field method for reliably measuring muscle build-up in horses.

5. Conclusions

General health status was found to be better in RS horses kept in group housing, contradicting fears about housing horses in groups. Hence, keeping RS horses in groups seems to be a feasible solution and our data indicate that group housing may increase horse welfare (measured as health status).

To further improve RS horse welfare, and probably also the economics and performance of the horses, an independent professional should perform body condition scoring and review feed rations regularly.

Supplementary Materials: Table S1 Clincial examinations per RS, Table S2 Prop of horses in clin ex, Table S3 working hours, Table S4 Self-rep inj & found risks, Table S5 water and time.

Author Contributions: Conceptualization: P.A., H.S. Data curation, J.Y., J.C.R.T., J.L., A.P. and H.S.; Formal analysis, J.Y.; Funding acquisition, J.Y., P.A., and H.S.; Investigation, J.Y., J.C.R.T., J.L., A.P., P.A. and H.S.; Methodology, J.Y., J.C.R.T., J.Y., A.P., P.A. and H.S.; Project administration, J.Y., J.L., A.P., P.A. and H.S.; Resources, J.Y.; Software, J.Y.; Supervision, J.Y., P.A., and H.S.; Visualization, J.Y., A.P., and H.S.; Writing – original draft, J.Y., J.L. and P.A.; Writing – review & editing, J.Y., J.C.R.T., J.L., A.P., P.A. and H.S.

Acknowledgments: We are very grateful to the 16 RS that participated in this study. Staff and horses took care of us and assisted in all conceivable ways while we were collecting data. We would also like to thank Maria Andersson for substantial contributions to the idea and design of this project, Astrid Borg for method development of the clinical examinations and Agneta Sandberg, at the Swedish Trotting Association, for help in developing the injury risk protocol. We also thank Mary McAfee for checking the language of the manuscript and two unknown referees for constructive comments.

Conflicts of Interest: The authors declare no conflict of interest. The funders had no role in the design of the study; in the collection, analyses, or interpretation of data; in the writing of the manuscript, or in the decision to publish the results.

Appendix A

Table A1. TJK: Protocol used in clinical examinations.

Clinical Examination Point	Name and Age of the Horse	Gender and Breed	Comments /Description Take Pictures
General condition		Normal / Deviating	
General impression at physical examination		(avoiding/ aggressive/ neutral/ seeking contact)	Aggressive horses will not be examined!
Rug		No / Yes	Torn and broken
Lesions from the equipment		(none/ fur coat wear, wounds)	
Lymph nodes			Where on the body
Cleanliness, body			Mud or manure
Cleanliness, legs		1 = fetlock 2 = cannon bone	Mud or manure
Thermal comfort		If rugged: Directly after taking of the rug (shivering with cold, neutral, sweating)	Note if sweating or shivering
Rectal temperature			
Hart rate		0 = 28–40, 1 = above	
Breathing frequency		0 = 8–16, 1 = above	
Cough reflex		0 = negative, 1 = positive	
Eyes		(non, 1 = discharge or protein lump and lines down the cheek, 2 = pus)	
Nasal discharge		(none, clear running, coloured or thick)	
Commissures of the lips/Signs of injuries caused by bitting		(none, wear, wounds)	Not noting healed wounds
Bars of the mouth		1 = old abrasions 2 = wound	
Faeces		Normal or loose	
Henneke BCS			
Mane and tail		(no itching, some itching visible, no hair/ rash)	
Coat		(normal, partly ragged or dull, dull and ragged/long)	Note if shorn /partly shorn
Skin status		(normal, visible furless spots/ crusts/ dandruff, serious skin issues on large parts of the body)	Note where on the body

Table A1. *Cont.*

Clinical Examination Point	Name and Age of the Horse	Gender and Breed	Comments /Description Take Pictures
Wounds		Note number, where on the body, severity, if possible cause of the wounds	
Exterior deviations			Only sever deviations
Hoof status		Normal shape, different size, injuries, thrush, wear,	
Shoe status		Yes / No	Half or full, Not if shoeing is good or not
Legs LF, LR, RF, RR, below carpus		Extosis, 1 = wind galls normal-mild, moderate, 2 = severe, involving	After veterinarian evaluation 1 = old abrasions
Lameness control; walk and trot		(none, arhythmical, mild lameness, severe lameness)	

Appendix B

Table A2. Check list – Risks of injury. Walk systematically through all parts of the RS where the horses spend time. All risks are photographed.

	Risk Present or Not	Comments
Ceiling, upper part of box doors, other parts of buildings where horses can hit their heads		General ceiling lower than 1.5x withers (2.2 m)
Width of stable aisle		Narrower than 2.5 m
Width of doors		measure
Width and height entrance lying hall/ shed		measure
Slippery floor		
Narrow openings where head/ hoof can get stuck		In relation to the size of the horse
Bars that horses can reach		Note if weak
Sharp hooks		
Weak or sharp box walls		
Holes in walls		
Windows construction and protection		
Rug hanger that the horses can reach		
Crib		
Haynet (indoors)		
Other feed arrangement		e.g., for salt stone
Fire protection		
Electrical cords can be reached		
Can horses access feed storage		

Table A2. *Cont.*

	Risk Present or Not	Comments
Can horses access tools		
Ventilation		
Fencing – give thorough description		
Loos fencing wire		
Horses can access plastic		
Objects were horses can get stuck		
Haynet (outdoors)		
Mud in the paddock		
Rug on horses in paddock		
Halter on horses in paddock		
Slippery ground in paddock		
Slippery walk to and from paddock		
Mould		Where
Dust		
Objects lying/ standing around		
Tidiness		

References

1. Broom, D. Animal welfare: concepts and measurements. *J. of Anim. Sci.* **1991**, *669*, 4167–4175. [CrossRef]
2. McGreevy, P. The advent of equitation science. *Vet. J.* **2007**, *174*, 492–500. [CrossRef] [PubMed]
3. Horseman, S.; Buller, H.; Mullan, S.; Whay, H.R. Current Welfare Problems Facing Horses in Great Britain as Identified by Equine Stakeholders. *PLoS ONE* **2016**, *11*, e0160269. [CrossRef]
4. McGreevy, P. *Equine Behaviour. A Guide for Veterinarians and Equine Scientists*, 2nd ed.; Saunders Elsevier: Edinburgh, UK, 2012; ISBN 978-0-7020-4337-6.
5. Swedish Board of Agriculture. *Report on Housing of Horses in Sweden*; Swedish Board of Agriculture: Jönköping, Sweden, 2016. Available online: https://webbutiken.jordbruksverket.se/sv/artiklar/ra1812.html (accessed on 18 March 2018).
6. Swedish Board of Agriculture. *Legislation for the Protection of Horses*; Swedish Board of Agriculture: Jönköping, Sweden, 2007; p. 13.
7. Keeling, L.J.; Bøe, K.E.; Christensen, J.W.; Hyyppä, S.; Jansson, H.; Jørgensen, G.H.M.; Ladewig, J.; Mejdell, C.M.; Särkijärvi, S.; Søndergaard, E.; et al. Injury incidence, reactivity and ease of handling of horses kept in groups: A matched case control study in four Nordic countries. *Appl. Anim. Behav. Sci.* **2016**, *185*, 59–65. [CrossRef]
8. Lesimple, C.; Poissonnet, A.; Hausberger, M. How to keep your horse safe? An epidemiological study about management practices. *Appl. Anim. Behav. Sci.* **2016**, *181*, 105–114. [CrossRef]
9. Martinson, K. Estimation of body weight and development of a body weight score for adult equids using morphometric measurements. *J. Anim. Sci.* **2014**, *92*, 2230–2238. [CrossRef] [PubMed]
10. Egenvall, A.; Lönell, C.; Roepstorff, L. Analysis of morbidity and mortality data in riding school horses, with special regard to locomotor problems. *Prev. Vet. Med.* **2009**, *88*, 193–204. [CrossRef] [PubMed]
11. Wyse, C.A.; McNie, K.A.; Tannahil, V.J.; Murray, J.K.; Love, S. Prevalence of obesity in riding horses in Scotland. *Vet. Rec.* **2008**, *162*, 590–591. [CrossRef] [PubMed]
12. Thatcher, C.D.; Pleasant, R.S.; Geor, R.J.; Elvinger, F. Prevalence of overconditioning in mature horses in Southwest Virginia during the summer. *J. Vet. Int. Med.* **2012**, *26*, 1413–1418. [CrossRef] [PubMed]
13. Hitchens, P.L.; Hultgren, J.; Frössling, J.; Emanuelson, U.; Keeling, L.J. Prevalence and risk factors for overweight horses at premises in Sweden assessed using official animal welfare control data. *Acta. Vet. Scand.* **2016**, *58*, 31–35. [CrossRef] [PubMed]

14. Connysson, M.; Essén-Gustavsson, B.; Lindberg, J.E.; Jansson, A. Effects of feed deprivation on Standardbred horses fed a forage-only diet and a 50:50 forage-oats diet. *Equine Vet. J.* **2010**, *42*, 335–340. [CrossRef] [PubMed]

15. Jansson, A.; Lindberg, J.E. A forage-only diet alters the metabolic response of horses in training. *Animals* **2012**, *6*, 1939–1946. [CrossRef] [PubMed]

16. Archer, D.C.; Proudman, C.J. Epidemiological cues to preventing colic. *Vet. J.* **2006**, *172*, 29–39. [CrossRef] [PubMed]

17. Egenvall, A.; Penell, J.; Bonett, B.N.; Blix, J.; Pringle, J. Demographics and costs of colic in Swedish horses. *J. Vet Int. Med.* **2008**, *22*, 1029–1037. [CrossRef] [PubMed]

18. Cohen, N.D.; Gibbs, P.G.; Woods, A.M. Dietary and other management factors associated with colic in horses. *J. Am. Vet. Med. Ass.* **1999**, *215*, 53–60.

19. Hudson, J.M.; Cohen, N.D.; Gibbs, P.G.; Thompson, J.A. Feeding practices associated with colic in horses. *J. Am. Vet. Med. Ass.* **2001**, *219*, 1419–1425. [CrossRef]

20. Hillyer, M.H.; Taylor, F.G.R.; Proudman, C.J.; Edwards, G.B.; Smith, J.E.; French, N.P. Case control study to identify risk factors for simple colonic obstruction and distension colic in horses. *Equine Vet. J.* **2002**, *34*, 455–463. [CrossRef] [PubMed]

21. Kaya, G.; Sommerfeld-Stur, I.; Iben, C. Risk factors of colic in horses in Austria. *J. Anim. Phys. Anim. Nutr.* **2009**, *93*, 339–349. [CrossRef] [PubMed]

22. Reeves, M.J.; Salman, M.D.; Smith, G. Risk factors for equine acute abdominal disease (colic): Results from a multi-center case-control study. *Prev. Vet. Med.* **1996**, *26*, 285–301. [CrossRef]

23. Proudman, C.J.; French, N.P.; Trees, A.J. Tapeworm infection is a significant risk factor for spasmodic colic and ileal impaction colic in the horse. *Equine Vet. J.* **1998**, *30*, 194–199. [CrossRef] [PubMed]

24. Back, H.; Nyman, A.; Lind, E.O. The association between *Anoplocephala perfoliata* and colic in Swedish horses-A case control study. *Vet. Parasit.* **2013**, *197*, 580–585. [CrossRef] [PubMed]

25. Corbett, C.J.; Love, S.; Moore, A.; Burden, F.A.; Matthews, J.B.; Denwood, M.J. The effectiveness of faecal removal methods of pasture management to control the cyathostomin burden of donkeys. *Parasites Vectors* **2014**, *7*, 48.

26. Wålinder, R.; Riihimäki, M.; Bohlin, S.; Högstedt, C.; Nordquist, T.; Raine, A.; Pringle, J.; Elfman, L. Installation of mechanical ventilation in a horse stable: effects on air quality and human and equine airways. *Environ. Health Prev. Med.* **2011**, *16*, 264–272.

27. Pinillos, R.G. *One Welfare – A Framework to Improve Animal Welfare and Human Wellbeing*; CAB International: Oxfordshire, UK, 2018; ISBN 9781786393845.

28. RRO – The Swedish Riding school association. A report on the riding schools in society. Supported by the Swedish Board of Agriculture. Ridskoleverksamheten i Sveriges kommuner *En rikstäckande kartläggning gentemot fritidsförvaltningar på uppdrag av Ridskolornas Riksorganisation med stöd av Jordbruksverket.* 2010. Available online: https://www.jordbruksverket.se/download/18.14121bbd12def92a91780005096/1370040654982/Augurs+FINAL+rapport+RRO+2010-11-21.pdf (accessed on 12 May 2016).

29. Viksten, S.M.; Visser, E.K.; Nyman, S.; Blokhuis, H.J. Developing a horse welfare assessment protocol. *Anim. Welf.* **2017**, *26*, 59–65. [CrossRef]

2

Common Feeding Practices Pose a Risk to the Welfare of Horses when Kept on Non-Edible Bedding

Miriam Baumgartner [1],*[ID], Theresa Boisson [1], Michael H. Erhard [2] and Margit H. Zeitler-Feicht [1]

[1] Ethology, Animal Husbandries and Animal Welfare Research Group, Chair of Organic Agriculture and Agronomy, TUM School of Life Sciences Weihenstephan, Technical University of Munich; Liesel Beckmann-Str. 2, 85354 Freising, Germany; boisson.theresa@gmail.com (T.B.); zeitler-feicht@wzw.tum.de (M.H.Z.-F.)

[2] Chair of Animal Welfare, Ethology, Animal Hygiene and Animal Husbandry, Department of Veterinary Sciences, Faculty of Veterinary Medicine, Ludwig-Maximilians-University Munich, Veterinärstr. 13, 80539 Munich, Germany; m.erhard@tierhyg.vetmed.uni-muenchen.de

* Correspondence: m.baumgartner@tum.de

Simple Summary: It is a basic high priority need of every horse to take in roughage continuously. In order to ensure the horses' behavioural, physical and mental welfare, any pause of feed intake should not last for more than 4 hours. However, this basic need is often neglected in practice. The aim of the presented study was to assess the welfare of horses that are fed restrictively (*non ad libitum*) and kept in individual housing systems. We analyzed whether the feed intake behaviour of horses on edible bedding differs from the one of horses on non-edible bedding. As a common practice, the individually stabled horses were fed roughage twice or thrice a day. Our results showed that with this restrictive feeding practice, the horses were not able to eat any roughage for approx. 9 h during the night. Horses on non-edible bedding altered their feed intake behaviour - i.e., they paused less often during their meals and at a later point in time than the horses on edible bedding. We conclude that special feeding patterns have to be implemented (e.g., automated forage feeding systems) to avoid any impairment of the horses' welfare if kept on non-edible bedding.

Abstract: During the evolution of the horse, an extended period of feed intake, spread over the entire 24-h period, determined the horses' behaviour and physiology. Horses will not interrupt their feed intake for more than 4 h, if they have a choice. The aim of the present study was to investigate in what way restrictive feeding practices (*non ad libitum*) affect the horses' natural feed intake behaviour. We observed the feed intake behaviour of 104 horses on edible ($n = 30$) and non-edible bedding ($n = 74$) on ten different farms. We assessed the duration of the forced nocturnal feed intake interruption of horses housed on shavings when no additional roughage was available. Furthermore, we comparatively examined the feed intake behaviour of horses housed on edible versus non-edible bedding. The daily restrictive feeding of roughage (2 times a day: $n = 8$; 3 times a day: $n = 2$), as it is common in individual housing systems, resulted in a nocturnal feed intake interruption of more than 4 hours for the majority (74.32%, 55/74) of the horses on shavings (8:50 ± 1:25 h, median: 8:45 h, minimum: 6:45 h, maximum: 13:23 h). In comparison to horses on straw, horses on shavings paused their feed intake less frequently and at a later latency. Furthermore, they spent less time on consuming the evening meal than horses on straw. Our results of the comparison of the feed-intake behaviour of horses on edible and non-edible bedding show that the horses' ethological feeding needs are not satisfied on non-edible bedding. If the horses accelerate their feed intake (also defined as "rebound effect"), this might indicate that the horses' welfare is compromised. We conclude that in addition to the body condition score, the longest duration of feed intake interruption (usually in the night) is an important welfare indicator of horses that have limited access to roughage.

Keywords: horse behaviour; feed intake pause; bedding; welfare indicator; feeding practices; roughage; horse welfare; individual housing system

1. Introduction

Under natural conditions horses spend most of their time foraging and grazing, approximately 12 to 16 h of the 24-h-period are is spent on the ingestion of food [1–8]. Approximately 60%–70% of the daytime and 30%-40% of the night time is spent on feed intake [7–11]. Even when stabled, horses which are fed *ad libitum* divide their feed into approximately 10 meals, comparable to free-ranging horses [1,4,11–18]. Neither during the day, nor at night do horses pause voluntarily for longer than 3 to 4 hours between meals, nor do they fast [1–3,6–8,11–13,15,18–22]. The interruptions in feed intake are mainly due to the horse's need to engage in other behaviours, e.g., resting, interacting with social partners and comfort activities. Resting periods alternate with feed intake [7,8,20–23]. Therefore, other motivations must overrule the horses´ constantly activated motivation to forage before horses interrupt feeding.

Contrary to popular belief, a feeling of satiation or myofibrillary fatigue of the masticatory muscles does not limit the horses in their intake of feed [2,7,8,20,24,25]. The ability to consume high amounts of roughage and thus the high motivation to eat is evolutionary. This is due to the fact that for thousands of years, the vast steppes of Eurasia have offered a wide plant diversity, yet scant vegetation. Horses are adapted to a diet that is rich in structural fibres and low in energy (i.e., rapidly hydrolyzable carbohydrates) [7,8,18,26]. Due to the horses´ evolutionary fitness benefit ("niche construction theory"), they have the ability to forage an unlimited amount of low energy and high fibre food (ultimate behavioural control mechanism) [27]. If there is an energy deficit, metabolic and gastrointestinal cues as well as external stimuli (e.g., food supply) will increase the horses´ motivation to eat [28]. It is obvious that only nursing foals limit their feed intake when enough energy has been taken in. In adult horses, this capability is diminished. The horse does not recognize when the energy intake exceeds the horse's demand, especially when eating energy-rich, low-fibre food [8,20]. Hence, a continuous intake of food is a high priority basic need for horses and that irrespective of the current energy demand. If horses are fed in individual housing systems according to common feeding practices, the horses' natural feed intake behavior can be compromised [5,7,8].

A large number of studies have shown that poor feeding practices can lead to health problems in the digestive tract of horses. In particular, long pauses of feed intake are associated with colics caused by dysfermentation [29–31], the development of stomach ulcers [32–38] and constipation [39,40]. According to Luthersson et al. [37], the risk of grade II gastric ulcers (equine gastric ulceration syndrome severity score ≥2) is significantly increased if the feed intake pause lasts for 6 h or longer. The explanation for this is a decreased production of saliva and a reduced buffering capacity in the stomach due to a lack of roughage [31].

In addition, if horses are restrictively fed with hay (*non ad libitum*) horses kept on shavings show abnormal repetitive behaviour more often than horses kept on straw [41,42]. According to Marsden [43], with horses that spend a shorter time on eating, there is a higher risk that they show abnormal behaviour for longer. Lack of roughage is considered as one of the main causes of behavioural disorders in horses [7,8,42–51]. However, a direct connection between the behavioural disorder "crib-biting" in horses and the occurrence of stomach ulcers has not been proven. This indicates that both diseases result from housing and management related stress and are not interdependent [52,53]. Consequently, horses have evolved in a way that requires continuous food intake, from an ethological (natural behaviour) and a physiological (digestive system) point of view. According to Fraser [54] three different, yet overlapping aspects need to be considered to evaluate animal welfare: basic health and functioning, natural living (behaviour) and affective states. If the horses´ foraging periods are too short or the feed intake pauses too long, the horses are not able to express their natural behaviour. According to the German "Guidelines for Good Animal Welfare Practice for the Keeping of Horses" of the German Federal Ministry of Food and Agriculture (GFFA, 55) a minimum feeding period of 12 h as well as a maximum of 4 hours of feed intake pauses is required to avoid compromises in animal welfare. Studies carried out more than 30 years ago have already shown that if horses are kept under common husbandry conditions with restrictive hay feeding on non-edible bedding, the time budget of the horses deviates considerably. Instead of the natural feeding period of approximately 12 h per day, the horses are sometimes only able to consume food for approximately 4 hours of the 24-h day [5]. Horses kept on non-edible bedding such as wood shavings are not able to obtain roughage apart from the hay meals that they are fed twice or thrice during daytime. Hence, there is a risk that the maximum feed intake pause of 4 hours will be exceeded [6,7,55]. This applies in particular to the nocturnal feed intake interruption. To date, no studies on feed intake pauses of horses kept in individual housing systems have been published. However, this is an important research topic because the individual housing system is currently the most common method of keeping horses in Europe. It is evident that straw, as a source of roughage, has benefits for the horses´ welfare. If straw is used as a bedding material, the horses engage with it in a higher frequency and duration. In particular, the horses eat the bedding and spend less time on inter alia standing motionlessly [56,57]. Since engagement is one of the most important functions that bedding material is supposed to fulfill, Werhahn et al. [56] recommend to use straw over straw pellets and shavings to fulfill behavioural needs. Moreover, the result of a current study is that roughage significantly influences the horses´ welfare if they are housed individually. If hay is fed restrictively, straw as bedding prevents dietary deprivations of horses [58].

It is still unknown whether the natural feed intake pauses of horses on non-edible bedding that are fed restricted amounts of hay are exceeded. The satisfaction of this basic need could potentially be restricted. The aim of the present study was to investigate in what way restrictive feeding practices alter the horses' natural feed intake behaviour and thereby compromise the horses´ welfare. In particular, the study's aim was to i) determine the duration of the nocturnal feed intake interruption of horses housed on shavings when no additional roughage was available and ii) examine the feed intake behaviour of horses housed on edible versus non-edible bedding comparatively. We mainly focused on the time the horses needed to finish the evening meal, the duration and frequency of pauses during feed intake and at what time a pause occurred first (latency) in general and per category according to their duration.

2. Materials and Methods

2.1. Animals

The study took place on 10 different farms in the greater Munich (Germany) area and included a total of 116 sport and leisure horses of different race, age and sex. Only horses that did not suffer from any gastrointestinal disease or teeth problems (according to the farm managers) were included in the study. Ponies were explicitly excluded. Hence, 104 of the 116 horses were included in the study. They were all crossbred horses. All of these selected horses had a normal nutritional status (body condition score 5 or 6 on a scale from 1 to 9 [59,60]). Of the 104 horses, 34 were housed in individual boxes without the possibility of putting their head outside (grilled partition between and in the front of the boxes), 25 were housed in individual boxes with the possibility of putting their head outside (open window toward the external environment) and 45 of the horses were housed in individual boxes having direct access to a small outside yard (without vegetation) of the approximate size of the box (12–15 m^2). All horses had at least the possibility of visual contact with conspecifics (limited tactile contact). During the study period, every horse that was kept in an individual housing system was allowed to graze for approximately 6 hours daily (4–8 h). At least 2 hours before the feeding of the evening meal all horses had to be back in their individual boxes. On every farm the horses were either stabled on edible or on non-edible bedding material.

2.2. Behaviour Observation

Different direct behaviour observations were performed on a different number of subsamples. The selection of subsamples was based on the following precondition: Horses stabled side by side on edible or non-edible bedding in boxes that are easily visible to the observer (64 out of 104 horses). These horses were continuously observed during the evening feeding ("continuous behaviour sampling", Table 1) [61]. The remaining 40 horses were all stabled on non-edible bedding and were sampled discontinuously from the time of the beginning to the end of the evening roughage meal. The observations were carried out between 2:30 p.m. and 6:00 p.m., depending on the farm, and lasted until midnight at the latest.

Table 1. Classification of horses per bedding type and observation method.

Category	CO-Horses [2]	DCO-Horses [3]	AR-Horses [4]	Total
Horses on straw	25	5	-	30
Horses on shavings	30 (25) [1]	4 (4)[1]	40 (26) [1]	74 (55) [1]
Total	55	9 (9)[1]	40 (26) [1]	104 (90) [1]

[1] Number of horses that finished their evening meal within the observation period [2] Continuously Observed horses; [3] DisContinuously Observed horses; [4] Additionally Recorded horses.

Subsequently, the voluntarily taken pause of feed intake will be defined as the "feed intake pause" and the forced interruption of feed intake due to restrictive roughage supply will be defined as the "feed intake interruption". We observed these horses continuously and determined the time it took them to finish their evening meal and for how long the feed intake interruption lasted during the night (55 Continuously Observed (CO-) horses and nine Discontinuously Observed (DCO-) horses = 64, Table 1). In addition, we continuously observed the same 55 out of these 64 horses to assess the duration of feed intake pauses during the evening meal (CO-horses, $n = 55$). It was impossible to continuously observe the remaining nine horses as they temporarily left the stable for equestrian use during the observation periods (DCO-horses, $n = 9$). For the latter, it was therefore only possible to determine the time they needed to finish the evening meal and the duration of the feed intake interruption during the night.

In addition, the time of beginning and end of the evening roughage meal were recorded for another 40 horses on non-edible bedding (e.g., shavings), which were not directly visible from the observation site (Additionally Recorded (AR-) horses to assess the duration of the nocturnal feed intake interruption, Table 1). We chose to proceed this way in order to include a higher number of horses on non-edible bedding to further strengthen the validity of our study on nocturnal feed intake interruptions. We checked upon the AR-horses regularly but just briefly, to record the time at which they had finished their evening meal. This required only a few seconds and we integrated the results into the overall study. Nine horses (DCO-horses) were not observed continuously, however, both groups of observed horses were roughly of the same sample size (relatively balanced number of horses per topic).

Of the continuously observed (CO-) horses, 30 horses were kept on non-edible bedding (shavings) and 25 horses on edible bedding (straw) (Tables 1 and 2). The latter had roughage *ad libitum* (hay, straw) at their disposal. The horses were able to eat straw whenever they wanted. The horses kept on non-edible bedding (shavings) did not have unlimited access to feed between roughage meals. All horses were fed with hay ($n = 9$ farms) or hay mixed with straw ($n = 1$ farm) either two ($n = 8$ farms) or three ($n = 2$ farms) times per day.

Table 2. Number of Continuously Observed (CO-) horses per farm and condition (bedding material: straw and shaving).

Farm	Horses on Straw Per Farm	Horses on Shavings Per Farm	CO-Horses in Total
1	4	6	10
2	2	5	7
3	3	1	4
4	3	2	5
5	3	3	6
6	3	6	9
7	7	2	9
8	0	1	1
9	0	3	3
10	0	1	1
Total	25	30	55
Min	0	1	1
Max	7	6	10
Mean	2.5	3	5.5
Median	3	2.5	5.5

The same observer observed all of the horses in the identical manner during one evening feeding. The observation was not repeated. The horses had the chance to accustom themselves with the observer for about 1 hour before the observation started. The horses in the study were used to humans and no longer noticed the observer once they had gotten accustomed to his presence. The observer was able to observe a group of 4 to 10 horses simultaneously from one observation point (CO- and DCO-horses). Since we only observed whether the horses took in food, or not, it was possible to distinguish the two activities in up to 10 horses while observing them at the same time (Table 2 as well as Tables S1–S5 in the Supplementary Materials).

2.3. Data Sampling

As per definition, the feed intake pause started when the horse had stopped foraging and chewing for at least one minute. The end of the feed intake pause was defined as the time when the horse started to eat again. We used the Microsoft Excel Programme 2016 to calculate the total number of minutes between these two events. Following the example of Krull [4], feed intake pauses were classified in three categories according to their duration: short feed intake pause = 1-10 min, medium feed intake pause = 11-30 min and long feed intake pause = 31 min and longer. For the purpose of our study the presence of isolated stalks of hay on the ground indicated that the horse had finished its evening meal. The time at which each horse finished its evening meal was recorded and defined as the start of the nocturnal feed intake interruption. Horses on straw or horses on non-edible bedding, which had access to straw stalks from the neighbouring box, occasionally started to forage straw after finishing the evening meal. This was not recorded separately. As per our definition, once the horses had finished their evening hay meal the nocturnal feeding interruption started. Each farm manager told us at what time the morning feeding would begin. We used this information as a basis to calculate the duration of the nocturnal feed intake interruption, which lasted from the end of the evening meal to the beginning of the morning feeding. The duration of the nocturnal feed intake interruption was only relevant for horses kept on non-edible bedding ($n = 74$ horses on shavings). We were also able to determine the duration of the nocturnal feed intake interruption for those horses that were temporarily taken out of the box for equestrian use, since all of them had finished their meal by the end of the observation period ($n = 4$ horses on shavings out of 9 Discontinuously observed (DCO-) horses, Table 1).

The observations ended with the end of the evening meal. All horses were observed until they finished their evening meal or until midnight when the stables were closed. The farm managers asked us to leave at midnight, also on those farms where the horses had not finished their evening meal by that time. For both, horses on straw and on shavings ($n = 55$, Continuously Observed (CO-) horses,

Tables 1 and 2), it was recorded when the first feed intake pause was taken in voluntarily (latency) as well as the frequency and the duration of the pauses. The term latency describes the time spent on foraging from the beginning of the evening feeding until the start of the first feed intake pause for each category as mentioned above (short, medium, and long feed intake pauses). Our aim was to find out in what way the duration of feed intake pauses differed when horses were kept on shavings versus horses that were kept on straw. For this purpose we additionally measured the frequency of feed intake pauses for each feeding pause category ($n = 55$ CO-horses, Table 2). The horses whose natural feed intake behaviour could have been altered due to equestrian use and which were thus forced to pause their feed intake during the observation period were excluded ($n = 9$ DCO-horses). We defined the total time until the end of the evening meal as the time, which the 64 horses spent on eating their evening meal or as the time in which hay was available (all CO- and DCO-horses). Every horse started the evening meal at an individual point of time and for all horses the evening meal ended when no more hay was available. As a consequence, the total time the horses spent on eating the evening meal includes the time they spent on feed intake pauses. In the case of DCO-horses, we did not subtract the time the horses spent on equestrian use to not alter the realistic total time that food was available every day.

We feared that a weighing of each prepared evening hay meal might introduce an undesired selection bias of the feeding farmers or caretakers. The farmers were not provided with any details about the study. However had they known that the weight of hay was of relevance for us, they might have altered the amount of hay (deliberately or undeliberately). The farmers or caretakers did not weigh the evening hay meal, thus we also refrained from doing so. For this reason, we did not determine the exact feed intake rate of the horses in the present study.

2.4. Statistical Analysis

Descriptive statistics were expressed as mean, standard error of the mean, median, minimum and maximum. Tables and graphs were created in Microsoft Excel Programme 2016. We performed the statistical analysis using Generalized Linear Mixed Models (GLMM). For this, the program R was used (R Studio Version 1.0.143; R software version 3.5.1, R Development Core Team, Vienna, Austria, 2018, package "glm2") [62]. Six separate models with different dependent variables were calculated: 1) total time for finishing the evening meal, 2) duration of nocturnal feed intake interruption, 3) frequency of feed intake pauses, 4) latency until the first feed intake pause, 5) number of horses that took no feed intake pause in relation to all observed (CO and DCO)-horses per farm, 6) duration of feed intake pause. Outcome variables were treated as continuous variables. We considered the bedding material (explanatory variable) as a categorical factor (straw versus shavings). Explanatory variable was the type of bedding in four models. The farm was specified as a random effect in all models except for those two that related to the effect of the farm as such. It was treated as an explanatory variable in the model in which we tested the effect of the farm on the duration of the nocturnal feed intake interruption and the number of horses that took no feed intake pause in relation to all observed horses per farm. Thus, an effect of feeding practices of the different farms on the nocturnal feed intake interruption was analyzed. The significance level was set at 0.05.

2.5. Ethics Statement

This study was non-invasive. It consisted in observing the horses under their current conditions of life. No specific treatments or interventions were done on the animals. The study complies with the Guidelines for Ethical Treatment of Animals in Applied Animal Behaviour and Welfare Research (ISAE Ethics Committee, 2017).

3. Results

3.1. Total Time for Finishing the Evening Meal

The evening meal started with the provision of roughage, which was delivered between 3:30 p.m. and 7:00 p.m., depending on the farm. The total time horses needed for finishing the evening meal (CO- and DCO-horses) including feed intake pauses and interruptions for equestrian use was 286 ± 70 min (4:48 h; median: 307 min; minimum: 100 min; maximum: 420 min, $n = 51$ horses). We only included those horses into the analysis, which had finished their evening meal by the end of the observation period (midnight). This applied to the majority of the groups on both bedding materials ($n = 51/64$ CO- and DCO-horses on eight farms, 79.69%). The type of bedding had an influence on the total time for finishing the evening meal ($p = 0.04$, Figure 1). The farm itself was irrelevant for the present study as in all cases the farm´s feeding practices did not have an influence on the total time the horses needed to finish the evening meal on both bedding materials ($p = 0.13$).

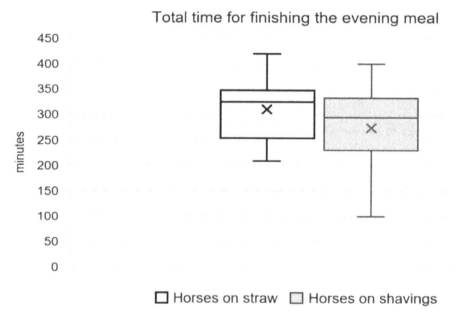

Figure 1. Total time for finishing the evening meal of horses on straw in comparison to horses on shavings, which finished their evening meal within the observation period (79.69%; $n = 51$ out of 64 horses; median: 325 min of $n = 17$ horses on straw; median: 307 min of $n = 34$ horses on shavings, $p = 0.04$).

3.2. Duration of the Nocturnal Feed Intake Interruption

In contrast to the horses on straw, which had permanent access to roughage since they were kept on edible bedding, the majority of the horses on shavings (74.32%, $n = 25$ CO + 4 DCO-horses + 26 AR-horses = 55 of 74 horses) were forced to interrupt their feed intake during the night for more than 4 hours. The nocturnal feed intake interruption was defined as the period between the end of the evening meal and the beginning of morning feed. The nocturnal feed intake interruption lasted on average 530 ± 85 min (8:50 ± 1:25 h, median: 525 min, minimum: 405 min, maximum: 803 min, Figure 2). The longest feed intake interruption was observed in a horse on shavings, which was fed at 3.30 p.m. and had eaten up after 127 min (2:07 h)—hence it had no roughage available. Thus, this horse was forced to interrupt its feed intake during the night for 13:24 h (803 min). The morning feeding started between 05:45 and 07:15 a.m., depending on the farm.

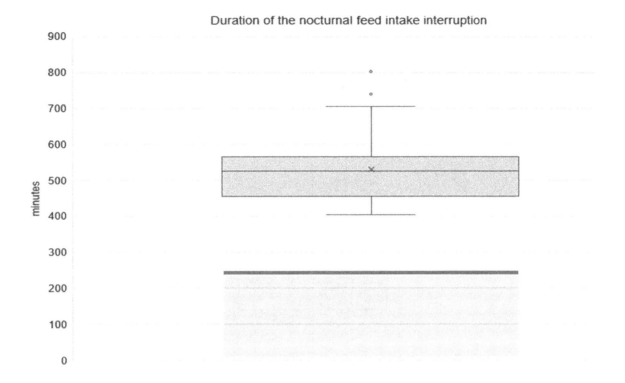

Figure 2. Duration of the nocturnal feed intake interruption of horses on shavings, which had finished their evening meal within the observation period (74.32%; $n = 55$ horses on shavings out of 74 horses on shaving in total). Green box: period without food supply which is still appropriate in terms of welfare; red line: tolerable maximum duration of feed intake interruptions at 240 min (4 hours) according to the German welfare standard [55]).

In 2 of the 10 farms, all seven of the seven CO-horses still had roughage available at the end of the observation period (midnight) (three of those horses on shavings) irrespective of the type of bedding they were housed on. Of the CO- and DCO-horses on shavings ($n = 34$, $n = 8/10$ farms), horses on shavings on three different farms did not finish their evening meal before the observation period ended (midnight), but still had hay available (estimated value: haystack for approximately 1:30 h of feed intake time). Thus, 85.29% ($n = 29/34$) of the sample of CO- and DCO-horses housed on shavings were forced to interrupt their feed intake for more than 4 hours in the night. In the sample of AR-horses on shavings, 65.00% (26 out of 40) had finished their entire roughage meal by midnight. The farms tended to influence the duration of the nocturnal feed intake interruption, but not significantly ($p = 0.06$). This is an indication that the feeding practices on the different farms (time of the evening feeding and amount of provided hay) varied.

3.3. Frequency of Feed Intake Pauses and Latency until the First Feed Intake Pause

Horses on straw paused their feed intake more often than horses on shavings ($n = 25$ horses, mean: 4.6 ± 3.7; median: 4.0 compared to n = 30 horses, mean: 2.6 ± 2.2; median 3.0; $p = 0.01$; Figure 3). The farms' feeding practice tended to have an effect on the frequency of the feed intake pauses ($p = 0.08$).

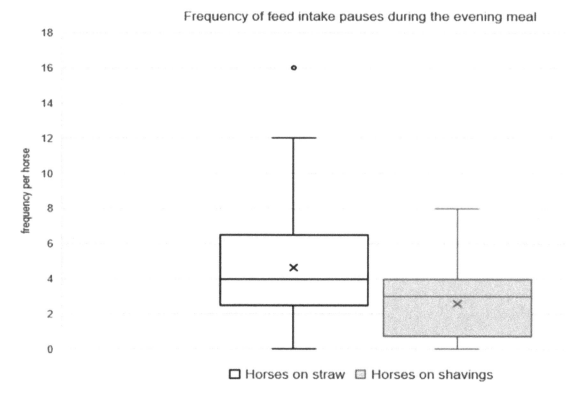

Figure 3. Frequency of feed intake pauses during the evening meal of horses on straw compared to horses on shavings ($n = 55$ horses; median: 4.0 of $n = 25$ horses on straw; median: 3.0 of $n = 30$ horses on shavings, $p = 0.01$)

We calculated the latency until the first feed intake pause irrespective of the pause category. The latency of horses on straw was observed at the same time as for the horses on shavings ($n = 25$ horses, first feed intake pause according to median: 104.00 min, mean: 125 ± 91 min versus n = 30 horses, first feed intake pause according to median: 134 min, mean: 153 ± 81 min, n.s. with $p = 0.22$; no influence of the farm $p = 0.42$). However, two horses on straw (8.00%, $n = 2/25$) of the CO-horses did not pause during the course of the evening meal, whereas seven horses on shavings (23.33%, $n = 7/30$, n.s. with $p = 0.12$) did not take a single feed intake pause. These horses on shavings continued eating hay without any interruption or pause until they were forced to pause when they had finished their evening meal.

3.4. Duration of Feed Intake Pause

A similar situation was observed with regards to the duration of the feed intake pause. Horses on straw paused almost twice as long as horses on shavings ($n = 25$ horses, median: 15.0 min, mean: 14 ± 9 min versus $n = 30$ horses, median: 7 min, mean: 11 ± 12 min, n.s. with $p = 0.36$, Figure 4). The farm, specified as a random effect, influenced the duration of the feed intake pause of the CO-horses on the two bedding variants ($p = 0.04$).

All CO-horses paused their feed intake rather for a short amount of time (a few minutes) than for a longer duration (short pauses: 48% of all categories of pauses for horses on straw and 54% of all categories of pauses for horses on shavings, Table 3). Horses on shavings took less feed intake pauses of a medium (11-30 min) duration in comparison to horses on straw (29% medium pauses of horses on shavings versus 37% medium pauses of horses on straw). Only very few of the horses took long (>30 min) pauses (15%–17% of the horses on straw and horses on shavings, n = 46 horses). Horses on straw took both, short (1–10 min) pauses (after 108 ± 53 min) and medium pauses (after 121 ± 71 min) at an earlier latency than the horses on shavings, irrespective of whether their pause was short, long, or medium. The horses on shavings paused feed intake at the earliest 2 h after the start of the meal

(128 ± 48 min). All of the horses on both types of bedding paused their feed intake for more than 30 min (long pauses) only after a latency of 3:24 h.

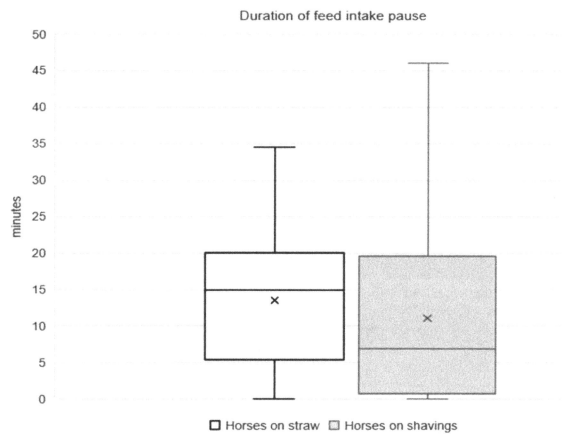

Figure 4. Duration of feed intake pause during the evening meal of horses on straw compared to horses on shavings (feed of hay or a mix of hay and straw to n = 55 horses; median: 15 min of n = 25 horses on straw; median: 7 min of n = 30 horses on shavings, n.s. with p = 0.36)

Table 3. Latency from the start of the evening feeding until the first pause divided into three categories according to the duration of the feed intake pauses (mean ± SEM, in minutes) and frequency of pauses per each category for horses on straw and shavings (in %).

Category	Short Pauses (1–10 Min)		Medium Pauses (11–30 Min)		Long Pauses (> 30 Min)	
Sample Size (Number of Horses)	55		55		46 [1]	
Horses on straw	48% [1]	108 ± 53	37% [1]	121 ± 71	15% [1]	204 ± 67
Horses on shavings	54% [1]	128 ± 48	29% [1]	168 ± 79	17% [1]	204 ± 86

[1] n = 9 horses that did not take long pauses were removed from the calculation of frequency of pauses per category (in %).

4. Discussion

Horses eat about 10 (7–13) meals in regular intervals within 24 h [4,7,17,55]. They hardly ever pause their feed intake for several hours [6–8,21,22,55]. Horses usually spend no hours without feeding [8,22]. They rarely stop foraging for longer than 3 hours (2 ± 1:18 h) [see for review 31] with 4 hours being the rare maximum [6–8,21,22,55]. The 24-h time budget of horses kept in individual housing systems on non-edible bedding differs greatly from that of free-ranging horses [1,5]. Kiley-Worthington [5] observed that considerable deviations in the time budget of the horses occur when housed in conditions with restrictive hay feeding and non-edible bedding. Instead of the natural feed intake duration of 12 - 16 h, she observed that an approximate feed intake of 4 hours during the 24-h day is possible at times. Especially if the feeding is restrictive and horses are housed on non-edible bedding, there is a higher risk that the maximum feed intake pause of 4 hours is exceeded [6,7,22,55]. The importance of continuous roughage intake for the physical and mental health of horses is often underestimated in practice [7,8]. The digestive system of the horses is in construction and function adapted to long feeding periods and a continuous feed intake, distributed over the 24-h day. Any major deviation from this disrupts the digestive process [36] and leads to behavioural disorders and even stereotypies (e.g., wood-chewing, ingesting wood-shavings and performing coprophagy, crib-biting, windsucking, weaving) [7,44,56,58]. A current study confirms alterations in the behaviour of horses housed in individual housing systems with restrictive feeding and non-edible-bedding. In particular, horses kept on straw express less aggressive behaviour compared to those living on non-edible bedding [58]. In our observations during the present study we noticed several times that horses on shavings showed abnormal behaviour and were aggressive towards horses on straw. This is in line with recent studies and might be indicative of the emotional state of frustration of the horses on non-edible bedding (frustration-related behaviour). Since only feed intake behaviour was the focus of the present study, no conclusions can be drawn from the observations mentioned.

We included 104 horses on 10 farms in our study of which the majority had a nocturnal feed intake interruption of 8:50 ± 1:25 h when kept on shavings. This clearly exceeds the required maximum of 4 hours according to the Guidelines to Good Animal Welfare Practice for the Keeping of Horses [55]. A horse would not take a voluntary feed intake pause of approx. 9 h because of its almost permanent motivation to forage. Therefore, feed-intake pauses of such length are not in line with the natural feeding behaviour of horses [7,22]. According to Wackenhut [63] and Mellor et al. [64] the proportion of horses kept individually on shavings in Germany and England is approximately 35%, which underlines how important it is to focus on the potential risks for animal welfare. Three basic requirements have to be fulfilled to ensure animal welfare: basic health and functioning, natural living (behaviour) and affective states [54]. If the horses´ feed intake pauses are too long, the horses are not able to act out their natural behaviour, which potentially compromises their welfare [54]. Furthermore, health problems in the digestive tract, i.e., grade II gastric ulcers, occur considerably more often if feed intake pauses last for more than 6 hours [37].

Only on two farms all of the horses on non-edible bedding still had roughage available at midnight. If the horses on these two farms would carefully divide the remaining quantity of roughage (an amount for approximately 1:30 h of feed intake time), they would have been able to take a feed intake pause of a maximum of 4 hours [6,55]. However, this could not be reliably assessed, as the observations had to stop at midnight for operational constraints. If the nocturnal feed intake interruption of all horses on non-edible bedding should be determined, the observation would have had to be continued during the nighttime, because 25.68% of horses did not finished their evening meal by midnight ($n = 19/74$).

All of the horses were allowed to graze daily. The grazing period ended at least 2 h prior to the evening feeding. Hence, it seems that the availability of straw predominantly determines whether the feed intake behaviour of horses on non-edible bedding differs from that on edible bedding (less time spent on the evening meal and fewer feed intake pauses). We observed that one horse bedded on straw preferred to forage straw instead of hay although hay was still available. If such a preference for straw would have been observed in more horses, we would have included it as a possible influence factor on the longer overall feeding time for the evening meals of the horses on straw in comparison to the horses on shavings. Since this happened only in one horse and only for a limited period, we can exclude this as a bias in the results.

Since we tried to bias the feeding routine as little as possible and did therefore not measure the specific hay amount for the calculation of the feed intake rate, we can only draw an indirect conclusion on the accelerated feed intake rate from the horses' modified feed intake behaviour on non-edible bedding. It can be assumed that horses on shavings ate their meal at a much faster rate, since it took them less time to finish the evening meal than the horses housed on straw. In addition, the former took significantly fewer feed intake pauses and only at a later latency. This indicates to an accumulated need deficit that could possibly lead to a "rebound effect" in horses on non-edible bedding. Accelerated feed intake due to lack of roughage or shortened grazing periods is in line with previous studies [8,65,66]. It has already been shown that horses suffer from a "rebound effect" as mentioned above when they are unable to act out their natural behaviour such as locomotion or resting [67–70]. Altered feed intake behaviour of horses on shavings as a consequence of restrictive roughage supply indicates an impaired welfare.

The treatment groups lacked uniformity, which could be considered as a restraining factor for the study. Our goal was to include a representative number of horses on ten different farms in order to obtain valid results concerning feeding practices for individually housed horses. We did not preselect the horses in advance, which might have biased the results. We recorded the end of the evening meal and the beginning of the morning feeding for 40 additional horses on non-edible bedding and were thus able to increase the sample size for the present study. Moreover, it was inevitable that horses were taken away for equestrian use during the observation periods. For this reason, we separated the horses in two groups: One continuously observed group and one discontinuously observed group. The confounding factor "equestrian use" applied to the latter. We decided that the temporary absence of the horse during the evening meal might have an influence on the natural feed intake behaviour, especially on the feed intake pauses. We excluded these horses (n = 9) in the analysis of the feed intake behaviour. For the stated reasons, not all hypotheses could be assessed on the basis of the exact same sample size of horses. Current welfare assessment protocols for horses measure the welfare criterion of "good feeding" or the welfare parameter of "freedom from hunger" only against the indicator "body condition score" [71–73]. However, this indicator can only be used to evaluate if the horses´ nutritional status is adequate. To date, current welfare assessment protocols for horses [54,55,71,73] do not take the indicators into account that relate to the behavioural and physiological aspect of appropriate feeding. In order to respect the psychological and physical well-being of the horses, one has to be aware of the fact that the feed intake behaviour is a basic high priority need of the horse. Excessive feed intake pauses as found in the present study represent a risk of pain, suffering and harm [5,18,21,33,37,42,51]. For this reason, a German court confirmed in its judgement (2019, file number RN 4 K 17.1298, Regensburg Administrative Court) that it has to be ensured that the maximum duration of feed intake pauses is 4 hours and that the duration of feed intake must at least be 12 h.

5. Conclusions

We conclude from our study that common feeding practices (two or three roughage meals during daytime) impair the horses´ welfare if they are kept on non-edible bedding. The reason for this is that in such housing systems the majority of the horses pause their feed intake for too long during the night. The average nocturnal feed intake interruption of horses observed in individual housing system on non-edible bedding was 9 h. Thus, the maximum tolerable feed intake pause of 4 hours is exceeded. Consequently, the well-being of the horses might be at risk. The changed feed intake behaviour of the horses allows us to draw a similar conclusion. The horses on non-edible bedding finished their hay meal faster and took fewer pauses during the meal than horses on edible bedding ("rebound effect"). If the basic high priority need of the horse to continuously engage in feed intake is neglected and the horses' access to roughage (straw) is limited, the horses will accelerate their feed intake of the roughage and eat almost non-stop. Altered feed intake behaviour indicates an impaired welfare.

However, further studies should be carried out in order to assess additional behavioural and physiological effects of restrictive feeding practices when there is no additional straw available. Video surveillance or longer observation periods could have helped to find out if all horses on non-edible bedding were affected by prolonged feed intake interruptions. Nevertheless, restrictive feeding of roughage for horses kept on non-edible bedding such as shavings or straw pellets can be regarded as inadequate. Novel feeding techniques (e.g., supply of roughage via forage feeding systems such as automated roughage feeders, feeding racks, time-controlled automatic feeders) need to be implemented and could satisfy the behavioural needs. Future welfare assessment should take the duration of the feed intake pauses into account. The "Weihenstephan's evaluation system for the assessment of animal welfare and the environmental impact of horse husbandry", which is currently under development, shall be used as a scientific consultation tool in Germany in the future. The tool includes the holistic feeding aspects of horses.

Supplementary Materials: Table S1: Table S1: Total time for finishing the evening meal dependent on farm and bedding; Table S2: Total time for finishing the evening meal dependent on beddingfor feed intake behaviour analysis. Table S3: "Horses on shavings", which ate up their evening meal within the observation period (Calculation of the duration of the nocturnal feed intake interruption; Table S4: Frequency of pauses, duration of pauses and latency until first feed intake pause during evening meal; Table S5: Number of feed intake pauses dependent on bedding.

Author Contributions: Conceptualized the study, M.B. and M.H.Z.-F.; carried out the practical data collection at farms, T.B.; performed the formal and statistical analyses, M.B.; drafted the manuscript, M.B.; revised the paper, M.B., T.B., M.H.E. and M.Z.-F.; acquired fundings, M.H.Z.-F. and M.B. All authors read and approved the final manuscript.

Acknowledgments: We would like to thank Irina Sasse for her assistance in translation, Bettina Gohm for a final spelling revision, and Helen Louton for very helpful comments on an earlier version of this manuscript. We would also like to thank the anonymous reviewers and the editor for their constructive comments to prior drafts of this manuscript.

References

1. Duncan, P. Timebudgets of camargue horses. *Behaviour* **1980**, *72*, 26–49. [CrossRef]
2. Mayer, E.; Duncan, P. Temporal Patterns of Feeding Behaviour in Free-Ranging Horses. *Behaviour* **1986**, *96*, 105–129.
3. Ihle, P. Ethological study on the daily routine of horses dependent on the type of husbandry. *Diss. Med. Vet.* **1984**.
4. Krull, H.D. Studies on the intake and digestibility of forage in horses. *Diss. Med. Vet.* **1984**.
5. Kiley-Worthington, M. *The Behaviour of Horses in Relation to Management and Training*; Allen & Co. Ltd.: London, England, 1987.
6. Davidson, N.; Harris, P. Nutrition and Welfare. In *The Welfare of Horses*; Waran, N., Ed.; Springer: Berlin/Heidelberg, Germany, 2007.
7. Zeitler-Feicht, M.H. *Horse Behaviour Explained: Origins, Treatment and Prevention of Problems*, 1st ed; Manson Publishing Ltd.: London, UK, 2004.
8. Ellis, A.D. Biological basis of behavior in relation to nutrition and feed intake in horses. In Proceedings of the EAAP Sci. Ser. 5th European Workshop on Equine Nutrition, Cirencester, UK, 19–22 September 2010; pp. 53–74.
9. Tyler, S.J. The behaviour and social organisation of New Forest Ponies. *Anim. Behav. Monogr.* **1972**, *5*, 85–196. [CrossRef]
10. Ruckebusch, Y.; Vigroux, P.; Candau, M. Analyse du Comportements Alimentaire Chez Les Equids. *CR Journée d'Etude Cereopa, Paris* **1976**, 69–72.
11. Hallam, S.; Campbell, E.P.; Qazamel, M.; Owen, H.; Ellis, A.D. Effects of traditional versus novel feeding management on 24 h time budget of stabled horses. In *Forages and Grazing in Horse Nutrition*; Saastamoinen, M., Fradinho, M.J., Santos, A.S., Miraglia, N., Eds.; Wageningen Academic Publishers: Wageningen, The Netherlands, 2012.
12. Ellis, A.D.; Redgate, S.; Zinchenko, S.; Owen, H.; Barfoot, C.; Harris, P. The effect of presenting forage in multi-layered haynets and at multiple sites on night time budgets of stabled horses. *Appl. Anim. Behav. Sci.* **2015**, *171*, 108–116. [CrossRef]
13. Boyd, L.E.; Carbonaro, D.A.; Houpt, K.A. The 24-h time budget of Przewalski horses. *Appl. Anim. Behav. Sci.* **1988**, *21*, 5–17. [CrossRef]

14. Berger, A.K.; Scheibe, K.; Eichhorn, K.; Scheibe, A.; Streich, J. Diurnal and ultradian rhythmus of behaviour in a mare group of Przewalski horses (Equus ferus przewalski) measured through one year under semi-reserve conditions. *Appl. Anim. Behav. Sci.* **1999**, *64*, 1–17. [CrossRef]

15. Van Dierendonk, M.C.; Bandi, N.; Batdorj, D.; Dugerlham, S.; Munkhtsog, B. Behavioural observations of reintroduced Takhi or Przewalski horses (Equus ferus przewalski) in Mongolia. *Appl. Anim. Behav. Sci.* **1996**, *50*, 95–114. [CrossRef]

16. Souris, A.; Kaszensky, P.; Julliard, R.; Walzer, C. Time budget-, behavioral synchrony- and body score developement of a newly released Przewalski's horse group (Equus ferus przewalski) in the Great Gobi B strictly protected area in SW Mongolia. *Appl. Anim. Behav. Sci.* **2007**, *107*, 307–321. [CrossRef] [PubMed]

17. Sweeting, M.P.; Houpt, C.; Houpt, K.A. Social facilation of feeding and time budgets in stabled ponies. *J. Anim. Sci.* **1985**, *60*, 369–371. [CrossRef] [PubMed]

18. Zeitler-Feicht, M.H. Feeding of horses considering behavioural aspects. In Tagungsbericht der Deutschen Veterinärmedizinischen Gesellschaft e.V. (DVG); Fachgruppe Angewandte Ethologie. Available online: https://www.dvg.net/index.php?id=2111&contUid=0 (accessed on 24 February 2020).

19. Edouard, N.; Fleurance, G.; Dumont, B.; Baumont, R.; Duncan, P. Does sward height affect the choice of feeding sites and voluntary intake in horses? *Appl. Anim. Behav. Sci.* **2009**, *119*, 219–228. [CrossRef]

20. Ralston, S.L. Controls of feeding in horses. *J. Anim. Sci.* **1984**, *59*, 1354–1361. [CrossRef]

21. Harris, P.A. Impact of Nutrition and Feeding practices on equines, their behaviour and welfare. In *Horse Behaviour and Welfare*; Wageningen Academic Publishers: Wageningen, The Netherlands, 2007.

22. Mills, D.; Redgate, S. Behaviour of horses. In *The Ethology of Domestic Animals: An Introductory Text*, 3rd ed.; Jensen, P., Ed.; CABI: Wallingford, UK, 2017.

23. Ninomiya, S.; Shusuke, S.; Kusunose, R.; Mitumasu, T.; Obara, Y. A note on a behavioural indicator of satisfaction in stabled horses. *Appl. Anim. Behav. Sci.* **2007**, *106*, 184–189. [CrossRef]

24. Hoh, J.F.Y. "Superfast" or masticatory myosin and the evolution of jaw-closing muscles of vertebrates. *J. Exp. Biol.* **2002**, *205*, 2203–2210.

25. Eizema, K.; Van der Burg, M.; De Jonge, H.W.; Dingboom, E.G.; Weijs, W.A.; Everts, M.E. Myosin heavy chain isoforms in equine gluteus medius muscle: Comparison of mRNA and protein expression profiles. *J. Histochem. Cytochem.* **2005**, *53*, 1383–1390. [CrossRef]

26. Coenen, M.; Vervuert, I. A minimum of roughage and a maximum of starch – necessary benchmarks for equine diets. *Pferdeheilkunde Equine Med.* **2010**, *26*, 147–151. [CrossRef]

27. Gygax, L. Wanting, liking and welfare: The role of affective states in proximate control of behaviour in vertebrates. *Ethology* **2017**, *123*, 689–704. [CrossRef]

28. Ralston, S.; Baile, C.A. Factors in the Control of Feed Intake of Horses and Ponies. *Neurosci. Biobehav. Rev.* **1983**, *7*, 465–470. [CrossRef]

29. Medina, B.; Girard, I.D.; Jacotot, E.; Julliand, V. Effect of preparation of Saccharomyces cerevisiae on microbial profiles and fermentation patterns in the large intestine of horses fed a high fiber or a high starch diet. *J. Anim. Sci.* **2002**, *80*, 2600–2609. [CrossRef] [PubMed]

30. Williamson, A.; Rogers, C.W.; Firth, E.C. A survey of feeding, management and faecal pH of Thoroughbred racehorses in the North Island of New Zealand. *N. Z. Vet. J.* **2007**, *55*, 337–341. [CrossRef] [PubMed]

31. Harris, P.A.; Ellis, A.D.; Fradinho, M.J.; Jansson, A.; Julliand, V.; Luthersson, N.; Santos, A.S.; Vervuert, I. Review: Feeding conserved forage to horses: Recent advances and recommendations. *Animal* **2017**, *11*, 958–967. [CrossRef]

32. Hammond, C.J. Gastric ulceration in mature Thoroughbred horses. *Equine Vet. J.* **1986**, *18*, 284–287. [CrossRef] [PubMed]

33. Nadeau, J.A.; Andrews, F.M.; Mathew, A.G.; Argenzio, R.A.; Blackford, J.T.; Sohtell, M.; Saxton, A.M. Evaluation of diet as a cause of gastric ulcers in horses. *Am. J. vet. Res.* **2000**, *61*, 784–790. [CrossRef]

34. Nadeau, J.A.; Andrews, F.M. Equine gastric ulcer syndrome: The continuing conundrum. *Equine Vet. J.* **2009**, *41*, 611–615. [CrossRef]

35. Luthersson, N.; Nadeau, J.A. Gastric ulceration. In *Equine Applied and Clinical Nutrition*, 1st ed.; Geor, J.R., Harris, P.A., Coenen, M., Eds.; Elsevier Ltd: Amsterdam, The Netherlands, 2013.

36. Luthersson, N.; Nielsen, K.H.; Harris, P.; Parkin, T.D.H. The prevalence and anatomical distribution of equine gastric ulceration syndrome (EGUS) in 201 horses in Denmark. *Equine Vet. J.* **2009**, *41*, 619–624. [CrossRef]

37. Luthersson, N.; Nielsen, K.H.; Harris, P.; Parkin, T.D.H. Risk factors associated with equine gastric ulceration syndrome (EGUS) in 201 horses in Denmark. *Equine Vet. J.* **2009**, *41*, 625–630. [CrossRef]

38. Murray, M.J.; Grady, T.C. The effect of a pectin-lecithin complex on prevention of gastric mucosal lesions induced by feed deprivation in ponies. *Equine Vet. J.* **2002**, *34*, 195–198. [CrossRef]

39. Gieselmann, A. Nutritive anamnesis of colic cases of the horse. *Diss. Med. Vet.* **1994**.

40. Coenen, M. Nutrition and Colic. *Pferdeheilkunde Equine Med.* **2013**, *29*, 176–182. [CrossRef]

41. Baumgartner, M.; Gandorfer, J.; Reiter, K.; Zeitler-Feicht, M.H. Abnormal behaviour of individually stabled horses dependent on situation and bedding material. Aktuelle Arbeiten zur artgemäßen Tierhaltung. *KTBL Schr.* **2015**, *510*, 190–192.

42. Lesimple, C.; Poissonnet, A.; Hausberger, M. How to keep your horse safe? An epidemiological study about management practices. *Appl. Anim. Behav. Sci.* **2016**, *181*, 105–114. [CrossRef]

43. Marsden, M.D. Feeding practices have greater effect than housing practices on behaviour and welfare of horse. In *Proc. of the 4th Internat. Symp. on Livestock and Environment*; University of Warwick: Coventry, UK, 1993; pp. 314–318.

44. Sweeting, M.P.; Houpt, K.A. Water Consumption and Time Budgets of Stabled Pony (Equus caballus) Geldings. *Appl. Anim. Behav. Sci.* **1987**, *17*, 1–7. [CrossRef]

45. Borroni, A.; Canali, E. Behavioural Problems in Thoroughbred Horses Reared in Italy. Available online: https://www.semanticscholar.org/paper/Behavioural-problems-in-thoroughbred-horses-reared-Canali-Borroni/1a28a0adf44dbea9c9c6c94c5cad505624b4bab3 (accessed on 24 February 2020).

46. Luescher, U.A.; McKeown, D.B.; Dean, H. A cross-sectional study on compulsive behaviour (satble vices) in horses. *Equine Vet. J.* **1998**, *27*, 14–18.

47. Broom, D.; Kennedy, M.J. Stereotypies in horses: Their relevance to welfare and causation. *Equine Vet. Educ.* **1993**, *5*, 151–154. [CrossRef]

48. McGreevy, P.D.; Cripps, P.; French, N.; Green, L.; Nicol, C.L. Management factors associated with stereotypic and redirected behaviour in the thoroughbred horse. *Equine Vet. J.* **1995**, *27*, 86–95. [CrossRef]

49. Ellis, A.D.; Visser, C.K.; Van Reenen, C.G. Effect of a high concentrate versus high fibre diet on behaviour and welfare of horses. In Proceedings of the 40th International Congress of the ISAE, Bristol, UK, 8–12 August 2006.

50. Parker, M.; Goodwin, D.; Redheard, E.S. Survey of breeders' management of horses in Europe, North America and Australia: Comparison of factors associated with the development of abnormal behaviour. *Appl. Anim. Behav. Sci.* **2008**, *114*, 206–215. [CrossRef]

51. Sarrafchi, A.; Blokhuis, H.J. Equine stereotypic behaviors: Causation, occurrence, and prevention. *J. Vet. Behav.* **2013**, *8*, 386–394. [CrossRef]

52. Wickens, C.L.; McCall, C.A.; Bursian, S.; Hanson, R.; Heleski, C.R.; Liesman, J.S.; McEhenney, W.H.; Trotter, N.L. Assessment of gastric ulceration and gastrin response in horses with history of crib-biting. *J. Equine Vet. Sci.* **2013**, *33*, 739–745. [CrossRef]

53. Daniels, S.P.; Scott, L.; De Lavis, I.; Linekara, A.; Hemmings, A.J. Crib Biting and Equine Gastric Ulceration Syndrome: Do horses that display oral stereotypies have altered gastric anatomy and physiology? *J. Vet. Behav.* **2019**, *30*, 110–113. [CrossRef]

54. Fraser, D. Understanding animal welfare. *Acta Vet. Scand.* **2008**, *50*. [CrossRef] [PubMed]

55. GFFA guidelines. Guidelines for Good Animal Welfare Practice for the Keeping of Horses (Leitlinien zur Beurteilung von Pferdehaltungen unter Tierschutzgesichtspunkten). In *Expert Group Horse Husbandries in Accordance with Animal Welfare (Sachverständigengruppe Tierschutzgerechte Pferdehaltung)*, 2nd ed.; German Federal Ministry of Food and Agriculture (Bundesministerium für Ernährung und Landwirtschaft): Berlin, Germany, 2009.

56. Werhahn, I.; Hessel, E.F.; Bachhausen, I.; Van der Weghe, H.F.A. Effects of Different Bedding Materials on the Behavior of Horses Housed in Single Stalls. *J. Equine Vet. Sci.* **2010**, *30*, 425–431. [CrossRef]

57. Greening, L.; Shenton, V.; Wilcockson, K.; Swanson, J. Investigating duration of nocturnal ingestive and sleep behaviors of horses bedded on straw versus shavings. *J. Vet. Behav.* **2013**, *8*, 82–86. [CrossRef]

58. Ruet, A.; Lemarchand, J.; Parias, C.; Mach, N.; Moisan, M.-P.; Foury, A.; Briant, C.; Lansade, L. Housing Horses in Individual Boxes Is a Challenge with Regard to Welfare. *Animals* **2019**, *9*, 621. [CrossRef] [PubMed]

59. Henneke, D.R.; Potter, G.D.; Kreider, J.L.; Yeates, B.F. Relationship between condition score, physical measurements and body fat percentage in mares. *Equine Vet. J.* **1983**, *15*, 371–372. [CrossRef] [PubMed]

60. Kienzle, E.; Schramme, S. Body Condition Scoring and prediction of body weight in adult Warm blooded horses. *Pferdeheilkunde Equine Med.* **2004**, *20*, 517–524. [CrossRef]

61. Martin, P.; Bateson, P. *Measuring behaviour—An Introductory Guide*, 3rd ed.; Cambridge University Press: Cambridge, UK, 2007.

62. R Core Team. *R A Language and Environment for Statistical Computing*; R Foundation for Statistical Computing: Vienna, Austria, 2018.

63. Wackenhut, K.S. Inquiries to the housing of high performance horses with a special consideration of the guidelines for the judgement of horse-keeping with the point of view of animal protection. *Diss. Med. Vet.* **1994**.

64. Mellor, D.J.; Love, S.; Walker, R.; Gettinby, G.; Reid, S.W.J. Sentinel practice-based survey of the management and health of horses in northern Britain. *Vet. Rec.* **2001**, *149*, 417–423. [CrossRef]

65. Longland, A.C.; Ince, J.; Newbold, J.C.; Barfoot, C.; Harris, P.A. Pasture Intake in Ponies with and without a Grazing Muzzle and Over Time. Available online: https://www.horsetalk.co.nz/2016/04/14/grass-length-grazing-muzzles-horses-hacked-off/ (accessed on 24 February 2020).

66. Glunk, E.C.; Shannon, M.S.; Pratt-Phillips, E.; Siciliano, P.D. Effect of Restricted Pasture Access on Pasture Dry Matter Intake Rate, Dietary Energy Intake, and Faecal pH in Horses. *J. of Equine Vet. Sci.* **2013**, *33*, 421–426. [CrossRef]

67. Dallaire, A.; Ruckebusch, Y. Sleep and Wakefulness in Housed Pony under Different Dietry Conditions. *Canad. J. Comp. Med. Rev.* **1974**, *38*, 65–71.

68. Freire, R.; Buckley, P.; Cooper, J.J. Effects of different forms of exercise on post inhibitory rebound and unwanted behaviour in stabled horses. *Equine Vet. J.* **2009**, *41*, 487–492. [CrossRef] [PubMed]

69. Chaplin, S.J.; Gretgrix, L. Effect of housing conditions on activity and lying behaviour of horses. *Animal* **2010**, *4*, 792–795. [CrossRef]

70. Lesimple, C.; Fureix, C.; LeScolan, N.; Richard-Yris, M.A.; Hausberger, M. Housing conditions and breed are associated with emotionality and cognitive abilities in riding school horses. *Appl. Anim. Behav. Sci.* **2011**, *129*, 92–99. [CrossRef]

71. Dalla Costa, E.; Dai, F.; Lebelt, D.; Scholz, P.; Barbieri, S.; Canali, E.; Zanella, A.J.; Minero, M. Welfare assessment of horses: The AWIN approach. *Anim. Welf.* **2016**, *25*, 481–488. [CrossRef]

72. Popescu, S.; Diugan, E.A. The relationship between the welfare quality and stress index in working and breeding horses. *Res. Vet. Sci.* **2017**, *115*, 442–450. [CrossRef]

73. Popescu, S.; Lazar, E.A.; Borda, C.; Niculae, M.; Sandru, C.D.; Spinu, M. Welfare Quality of Breeding Horses Under Different Housing Conditions. *Animals* **2019**, *9*, 81. [CrossRef]

3

Garlic (*Allium Sativum*) Supplementation Improves Respiratory Health but has Increased Risk of Lower Hematologic Values in Horses

Markku Saastamoinen [1],*[ID]**, Susanna Särkijärvi** [1] **and Seppo Hyyppä** [2]

[1]   Natural Resources Institute Finland (Luke), Production Systems, Tietotie 2, 31600 Jokioinen, Finland; susanna.sarkijarvi@luke.fi

[2]   Ypäjä Equine College, Opistontie 9, 32100 Ypäjä, Finland; seppo.hyyppa@hevosopisto.fi

*   Correspondence: markku.saastamoinen@luke.fi

Simple Summary: The hypotheses of this study were that garlic supplementation may help to clear mucus in the airways, but also causes declining hematologic values in prolonged feeding. The results show that long-term supplementation of dried garlic on the level of 32 mg/kg BW improved respiratory health in terms of reduced amount of tracheal symptoms and accumulation of tracheal exudates. However, the garlic supplemented horses showed slightly declining hemoglobin (Hb), hematocrit (HcT) and red blood cells (RBC) values.

Abstract: Garlic (*Allium sativum*) is claimed to have numerous beneficial properties to the health of humans and animals. It is commonly used for example to treat respiratory diseases and infections in horses' lungs. However, in addition to its possible positive influences, garlic may also have adverse health effects. The hypotheses of this study were that garlic supplementation may help to clear mucus in the airways, but also causes declining hematologic values in prolonged feeding. To our knowledge, this is the first organized study in controlled conditions to show the health effects of garlic supplementation for horses so far. The results show that long-term supplementation of dried garlic on the level of 32 mg/kg BW seemed to reduce the amount of tracheal symptoms and accumulation of tracheal exudates. Additionally, the number of neutrophil cells in the tracheal mucus was numerically smaller in the garlic supplemented horses. However, the garlic supplemented horses showed slightly declining Hb, HcT and RBC values during an 83-day study period. Consequently, it is possible that even low garlic supplementation levels can be detrimental to the horse's hematology when the supplementation period is long.

Keywords: horse; nutrition; respiratory health; dietary supplementation; hematology values

1. Introduction

Garlic (*Allium sativum*) has been used in the diets of humans and animals for centuries because of its believed positive health effects, which contain many active components [1]. In horse nutrition and care, garlic is typically used to treat respiratory diseases and infections in their lungs, and to provide relief from the symptoms of coughs. In horse stables, the hygienic quality of the air may be poor and airborne dust concentrations may reach levels that are detrimental to horses' respiratory health [2,3]. The main sources of respirable particles and allergens causing respiratory symptoms are forages and bedding materials [4,5]. The highest dust measurements are observed in winter when the stable doors are closed [6].

The onion family (*Allium* species) is rich in an active component of organosulfur compounds [7] that are associated with the above-mentioned beneficial properties, but, on the other hand, are also reviewed to be associated with toxicosis in mammals [8]. One of those toxins is N-propyl-disulfide, that alters the enzyme glucose-6-phosphate dehydrogenase in red blood cells. This interferes with the cells' ability to prevent oxidative damage to hemoglobin [9]. Ingestion of onions can cause hemolytic anemia in horses [10]. The toxic effects and the mechanism are reviewed by Hutchison [11] and Pearson et al. [12]. There are scientific papers presenting and reviewing studies on adverse effects of garlic in humans and various animal species, i.e. horses, cattle, birds, rats and dogs [8,11,13–15].

Garlic is claimed to have also many other positive effects and is, consequently, often included as a supplemental feed to the diets of horses [7,16]. The benefits are, for example, antidiabetic and antimicrobial as well as antiparasitic properties [17–19]. Garlic is also commonly used as an insect repellent [13]. Studies on influences of garlic in horse nutrition are, however, scarce, and the dosage for beneficial effects is not known. Furthermore, there is little information on possible adverse health effects to determine the safe use of garlic for horses. Bergero and Valle [20] concluded that the traditional use of herbs is not always properly based on dosages, and moreover, safety is not automatically provided. Supplements considered safe in humans and other species are not always safe in horses. The authors of a recent study [15] suggested that the usage of garlic as a feed additive should be monitored carefully because of the detrimental effects of overdosing.

There may be a risk of anemia when certain sizes of garlic doses are fed to horses for a long period of time. Pearson et al. [12] reported that a daily dose of dried garlic over 200 mg/kg BW developed indications of Heinz body anemia, and concluded that the potential for garlic toxicosis is present when horses are chronically fed with garlic. However, they had only two supplemented horses in their study. Based on this limited data, the report of The National Academy of Sciences [21] in the U.S. gives presumed and historical safe intakes of 90 and 15 milligrams per kilogram of body weight, and concluded that the threshold level above which the risk of an adverse event will increase significantly is likely to be between 15 and 200 mg/kg BW of dried garlic, potentially depending on the health and oxidative status of the individual horse involved.

It is possible that even low supplementation levels may be detrimental when the period of supplementation is long in duration. Elghandour et al. [15] suggested that herbal supplements not tested in horses have to be evaluated to verify the possible negative side effects, followed by standardization of the dosage. Consequently, data is needed to provide more precise dosing information to obtain the beneficial effects and, on the other hand, to avoid the detrimental effects of garlic.

The aim of this study was to evaluate the possible positive influence on airway health, as well as possible adverse health effects of garlic supplementation in horses. Our hypotheses were that garlic supplementation may help to clear mucus in the airways, but may also result in decreased hematologic values in prolonged feeding. To our knowledge, there are no studies with this or a larger number of horses carried out in controlled circumstances so far.

2. Materials and Methods

2.1. Geographical Area and Climate Conditions

The experiment was conducted in the facilities of MTT Agrifood Research Finland (currently Natural Resources Institute Luke) in the south western part of Finland (latitude 60°) during early and mid-winter (November to January) climatic conditions. The average outdoor temperatures were $-1.0\,°C$ (-6–$3\,°C$), $-6.0\,°C$ (-21–$1\,°C$) and $-4.6\,°C$ (-24–$1\,°C$) in November, December and January, respectively. When compared to the long-term weather statistics of the Finnish Meteorological Institute (Climate guide.fi), the temperatures were within the long-term monthly averages.

2.2. Horses and Housing Conditions

Twelve Finnhorse mares (aged 5 to 17 years, weighing 595 to 710 kg) were housed in a box stable with an automatic ventilation system in individual stalls (3 m × 3 m) and peat as bedding. The peat was manufactured for use as bedding in horse stalls (Vapo Ltd., Jyväskylä, Finland), and was chosen to be used as bedding because of its beneficial effects on stable air quality [22]. The horses had been stabled in the same stable since they were taken indoors from pastures at the beginning of September. The stalls were manually cleaned daily between 8 and 12 a.m. when the horses were in outdoor paddocks. All feces and wet material were removed and new bedding material was added.

The experimental design was a randomized block design with repeated measurements. After the first endoscopy, the horses were formed into pairs based on matched health status and upper respiratory tract characteristics as determined by the endoscopy (symptomatic similarity), as reported in our earlier study [22]. The two horses of each pair were then allotted to an experimental group and a control group and placed in the stable so that the stable air conditions were as equal as possible for all pairs of horses (for both groups). The horses had free daily exercise in paddocks in groups (grouped by experimental groups) for four hours, and for one hour of riding or driving. The stable temperatures and humidity, as well as outdoor temperatures and weather conditions were followed and recorded daily at 8 a.m.

2.3. Experimental Feeding

The horses were individually fed three times per day (morning, noon, and evening) with a hay (871 g DM/kg) and oat (883 g DM/kg) diet supplemented with a linseed-molasses-beet pulp mixture (905 g DM/kg, Neomed Ltd, Somero, Finland) at maintenance energy level [23], with a forage-to-concentrate ratio of 80:20. The hay was produced by MTT and was artificially cured (barn drying) to ensure good hygienic and nutritional quality. Based on the chemical analysis, it was of medium nutritional quality [24], and it fulfilled the criteria of good quality dried hay (sensory evaluation of dust, smell, color carried out by experienced researchers). The oats and concentrate mixture were moistened with water to minimize the release of airborne particles from the feeds. In addition, the experimental group (one horse of each matched pair) was supplemented with 20 g of dried garlic flakes (moistened before feeding), corresponding to 0.2 % of the DM intake and 32 mg/kg BW, which is within the safety limits given by The National Academy of Sciences [21]. The supplementation continued for a total of 83 days. The diets were also balanced with a commercial mineral-vitamin mixture (Suomen Rehu Ltd., Seinäjoki, Finland). The palatability of the feeds fed to horses was good, and the horses ate all the hay, concentrates (oats, mixed feed) and the garlic supplementation offered to them, with good appetite. Thus, no feed residuals existed.

2.4. Respiratory Tract Examination and Blood Sampling

Upper respiratory tract (ethmoidal region, pharyngeal openings of guttural pouches, soft palate, larynx and trachea) examination by endoscopy was performed three times during the study (days 0, 41 and 83), and the findings were recorded (found = 1; not found = 0). Tracheobronchial aspirates were drawn at the time of the endoscopy and cytological and a bacteriological (neutrophil cells) evaluation was conducted from the tracheal mucus. The classification of the neutrophil cells in bronchoalveoral smear samples was as follows: none or some single cells (–); single cells and few small pools of cells (+); several large pools of cells (++); abundant pools of cells (+++); and an extreme abundance of cells (++++).

Blood samples from the jugular vein, as measures of health and wellbeing of the horses, were collected at the same interval as the endoscopic examinations were done. The blood analysis consisted of white blood cells (WBC), red blood cells (RBC), mean cell corpuscular volume (MCV), hematocrit (HcT) and hemoglobin (Hb) contents. The white blood cells were differentiated. All samples were analyzed in the MTT clinical laboratory. The endoscopy examination was done and all samples from the horses were collected by a veterinarian researcher.

In animal handling and sample collection, the European Union directives (1999/575/EU; 2007/526/EU) and national animal welfare and ethical legislation based on the directives above and set by the Ministry of Agriculture and Forestry of Finland were followed carefully. The experimental procedures were evaluated and approved by The Animal Care Committee of MTT Permit 43/2000 before the study was commenced.

2.5. Statistical Analysis

The information of the first endoscopy (day 0) was excluded from the data because it was included in the animal pair-variable in the model. The data (samples from horses) were analyzed with a linear model applying the MIXED procedure of the SAS system (SAS Institute Inc., Cary, NC, USA) with the following statistical model: $Y_{ijk} = \mu + p_i + b_j + (p \times b)_{ij} + t_k + (p \times t)_{jk} + (b \times t)_{jk} + e_{ijk}$, where Y_{ijk} is the observation, μ is the overall mean, p_i is the random effect of ith animal pair ($i = 1 \ldots 6$), b_j is the fixed effect of jth feeding ($j = 1 \ldots 2$), t_k is the fixed effect of the time period ($k = 2$ or 3), and e_{ijk} is the normally distributed error with a mean of 0 and variance of δ^2. The terms $(p \times b)_{ij}$, $(p \times t)_{jk}$ and $(b \times t)_{jk}$ are compound factor effects. The differences were tested with a Tukey's test. Categorical variables (neutrophil cells in tracheal mucus) and 0/1-variables were not tested statistically, but were presented descriptively, because of the small number of observations and their subjective scoring making them less informative.

3. Results and Discussion

3.1. Housing Conditions

The average temperatures and humidity of the stable air in November, December and January are presented in Table 1. These temperatures were mainly within the target indoor temperature range (8–12 °C) in Finland, corresponding also to those reported for the winter months in the same stables in a previous study dealing with stable air quality [22]. However, on three days in December and one day in January, rather low stable temperatures (3–4 °C) were observed. On those days the outdoor temperatures were at their lowest (−20 °C or below). The stable air humidity was at the lowest levels naturally during the days when the outdoor temperatures were the lowest.

Table 1. Monthly averages of the stable air temperature and humidity.

Housing Conditions	November	December	January
Temperature °C			
Mean	9.6	7.3	10.7
Range	8–11	3–10	4–12
Humidity %			
Mean	71.5	65.1	66.8
Range	58–82	46–82	43–85

3.2. Respiratory Health

Data on the neutrophil cells in the tracheal mucus (in terms of categorical variables), and the endoscopy examination findings (0/1-variables) were not tested statistically because of the small number of observations and their subjective scoring, thus being less informative. The garlic supplementation seemed to reduce the tracheal exudate score based on the endoscopy examination (Table 2). The clinical signs disappeared in three of the six horses that were given garlic supplements, while one horse remained without any signs during the study period. Clinical signs remained in two horses from the garlic supplemented group. Concerning the control horses, they remained in three horses, fluctuated in two horses and disappeared in one horse. The present results are supported by Pearson [5], who reported a significant decrease in the respiratory rate in horses supplemented with garlic.

Table 2. Numbers of horses with endoscopy examination findings in the two groups.

Experimental Group; Horse	Initial (Day 0)	Day 41	Day 83	Control Group; Horse	Initial (Day 0)	Day 41	Day 83
1	1	0	1	7	1	1	1
2	1	0	0	8	0	1	0
3	1	1	0	9	1	0	0
4	0	1	0	10	1	1	1
5	1	1	1	11	1	0	1
6	1	0	0	12	0	0	0
Total	5	3	2	Total	4	3	3

1 = findings; 0 = no findings.

The tracheobronchial aspirates obtained during the endoscopy contained either scarce or moderate numbers of neutrophils (Table 3). The number of neutrophils in the tracheal mucus was smaller among the group of garlic supplemented horses on day 83. During the study course (days 41 and 83), a large number of neutrophil cells (+++) was found only in two samples in the supplemented group, but in four samples of the six horses in the control group. The neutrophils in the tracheal mucus of the control horses remained high or increased in three individuals during the study, but decreased in two of the supplemented horses (from +++ to +), and in one of them the number increased (from ++ to +++). An elevated number of neutrophils or the detection of Curschaman's spirals is suggested to correlate with COPD symptoms [25]. However, no horses were diagnosed with COPD in the present study. Nor were any clinical signs of declined health status observed.

Table 3. Incidence of neutrophil cells in tracheal mucus in individual horses.

Experimental Group; Horse	Initial (Day 0)	Day 41	Day 83	Control Group; Horse	Initial (Day 0)	Day 41	Day 83
1	++	++	+++	7	+	+++	++
2	–	(+)	–	8	–	–	++
3	+++	++	+	9	++	–	(+)
4	(+)	–	–	10	+++	+++	+++
5	+	+++	+	11	+	+	+++
6	–	–	(+)	12	–	(+)	–

– = none or some single cells; + = single cells and few small pools of cells; ++ = several large pools of cells; +++ = abundant pools of cells.

The differences between the groups were small, and because of the small data set and subjective evaluation of the endoscopy findings and incidence of the neutrophil cells, drawing reliable conclusions is not possible. In addition, we used bedding material (peat) low in respirable particles and hay with high hygienic quality. In practical conditions, there are several factors affecting the respiratory health of horses [26]. It is most important to use alternative bedding materials and feeds of high hygienic quality which are free from dust (molds, respirable particles) to reduce airborne dust and aeroallergens in stables [5,22].

At the beginning of the trial, the horses had spent about two months (September–October) in the stable with outdoor exercise sessions as described earlier. The first endoscopic examination at the beginning of the experiment revealed that 9 out of the 12 horses had signs of respiratory symptoms. Thus, moving the horses from the pasture into indoor housing at the beginning of September appeared to expose the horses to respiratory disease because of the air quality in the stable. This is supported by Elfman et al. [6] who found (in Swedish weather conditions comparable to those in Finland) that dust and airborne bacteria levels increase in September compared to other seasons in their study. Also, our previous study [22] reported that respiratory symptoms increased during the study period (October – December) in quite similar environmental circumstances, but that peat as a bedding material might have positive effects on respiratory health because of its beneficial influence on the air quality of the stable.

3.3. Blood Analysis

There were no statistically significant differences in the blood parameters between the groups (Table 4). However, the garlic supplemented horses showed a slight declining trend in Hb, HcT and RBC values during the study course, but the control horses remain higher values. The means of days 41 and 83 as well as the final Hb, HcT and RBC values on the day 83 in the supplemented horses were numerically smaller compared to the control horses, and the mean final Hb value of the garlic supplemented horses was at the lowest limit of the normal range for Finnhorses [27,28], or even below it. Regarding the results of individual horses (Figure 1), the Hb and RBC values on both days 41 and 83 were lower than the initial value in four of the six garlic supplemented horses, and five of them showed lower values also on day 83 compared with the initial value. These findings may indicate slight anemia in the garlic supplemented horses.

Table 4. Hematology (mean, (s.d)) of the groups during the study period.

Hematology Parameter	Experimental Group			Control Group		
	Initial (Day 0)	Day 41	Day 83	Initial (Day 0)	Day 41	Day 83
Hb, g/L	130.5 (11.5)	128.3 (7.6)	118.0 (7.0)	130.7 (11.2)	135.5 (17.7)	127.3 (15.3)
HcT, %	36.9 (0.0)	35.4 (0.0)	32.3 (0.0)	36.4 (0.0)	37.2 (0.0)	34.9 (0.0)
RBC, $\times 10^{12}$/L	7.17 (0.7)	7.15 (0.6)	6.57 (0.6)	7.29 (0.6)	7.46 (1.0)	7.04 (0.8)
MCV, fl	50.0 (1.3)	49.5 (1.1)	49.2 (1.1)	50.1 (2.2)	50.0 (1.6)	49.7 (1.4)
WBC, $\times 10^9$/L	8.15 (1.3)	7.52 (0.7)	6.97 (0.7)	7.76 (1.3)	6.95 (0.9)	7.52 (1.3)

Hb = hemoglobin; HcT = hematocrit; RBC = red blood cells; MCV = mean cell corpuscular volume; WBC = white blood cells.

The other parameters were within the ranges reported for healthy Finnhorses [27–29], but the WBC value declined numerically in the garlic supplemented horses. It is not possible to draw any conclusion from this, but WBCs increase in the case of inflammatory diseases [30]. Thus, this result supports the speculation on the beneficial respiratory health impacts of garlic reported above. No differences in the differentiated white blood cells existed (data not shown).

Studies on the effects of garlic supplementation on blood chemistry in controlled environments are scarce. The supplemented amount of dried garlic in the present study (32 mg/kg BW) is within recommended limits, however it is clearly close to the lower limit (15 and 200 mg/kg BW of dried garlic) given by The National Academy of Sciences [21].

Pearson et al. [12] showed Heinz body anemia due to the garlic supplementation. Because Heinz bodies and bilirubin were not analyzed in our study, it is not possible to make any conclusions regarding the type of anemia. The above-mentioned authors [12] found that recovery from anemia was largely complete five weeks after the removal of garlic supplementation. The supplementation period in their study was 71 days, but they used only four horses (two supplemented with garlic). They concluded that the toxic effect in their study was caused, at least, by oxidative damage of RBCs (RBC was numerically lower for the garlic supplemented horses in the present study as well). Garlic extracts have also been reported to cause Heinz body anemia in dogs [31] and oxidation of RBC in sheep [32]. In horses urticaria associated with dry garlic feeding has also been reported [33]. Furthermore, the feeding of other *Allium* species (e.g., *A. cepa*) has been reported to result in decreased Hb and HcT levels in some other animal species (pigs, dogs, goats) [34–36]. Pierce et al. [37] reported anemia caused by wild onion poisoning in horses.

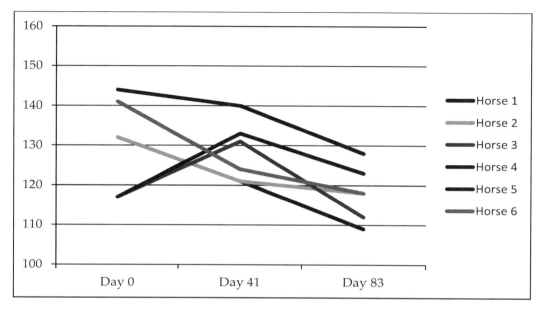

(**a.1**) Hemoglobin (g/l) in the garlic supplemented horses.

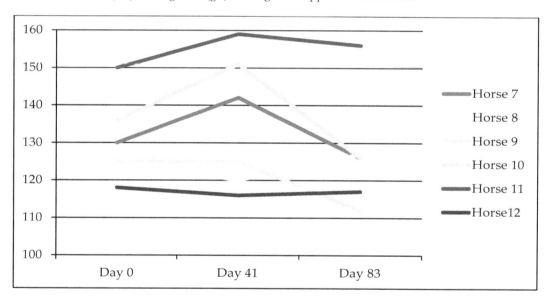

(**a.2**) Hemoglobin (g/l) in the control horses.

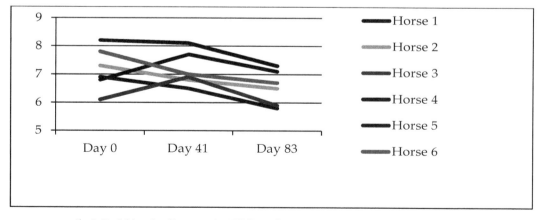

(**b.1**) Red blood cell count (× 10^{12}/l) in the garlic supplemented horses.

Figure 1. *Cont.*

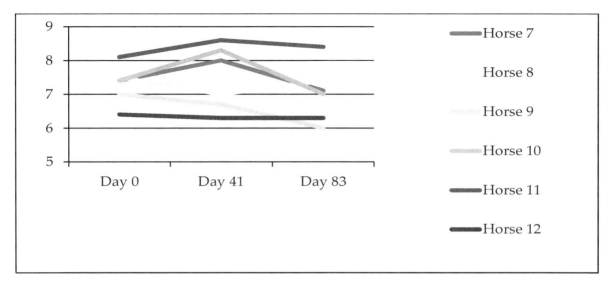

(**b.2**) Red blood cell count (× 10^{12}/l) in the control horses.

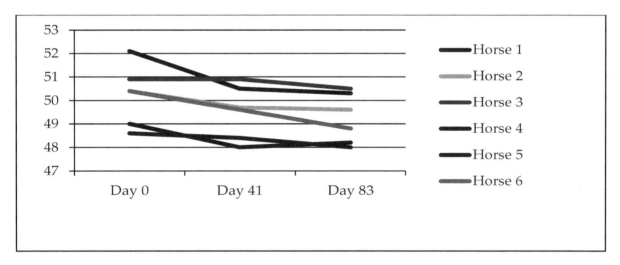

(**c.1**) Mean cell corpuscular volume (fl) in the garlic supplemented horses.

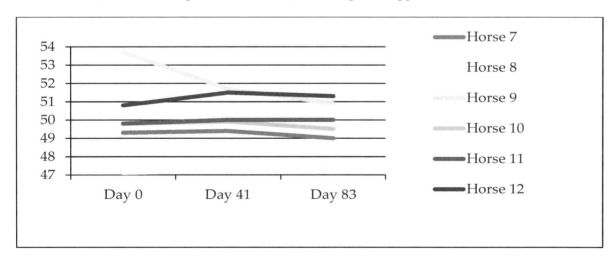

(**c.2**) Mean cell corpuscular volume (fl) in the control horses.

Figure 1. Hematology values of the horses in the garlic supplemented (**a.1–c.1**) and control (**a.2–c.2**) horses.

The decrease in Hb and other hematology values is more critical to oxidatively stressed hard-working horses than horses in light work [30]. The safe limit of garlic supplementation potentially depends on factors such as differences between individual horses, and the health status and exercise level of the horse. Low intake levels (15 mg/kg BW dried garlic) are unlikely to result in a risk of adverse effects in healthy, non-exercising, non-oxidatively stressed adult horses [21]. However, Bergero and Valle [20] pointed out that the form of garlic supplementation (dry, fresh, garlic oil, extract) may contain different substances with different biological effects.

4. Conclusions

To our knowledge, this is the first organized study in controlled conditions to show possible positive and negative health effects of garlic supplementation in horses. Although this study shows that dried garlic may help to remove tracheal mucus, it also points out that there may be a risk of adverse effects on hemoglobin levels and red blood cell amount if fed with garlic for long periods of time. The supplementation level (32 mg/kg BW) of dried garlic fed to horses seemed to reduce the tracheal symptoms and accumulation of tracheal exudates, but may also cause decreased hematologic values when fed continuously for a nearly three-month period. However, the small data set and subjective evaluation applied in the present study mean that the results need to be considered as preliminary results. Consequently, further research is needed to identify safe garlic doses and supplementation duration for horses, as well as to examine the positive and preventive health effects.

Author Contributions: Designing the experiment, M.S. and S.S.; methodology, investigation, S.S., S.H. and M.S.; data analysis, S.S.; writing—original draft preparation, M.S.

References

1.   Rahman, M.S. Allicin and other functional active components in garlic: Health benefits and bioavailability. *Int. J. Food Prop.* **2007**, *10*, 245–268. [CrossRef]
2.   McGorum, B.C.; Ellison, J.; Cullen, R.T. Total and respirable airborne dust endotoxin concentrations in three equine management systems. *Equine Vet. J.* **1998**, *30*, 430–434. [CrossRef] [PubMed]
3.   Berndt, A.; Derksen, F.J.; Robinson, N.E. Endotoxin concentrations within the breathing zone of horses are higher in stables than on pasture. *Vet. J.* **2010**, *183*, 54–57. [CrossRef] [PubMed]
4.   Raymond, S.L.; Curtis, E.F.; Winfield, L.M.; Clarke, A.F. A comparison of respirable particles associated with various forage products for horses. *Equine Pract.* **1997**, *19*, 23–26.
5.   Vandeput, S.; Istasse, L.; Nicks, B.; Lekeux, P. Airborne dust and aeroallergen concentrations in different sources of feed and bedding for horses. *Vet. Q.* **1997**, *19*, 154–158. [CrossRef] [PubMed]
6.   Elfman, L.; Wålinder, R.; Riihimäki, M.; Pringle, J. Air quality in horse stables. In *Chemistry, Emission Control, Radioactive Pollution and Indoor Air Quality*; Mazzeo, D., Ed.; Intech: Rijeka, Croatia, 2011; pp. 655–680.
7.   Munday, R.; Munday, C.M. Relative activities of organosulfur compounds derived from onions and garlic in increasing tissue activities of quinone reductase and glutathione transferase in rat tissues. *Nutr. Cancer* **2001**, *40*, 205–210. [CrossRef] [PubMed]
8.   Wade, L.L.; Newman, S.J. *Hemoglobinuric nephrosis* and *Hepastoplenic eryrhophagocytosis* in a dusky-headed conure (*Aratinga weddeli*) after ingestion of garlic (*Allium sativum*). *J. Avian Med. Surg.* **2004**, *18*, 155–161. [CrossRef]
9.   Ruwende, C.; Hill, A. Glucose-6-phosphate dehydrogenase deficiency and malaria. *J. Mol. Med.* **1998**, *76*, 581–588. [CrossRef]
10.   Harvey, J.W.; Racker, D. Experimental onion-induced hemolytic anemia in dogs. *Vet. Pathol.* **1985**, *22*, 387–392. [CrossRef]
11.   Hutchison, T.W.S. Onions as a cause of Heinz body anaemia and death in cattle. *Can. Vet. J.* **1977**, *18*, 358–360.

12. Pearson, W.; Boermans, H.J.; Bettler, W.J.; McBride, B.W.; Lindinger, M.I. Association of maximum voluntary dietary intake of freeze-dried garlic with Heinz body anemia in horses. *Am. J. Vet. Res.* **2005**, *66*, 457–465. [CrossRef] [PubMed]

13. Williams, C.A.; Lamprecht, E.D. Some commonly fed herbs and other functional foods in equine nutrition: A review. *Vet. J.* **2008**, *178*, 21–31. [CrossRef] [PubMed]

14. Borrelli, F.; Caspasso, R.; Izzo, A.A. Garlic (*Allium sativum* L.): Adverse effects and drug interactions in humans. *Mol. Nutr. Food Res.* **2007**, *51*, 1386–1397. [CrossRef] [PubMed]

15. Elghandour, M.M.Y.; Reddy, P.R.; Salem, A.Z.M.; Reddy, P.P.R.; Hyder, I.; Barbabosa-Pliego, A.; Yasawini, D. Plant bioactives and extracts as feed additives in Horse nutrition. *J. Equine Vet. Sci.* **2018**, *69*, 66–77. [CrossRef]

16. Pearson, W. Ethnoveterinary medicine: The Science of botanicals in equine health and disease. In Proceedings of the Second Annual European Equine Health and Nutrition Congress, Lelystad, The Netherlands, 19–20 March 2003; pp. 31–40.

17. Kook, S.; Gun-Hee, K.; Choi, K. The antidiabetic effect of onion and garlic in experimental diabetic rats: Meta-analysis. *J. Medic. Food* **2009**, *12*, 552–560. [CrossRef] [PubMed]

18. Soffar, S.A.; Mokhtar, G.M. Evaluation of the antiparasitic effect of aqueous garlic (*Allium sativum*) extract in *Hymenolepiasis nana* and *Giardia*. *J. Egypt. Soc. Parasitol.* **1991**, *21*, 497–502. [PubMed]

19. Cellini, L.; Di Campli, E.; Masuli, M. Inhibition of *Helicobacter pyroli* by garlic extract (*Allium sativum*) FEMS. *Immunol. Med. Microbiol.* **1996**, *13*, 273–277. [CrossRef] [PubMed]

20. Bergero, D.; Valle, E. A critical analysis on the use of herbs and herbal extracts in feeding sport horses. *Pferdeheilkunde* **2006**, *22*, 550–557. [CrossRef]

21. The National Academies Report in Brief. *Safety of Dietary Supplements for Horses, Dogs and Cats*; The National Academy of Sciences; The National Academies Press: Washington, DC, USA, 2008; 260p.

22. Saastamoinen, M.; Särkijärvi, S.; Hyyppä, S. Reducing respiratory health risks to horses and workers: A comparison of two stall bedding materials. *Animals* **2015**, *5*, 967–977. [CrossRef]

23. Luke. Feed Tables and Nutrition Recommendations. Natural Resources Institute Finland. 2018. Available online: https://portal.mtt.fi/portal/page/portal/Rehutaulukot/feed_tables_englishorhttp://urn.fi/URN: ISBN:978-952-326-054-2 (accessed on 15 October 2018).

24. Saastamoinen, M.; Hellämäki, M. Forage analysis as a basis of feeding of horses. In *Forages and Grazing in Horse Nutrition*; Saastamoinen, M., Fradinho, M.J., Santos, A.S., Miraglia, N., Eds.; EAAP Publication 132; Wageningen Academic Publishers: Wageningen, The Netherlands, 2012; pp. 304–314.

25. Clarke, E.G.C.; Clarke, M.L. *Garner's Veterinary Toxicology*, 3rd ed.; Ballieri & Tindal: London, UK, 1967.

26. Riihimäki, M.; Raine, A.; Elfman, L.; Pringle, J. Markers of respiratory inflammation in horses in relation to seasonal changes in air quality in a conventional racing stable. *Can. J. Vet. Res.* **2008**, *72*, 432–439.

27. Pösö, A.R.; Soveri, T.; Oksanen, H.E. The effect of exercise on blood parameters in Standardbred and Finnish-bred horses. *Acta Vet. Scand.* **1983**, *24*, 170–184. [PubMed]

28. Movet. Laboratory Handbook. Available online: www.movet.fi (accessed on 15 October 2018). (In Finnish)

29. Saastamoinen, M.T. Propionic acid treated grain (oats) in the diet of horses. *Agric. Sci. Finl* **1994**, *3*, 161–168. [CrossRef]

30. Lindner, A. *Laboratory Diagnosis for Sport Horses*; Wageningen Academic Publishers: Wageningen, The Netherlands, 1998; 64p.

31. Hu, Q.; Yang, Q.; Yamoto, O.; Yamasiki, M.; Meade, Y.; Yoshihara, T. Isolation and identification of organosulfur compounds oxidizing canine erythrocytes from garlic (*Allium sativum*). *J. Agric. Food. Chem.* **2002**, *50*, 1059–1062. [CrossRef] [PubMed]

32. Stevens, H. Suspected wild garlic poisoning in sheep. *Vet. Rec.* **1984**, *115*, 363. [CrossRef] [PubMed]

33. Miyazawa, K.; Ito, M.; Ohsaki, K. An equine case of urticaria associated with dry garlic feeding. *J. Vet. Med. Sci.* **1991**, *53*, 747–748. [CrossRef] [PubMed]

34. Ostrowska, E.; Nicholas, K.G.; Sterling, S.J.; Brendan, G.T.; Rodney, B.J.; Earling, D.R.; Jois, M.; Dunshea, F.R. Consumption of brown onions (*Allium cepa* var. *cavalier* and var. *density*) moderately modulates blood lipids, haematological and haemostatic variables in healthy pigs. *Br. J. Nutr.* **2004**, *91*, 211–218.

35. Ogawa, E.; Shinoki, T.; Akahori, F.; Masaka, T. Effect of onion ingestion on anti-oxidizing agents in dog erythrocytes. *Jpn. J. Vet. Sci.* **1986**, *48*, 685–691. [CrossRef]

36.   Heidarpour, M.; Fakrieh, M.; Aslani, M.R.; Mohri, M.; Keywanloo, M. Oxidative effects of long-term onion
       (*Allium cepa*) feeding on goat erythrocytes. *Comp. Clin. Pathol.* **2011**, *22*, 195–202. [CrossRef]
37.   Pierce, K.R.; Joyce, J.R.; England, R.B.; Jones, P. Acute hemolytic anemia caused by wild onion poisoning in
       horses. *Am. Vet. Med. Ass. J.* **1972**, *160*, 323–327.

4

Evaluation of the Accuracy of Horse Body Weight Estimation Methods

Wanda Górniak [1],*[ID], Martyna Wieliczko [1], Maria Soroko [2] and Mariusz Korczyński [1][ID]

[1] Department of Environment Hygiene and Animal Welfare, Wroclaw University of Environmental and Life Sciences, Chelmonskiego 38C, 51-630 Wroclaw, Poland; mbarabasc@gmail.com (M.W.); mariusz.korczynski@upwr.edu.pl (M.K.)

[2] Institute of Animal Breeding, Wroclaw University of Environmental and Life Sciences, Chelmonskiego 38C, 51-630 Wroclaw, Poland; maria.soroko@upwr.edu.pl

* Correspondence: wanda.gorniak@upwr.edu.pl

Simple Summary: Horse body weight estimation and monitoring of the weight variations are necessary to determine the amount of feed and feed additives for the proper functioning of the animal. Due to the cost and practical challenge of weighing horses on a large scale, several alternative methods for estimating the body weight of horses have been developed. One of them is to determine a horse's body weight using a formula. The aim of the study was to evaluate established formulae for estimating horse body weight from data gathered using measurement tape. The investigation was conducted on a group of 299 adult horses of five breeds: ponies, Polish Noble Half Breed, Silesian Breed, Wielkopolski Breed and Thoroughbred. For each horse, body measurements were performed and the actual body weight of the horses was measured with an electronic scales. The horse's body weight measurements were compared with the result of seven different formulae. It was found that the use of formulae for body weight estimation can be useful in determining feed dosages and additives, medicines, or deworming agents.

Abstract: Methods of estimating horse body weight using mathematical formulae have better accuracy than methods of reading body weight from measuring tape. The aim of the study was to evaluate established formulae for estimating horse body weight from data gathered using measurement tape. The research was conducted in a group of 299 adult horses and ponies of selected breeds: ponies ($n = 58$), Polish Noble Half Breed ($n = 150$), Silesian Breed ($n = 23$), Wielkopolski Breed ($n = 52$), and Thoroughbred ($n = 16$). Body measurements were performed on each horse using a measuring stick and tape. The actual body weight of the horses was measured with electronic scale. Statistical analysis was carried out separately for individual breeds of horses. In each of the research groups formulae were selected, the results of which were closest to the actual horse body weight readings. The use of formulae for body weight estimation can be useful in determining feed dosages and additives, medicines or deworming agents. Regular weight measurement is important for maintaining a healthy horse.

Keywords: horses; breed; body weight estimation; estimation formula

1. Introduction

Body weight (BW) measurement and its regular recording is important in assessing the health of a horse. In young horses, regular weight measurements are very important because they provide essential information on the proper development of the animal. Weight estimation and monitoring of changes are necessary to determine the amount of feed and feed additives for the proper functioning of the animal. Measurements also provide basic information for calculating the correct dosage of medication when treatment is required, or when a deworming preparation is administered. Excessive or inadequate dosing of these agents may result in treatment failure or drug resistance [1–3]. A sudden change in body weight is also an important indicator of a change in health [4,5], and an increase in the weight of grazing horses in spring is closely related to the occurrence of laminitis [6,7]. Regular weight measurement and adjustment of the feeding plan based on this information are therefore very important for the maintenance of a healthy horse.

Due to the cost and practical challenge of weighing horses on a large scale, several alternative methods for estimating the body weight of horses have been developed. One of them is to determine a horse's body weight using a formula that includes chest circumference and body length measurements. Previous research has focused on developing the most accurate equation model using this method [8–11]. In some studies, half the circumference of the chest (from the withers to the mid-abdominal line) was measured, and then the doubled value was used in the formula instead of measuring the entire circumference of the trunk. This minimizes the measurement error that may sometimes occur when the measuring tape is bent on the opposite side of the horse from the person performing the measurement [12]. In the studies conducted by Marcenac and Aublet [13], Ensminger [14], Carroll and Huntington [8], Jones et al. [10], and Martinson et al. [15], it was found that methods of estimating horse body weight using mathematical formulae have better accuracy than methods of reading BW from measuring tape, whose estimation of body weight compared to the actual weight may differ significantly [12,16]. The aim of the study was to evaluate established formulae for estimating horse body weight from data gathered using measurement tape.

Those models are developed empirically, and as such, their correspondence to the actual value of horse body weight is limited. Moreover, different breeds of horses present different body postures, which increases the possibility of errors. In consequence, the models are likely to have various reliability for different breeds. In the literature, little is known about the accuracy of the formulae, with respect to different breeds of horses. Therefore, there is a need to evaluate the error associated with employing the formulae for a different breed. The results of horse body weight, calculated by means of the formulae, were compared with the true mass of the horse measured on a weighing scale.

2. Materials and Methods

All horses qualified for the research were subjected to standard procedures without any harm or discomfort and therefore did not require the consent of the Local Ethical Commission for Animal Experiments at the Institute of Immunology and Experimental Therapy of the Polish Academy of Sciences in Wroclaw, Poland (Act of 15 January 2015 on protection animals used for scientific or educational purposes). Consent from all horse owners was obtained prior to the investigations.

2.1. Animals

The investigation was conducted on a group of 299 adult horses and ponies kept in five Horse Studs: Pepowo, Bonza, Jeziorki Osieczna, Klodzka Roza and Leka Mroczenska in Poland between July–August 2019. Five groups were specified: ponies ($n = 58$), Polish Noble Half Breed ($n = 150$), Silesian Breed ($n = 23$), Wielkopolski Breed ($n = 52$) and Thoroughbred ($n = 16$). Data were collected from horses that fulfilled the following conditions: age ≥ 3 years, BCS (Body Condition Score) of 5.0–5.5 [17,18] with 106 geldings, 155 nonpregnant mares, and 38 stallions represented. The BCS included six areas of the horse's body (neck, withers, back, tail, ribs and back) to classify body condition on a scale of 1 (weak/poor) to 9 (extremely fat). Each area of the body was scored separately and scores were averaged to represent overall body condition. All of the horses were barefoot on all of the hooves. Included animals were housed in stables overnight and during the day were turned out on a pasture, having access to hay and water ad libitum. Currently, bred horses of the Polish Noble Half Breed and Wielkopolski Breed are maintained in Poland as a type of sports horse in the disciplines of dressage and jumping, very often as a result of mating with German sports horses of the Hanoverian, Holsteiner, or Trakehner Breeds. Horses of the Silesian Breed are characterized by a massive, harmonious conformation and long and strong, muscular neck. Horses of this breed are used mainly in carriage driving. The studied group of ponies was used in sport in the disciplines of dressage and jumping. The height at the withers of the pony was in the range of 100–146 cm.

2.2. Data Collection

Measurements were collected in the stable corridor on a concrete floor with an even surface. Plastic measurement tape (Zoometric Tape, Hauptner, Dietlikon-Zürich, Switzerland) with maximum measurable length 250 cm, and a measuring stick (Aluminum Horse Height Measuring Stick, Busse, Lohne, Germany) with a range 100–180 cm, were used. Trained staff performed single measurements of the horse body using the methodology described in original methodology, i.e., girth circumference at the base of the mane hairs [15], girth circumference measured over the highest point of the withers [8,14], the circumference around the abdomen at the point of the umbilicus [10], the body length from the point of buttock (tuber ischium) to the point of shoulder (head of humerus) [8,15], the length from the point of buttock to elbow (olecranon) [10,14], height at the withers (height at the third thoracic vertebra), the circumference of the neck located halfway between the poll and withers [15]. All measurements were taken by the same two persons, while a colleague held the horse. The horse limbs were always positioned parallel to each other to minimize measurement error. The true bodyweight of the horses was measured using a portable electronic scale Rhewa 82 Alpha (Rhewa Waagen, Mettmann, Germany) with a weighting platform of maximum load capacity of 1000 kg and a stated accuracy 0.5 kg.

2.3. Data Analysis

Statistical analysis was carried out with the use of Statistica software (v. 13.3, StatSoft Inc., Tulsa, OK, USA). The analyses were made separately for individual breeds. The investigation would only be valid if the formulae presented here gave adequate results. Therefore, in order to verify the utility of the formulae, the chi-square test and Shapiro Wilk and Kolmogorov Smirnov tests were performed on the residuals. The selection of the statistical procedure was dependent on the number of horses in the group. For data samples, fewer than 30, the Shapiro Wilk and Kolmogorov Smirnov test were performed. For greater populations, these tests do not provide accurate and reliable results. Hence, the chi-square test was performed. In both cases, the residuals were subjected to the statistical evaluation where two hypotheses were established—i.e., residuals originate from a Gaussian distribution, or alternatively, the residuals do not originate from a Gaussian distribution. The predefined formula was considered as the model, and the measured value of horse mass determined with the aid of the scale was considered as an observed value. It appeared that in the case of each considered formula there was no basis to reject the null hypothesis. Hence, the residuals could be considered to originate from the Gaussian distribution in all cases. In this sense, all the formulae could be considered for further deliberation.

The results of horse body weight measurements were compared with the formulae in Table 1. For this purpose, the differences between the observed value (i.e., true horse body weight) and the estimated value (i.e., calculated from the formulae) were determined. Additionally, the correctness of the formulae was verified by means of the root mean square error (RMSE).

Table 1. Formulas for estimating horse body weight used in the experiment.

Reference	Application	Formula
Marcenac and Aublet [13]	adult horses	$=G \ (m)^3 \times 80$
Ensminger [14]	adult horses	$=[(G \ (in)^2 \times L2 \ (in)) + 22.7]/660$
Carroll and Huntington [8]	adult horses	$=(G \ (cm)^2 \times L \ (cm))/11877$
Jones et al. [10]	>2 year, 230 to 707 kg	$=(G2 \ (cm)^{1.78} \times L2 \ (cm)^{0.97})/3011$
Martinson et al. [15]	Arabian type horses	$=(G \ (cm)^{1.486} \times L \ (cm)^{0.554} \times H \ (cm)^{0.599} \times N \ (cm)^{0.173})/3596$
Martinson et al. [15]	ponies	$=(G \ (cm)^{1.486} \times L \ (cm)^{0.554} \times H \ (cm)^{0.599} \times N \ (cm)^{0.173})/3606$
Martinson et al. [15]	stock horses	$=(G \ (cm)^{1.486} \times L \ (cm)^{0.554} \times H \ (cm)^{0.599} \times N \ (cm)^{0.173})/3441$

G—girth circumference; G2—abdominal circumference on the navel; L—body length from shoulder to ischium; L2—length from elbow to ischium; H—height at withers; N—neck circumference.

3. Results

On the basis of the data obtained from the conducted measurements, the bodyweight of horses was estimated according to the individual formulae (Table 1). The selected formulae were applied to all breeds of horses, and not only to those groups for which the formulae were originally developed. The average value of mass calculated using all of the formulae, as well as the measured value of mass, along with the standard deviation, is shown in Table 2. Average differences between the actual body weight and the calculated weight, depending on the breed, are shown on individual graphs (Figures 1–5).

Table 2. Average mass of horses calculated with the aid of formulae.

Formula	Application	Ponies $n = 58$	Polish Noble Half Breed $n = 150$	Silesian Breed $n = 23$	Wielkopolski Breed $n = 52$	Thoroughbred $n = 16$
		$\bar{x}[kg]$ $\pm SD$	$\bar{x}[kg]$ $\pm SD$	$\bar{x}[kg]$ $\pm SD$	$\bar{x}[kg]$ $\pm SD$	$\bar{x}[kg]$ $\pm SD$
Marcenac and Aublet [13]	adult horses	322 106	566 76	584 112	566 65	523 39
Ensminger [14]	adult horses	264 86	480 58	499 96	487 56	441 37
Carroll and Huntington [8]	adult horses	291 98	529 65	549 103	538 81	489 40
Jones et al. [10]	>2 year, 230 to 707 kg	313 90	511 67	534 91	499 58	469 41
Martinson et al. [15]	Arabian type horses	312 ⊥ 97	558 ⊥ 57	571 ± 89	567 ± 64	514 ± 34
Martinson et al. [15]	ponies	311 ± 97	556 ± 37	569 ± 89	565 ± 64	513 ± 34
Martinson et al. [15]	stock horses	326 ± 02	583 ± 60	596 ± 93	592 ± 67	537 ± 35
Measured weight	N/A	306 ± 99	561 ± 63	588 ± 96	567 ± 60	501 ± 26

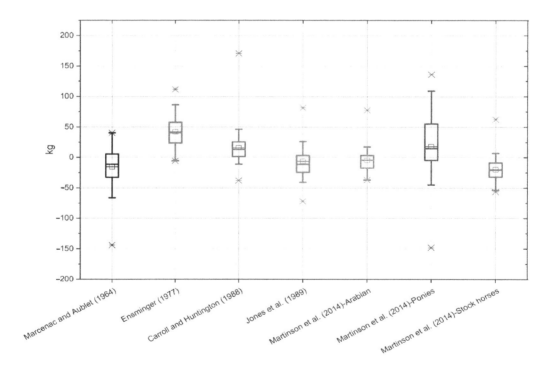

Figure 1. The box whiskers plot of the differences between real and estimated body weight for ponies.

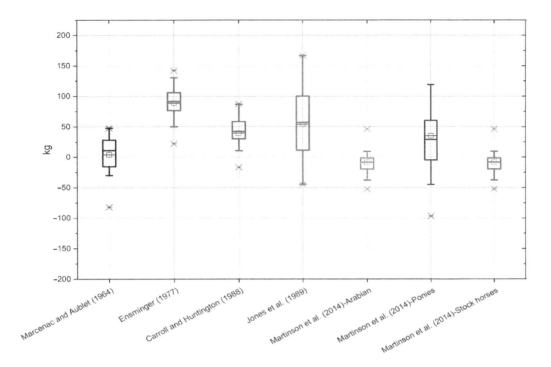

Figure 2. The box whiskers plot of the differences between real and estimated body weight for Silesian Breed.

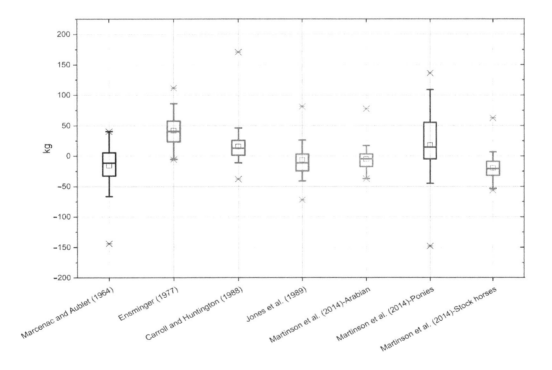

Figure 3. The box whiskers plot of the differences between real and estimated body weight for Polish Noble Half Breed.

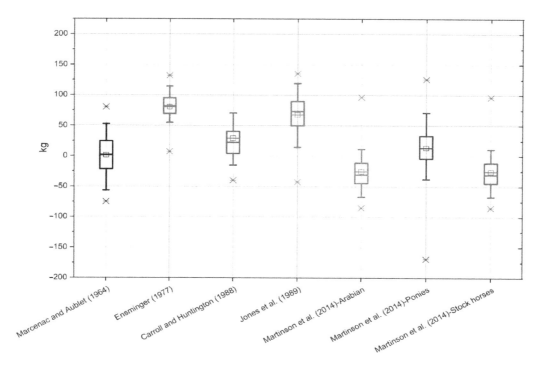

Figure 4. The box whiskers plot of the differences between real and estimated body weight for Wielkopolski Breed.

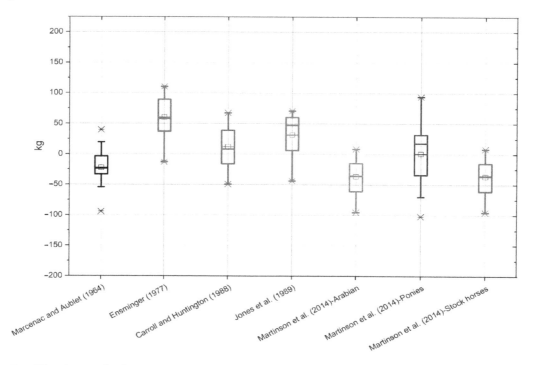

Figure 5. The box whiskers plot of the differences between real and estimated body weight for Thoroughbred.

In the group of ponies (Figure 1) the most accurate formula was Martinson et al. [15] for Arabian type horses, and the least precise were Ensminger [14] and Martinson et al. [15] for ponies.

In the case of horses of Silesian Breed (Figure 2) the results of Martinson et al. [15] for Arabian type horses and stock horses were closest to the true body weight. The least accurate were Ensminger [14] and Jones et al. [10]. For Polish Noble Half Breed (Figure 3) the most accurate was Martinson et al. [15] for Arabian type horses, whereas the furthest from the true body weight were the results from Ensminger [14]. In the group of Wielkopolski Breed horses (Figure 4) the Marcenac and Aublet [13] formula proved to be the most accurate, and the least accurate were the Ensminger [14] and Jones et al. [10] formulae. The most accurate formula in the group of Thoroughbreds (Figure 5) was the Carroll and Huntington formula [8], whereas the least accurate was the formula from Ensminger [14].

The analysis of formula matching on the basis of RMSE (Figure 6) indicated that the smallest matching error exists when the bodyweight of ponies, Silesian Breed and Polish Noble Half Breed was calculated using the formula from Martinson et al. [15] designated for Arabian type horse and stock horses (Table 1). For Thoroughbreds, the smallest RMSE error was for the Caroll and Huntington [8] formula, and for horses of the Wielkopolski Breed it was the Marcenac and Aublet [13] formula.

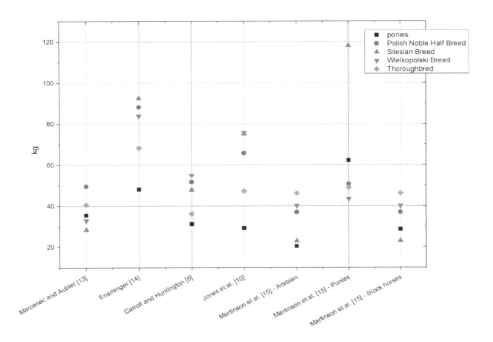

Figure 6. Root mean square error for individual horse groups and formulas.

4. Discussion

Research conducted on weight estimation methods by Ellis and Hollands [9] has shown that none of the methods employed to date can be recommended for the examined horse population. The research was conducted on a group of 600 horses of different breeds and ages in the UK. Bodyweight estimation from two measuring tapes, the Carroll and Huntington [8] formula, and visual bodyweight evaluation were subjected to comparison. The most accurate of the methods studied proved to be the estimation formula developed by Carroll and Huntington [8]; however, body weights significantly differed from the actual weights ($p < 0.001$). In previous studies, it was found that visual estimation of horse bodyweight was unreliable [19]. The research presented by Carroll and Huntington [8] was conducted mainly on Thoroughbreds and ponies, and suggested that the estimation formula was 90% accurate. In the experiment carried out on Hucul horses, it was shown that the most reliable procedure to estimate body weight was to use the Carroll and Huntington formula [8], which underestimated the actual body weight by an average of 7 kg, and the error of this method was 4.5% [20]. Our results show that the Carroll and Huntington [8] formula was closest to the actual body weight for Thoroughbreds.

Gharahveysi [21] conducted a study on 244 Iranian Arabian horses, demonstrating that there were no significant differences ($p > 0.05$) between the Ensminger [14] and Marcenac and Aublet [13] body weight estimation methods and a true weight, while the Jones et al. [10] formula showed a highly significant difference ($p < 0.01$). In our present study, it was found that the Marcenac and Aublet [13] formula was marked by a small relative error in Wielkopolski Breed and Silesian Breed groups. The Jones et al. [10] formula did not prove to be effective in any group.

Martinson et al. [15] examined 629 adult horses and ponies in order to develop an ideal weight estimation formula. The height at the withers and neck circumference were added to the estimation model, and all measurements could be changed by matching the additive model to the logarithmic scale of all variables. The equation obtained for body mass estimation manifested a very high correlation ($r^2 = 0.92$) with true horse weight. In the present research, this formula was the most accurate method for determining bodyweight in the group of ponies, Silesian Breed, and Polish Noble Half Breed. The high accuracy of calculations based on this formula may result from taking into account not only the length of the diagonal torso and chest circumference, but also the height at the withers and the circumference of the neck. The accumulation of fat deposits increases the circumference of the thorax, and primarily the circumference of the neck [22,23]. Catalano et al. [24] also stated that adding height at the withers and neck circumference significantly improved the accuracy of the equations for estimating BW draft and warmblood horses, especially for overweight or underweight horses. In another study, they examined Miniature, saddle-type, and Thoroughbred horses, and indicated that the adding breed type, height, neck circumference, body length, and girth circumference improve BW estimation [25]. Jensen et al. [26] assessed the accuracy of various formulas for estimating bodyweight of Icelandic and warmblood horses, and also assessed the relationship between the variables for cresty neck score, BCS, and plasma concentrations of insulin. Overall, the concordance correlation coefficient was high for most formulas, but complex using at least four morphometric measurements were more accurate. Plasma insulin levels were higher ($p < 0.001$) in Icelandic horses than in warmblood horses, which was reflected in higher body fat, suggesting differences in body condition score. In our investigation, ponies and Silesian Breed horses are highly predisposed to storing fat in their neck, similar to Icelandic horses, so formulas with more measurements were more accurate.

For a more accurate estimate, the body fat and muscularity of the horse should be considered. Horse weight can vary greatly depending on the quality of work performed; horses in sport training should, in principle, be heavier due to greater muscle development. One of such equations for estimating horse weight, which includes more measurements (i.e., cannon bone circumference, neck circumference at the base) and BCS (points) is the formula developed by Kienzle and Schramme [18]. Due to ongoing breeding advances and the mixing of different horse breeds, the phenotype of horses changes. The formulae for estimating body weight may consequently need to be readjusted accordingly.

5. Conclusions

The Martinson et al. [15] formula recommended for Arabian type horses was the most accurate in estimating body weight in the group of ponies, Silesian Breed, and Polish Noble Half Breed, which may result from a similar fat distribution pattern in these breeds. The Marcenac and Aublet [13] formula were marked by a small relative error in the group of horses of the Wielkopolski Breed. Currently, bred horses of the Wielkopolski Breed are maintained as sports horses, very often as a result of mating with German sports horses of the Hanoverian, Holsteiner, or Trakehner Breeds. The estimations most similar to true body weight in the group of Thoroughbred horses was based on the Carroll and Huntington [8] formula.

Author Contributions: Conceptualization, W.G. and M.W.; methodology, W.G.; validation, W.G., M.W. and M.S.; formal analysis, W.G.; investigation, M.W.; data curation, M.W.; writing—original draft preparation, W.G. and M.S.; writing—review and editing, M.S. and M.K.; visualization, W.G.; supervision, M.K.; project administration, W.G. All authors have read and agreed to the published version of the manuscript.

Acknowledgments: The authors would like to thank all horse owners and stable managers for their help with the study.

References

1. Kaushal, R.; Bates, D.W.; Landrigan, C.; McKenna, K.J.; Clapp, M.D.; Federico, F.; Goldmann, D.A. Medication errors and adverse drug events in pediatric inpatients. *J. Am. Med. Assoc.* **2001**, *285*, 2114–2120. [CrossRef] [PubMed]
2. Powers, J.H. Risk perception and inappropriate antimicrobial use: Yes, it can hurt. *Clin. Infect. Dis.* **2009**, *48*, 1350–1353. [CrossRef] [PubMed]
3. Runciman, B.; Walton, M. *Safety and Ethics in Healthcare: A Guide to Getting It Right*, 1st ed.; Ashgate Publishing, Ltd.: London, UK, 2007.
4. Alford, P.; Geller, S.; Richrdson, B.; Slater, M.; Honnas, C.; Foreman, J.; Robinson, J.; Messer, M.; Roberts, M.; Goble, D.; et al. A multicenter, matched case-control study of risk factors for equine laminitis. *Prev. Vet. Med.* **2001**, *49*, 209–222. [CrossRef]
5. Frank, N.; Elliot, S.; Brandt, L.; Keisler, D.H. Physical characteristics, blood hormone concentrations, and plasma lipid concentrations in obese horses with insulin resistance. *J. Am. Vet. Med. Assoc.* **2006**, *228*, 1383–1390. [CrossRef] [PubMed]
6. Giles, S.L.; Rands, S.A.; Nicol, C.J.; Harris, P.A. Obesity prevalence and associated risk factors in outdoor living domestic horses and ponies. *Peer J.* **2014**, *2*, e299. [CrossRef]
7. Wylie, C.E.; Collins, S.N.; Verheyen, K.L.; Newton, J.R. Risk factors for equine laminitis: A systematic review with quality appraisal of published evidence. *Vet. J.* **2012**, *193*, 58–66. [CrossRef]
8. Carroll, C.L.; Huntington, P.J. Body condition scoring and weight estimation of horses. *Equine Vet. J.* **1988**, *20*, 41–45. [CrossRef]
9. Ellis, J.; Hollands, T. Accuracy of different methods of estimating the weight of horses. *Vet. Rec.* **1998**, *143*, 335–336. [CrossRef]
10. Jones, R.S.; Lawrence, T.L.; Veevers, A.; Cleave, N.; Hall, J. Accuracy of prediction of the liveweight of horses from body measurements. *Vet. Rec.* **1989**, *125*, 549–553. [CrossRef]
11. Milner, J.; Hewitt, D. Weight of horses: Improved estimates based on girth and length. *Can. Vet. J.* **1969**, *10*, 314–316.
12. Kyung-Nyer, K. Equine Body Weight Estimation Using Three-Dimensional Images, Electronic, Scholarly Journal. Master's Thesis, Colorado State University, Fort Collins, CO, USA, 2015.
13. Marcenac, L.N.; Aublet, H. *Encyclopedie du Cheval*; Librairie Maloine: Paris, France, 1964.
14. Ensminger, M.E. *Horses and Horsemanship*, 5th ed.; Interstate Printers & Publishers: Danville, CA, USA, 1977.
15. Martinson, K.L.; Coleman, R.C.; Rendahl, A.K.; Fang, Z.; McCue, M.E. Estimation of body weight and development of a body weight score for adult equids using morphometric measurements. *J. Anim. Sci.* **2014**, *92*, 2230–2238. [CrossRef]
16. Hoffmann, G.; Bentke, A.; Rose-Meierhöfer, S.; Ammon, C.; Mazetti, P.; Hardarson, G.H. Estimation of the body weight of Icelandic horses. *J. Equine Vet. Sci.* **2013**, *33*, 893–895. [CrossRef]
17. Henneke, D.R.; Potter, G.D.; Kreider, J.L.; Yeates, B.F. Relationship between condition score, physical measurements and body fat percentage in mares. *Equine Vet. J.* **1983**, *15*, 371–372. [CrossRef] [PubMed]
18. Kienzle, E.; Schramme, S.C. Beurteilung des Ernährungszustandes mittels Body Condition Scores und Gewichtsschätzung beim adulten Warmblutpferd. *Pferdeheilkunde* **2004**, *6*, 517–524. [CrossRef]
19. Johnson, E.; Asquith, R.; Kivipelto, J. Accuracy of weight determination of equids by visual estimation. In Proceedings of the 11th ENPS, Stillwater, OK, USA, 18–20 May 1989.
20. Łuszczyński, J.; Michalak, J.; Pieszka, M. Assessment of methods for determining body weight based on biometric dimensions in Hucul horses. *Sci. Ann. Pol. Soc. Anim. Prod.* **2019**, *4*, 9–20. [CrossRef]
21. Gharahveysi, S. Compare of different formulas of estimating the weight of horses by the Iranian Arab Horse data. *J. Anim. Vet. Adv.* **2012**, *11*, 2429–2431. [CrossRef]
22. Carter, R.A.; Geor, R.J.; Staniar, W.B.; Cubitt, T.A.; Harris, P.A. Apparent adiposity assessed by standardised scoring systems and morphometric measurements in horses and ponies. *Vet. J.* **2009**, *179*, 204–210. [CrossRef]
23. Thatcher, C.D.; Pleasant, R.S.; Geor, R.J.; Elvinger, F. Prevalence of overconditioning in mature horses in southwest Virginia during the summer. *J. Vet. Intern. Med.* **2012**, *26*, 1413–1418. [CrossRef]
24. Catalano, D.N.; Coleman, R.J.; Hathaway, M.R.; McCue, M.E.; Rendahl, A.K.; Martinson, K.L. Estimation of actual and ideal bodyweight using morphometric measurements and owner guessed bodyweight of adult draft and warmblood horses. *J. Equine Vet. Sci.* **2016**, *39*, 38–43. [CrossRef]

25. Catalano, D.N.; Coleman, R.J.; Hathaway, M.R.; Neu, A.E.; Wagner, E.L.; Tyler, P.J.; McCue, M.E.; Martinson, K.L. Estimation of actual and ideal bodyweight using morphometric measurements of Miniature, saddle-type, and Thoroughbred horses. *J. Equine Vet. Sci.* **2019**, *78*, 117–122. [CrossRef]

26. Jensen, R.B.; Rockhold, L.L.; Tauson, A.H. Weight estimation and hormone concentrations related to body condition in Icelandic and Warmblood horses: A field study. *Acta Vet Scand.* **2019**, *61*, 63. [CrossRef] [PubMed]

5

Impact of Year-Round Grazing by Horses on Pasture Nutrient Dynamics and the Correlation with Pasture Nutrient Content and Fecal Nutrient Composition

Sara Ringmark [1,*], Anna Skarin [2] and Anna Jansson [1]

[1] Department of Anatomy, Physiology and Biochemistry, Swedish University of Agricultural Sciences, SE-75007 Uppsala, Sweden

[2] Department of Animal Nutrition and Management, Swedish University of Agricultural Sciences, SE-75007 Uppsala, Sweden

* Correspondence: sara.ringmark@slu.se.

Simple Summary: Horse grazing may benefit biodiversity. This study compared the effect of horses grazing year-round to that of mowing on pasture quality in a forest-grassland landscape in Sweden. Twelve Gotlandsruss stallions were kept in three enclosures (~0.35 horse/hectare) without supplementary feeding for 2.5 years. Each enclosure contained three exclosures where pasture was not grazed, but mown monthly. Horse grazing increased the diversity of pasture nutrient content. Moreover, energy and protein concentrations and grass availability increased in areas grazed by horses, but decreased where grass was mown. This indicates that year-round grazing can be used to increase biodiversity, a suggestion supported by botanical observations. Nutrient content in horses' droppings was found to correlate with nutrient content in pasture, so analysis of droppings may be used to roughly estimate the quality of pasture consumed by horses. Under the conditions studied, pasture protein content was sufficient to meet horse requirements year-round, while energy content and pasture availability may have been limited in winter. Monthly data presented here on the nutritive value of pasture can help guide the management of year-round grazing systems in the Nordic countries.

Abstract: Horse grazing may benefit biodiversity, but the impact of year-round grazing on nutrient dynamics has not been evaluated previously. This study compared pasture quality in a forest-grassland landscape grazed year-round by horses with that in exclosed mown areas. Twelve Gotlandsruss stallions were kept without supplementary feeding in three enclosures (~0.35 horse/ha) outside Uppsala, Sweden, from May 2014 to September 2016. Each enclosure contained three mown exclosures, where grass sward samples were collected monthly and analyzed for chemical composition and vegetation density. Fecal grab samples were collected and analyzed for crude protein (CP) and organic matter (OM) content. There were no differences in exclosure pasture energy or CP content between enclosures ($p > 0.05$). In grazed areas, there were differences in grass energy and CP content ($p > 0.05$) between enclosures. During the three summers studied, energy and CP content increased in the enclosures, but decreased in the exclosures. By the end, biomass content/ha was greater in the enclosures than in the exclosures. Fecal OM and CP content showed moderate to strong correlations with pasture nutrient content ($r = 0.3–0.8$, $p < 0.05$). Thus, in contrast to monthly mowing, horse grazing diversified pasture chemical composition and increased its nutritive value.

Keywords: pasture; horse nutrition; crude protein; exclosures

1. Introduction

Year-round grazing by cattle, sheep, and horses is common in many European countries, but not in Sweden. To our knowledge, the effects on pasture quality and quantity of keeping horses year-round on extensive grazing have not been evaluated previously in the Scandinavian countries. Reasons for this might include the comparatively short growing season, the need for shelter to meet animal welfare legislation, and expected low nutrient content of pasture during winter. Lack of validated methods for monitoring horse nutrient intake on pasture may be another reason. However, studies in Germany have shown that the nutrient content of pastures grazed year-round can meet or exceed the requirements of adult cattle and horses, even in winter [1]. In contrast, a study on year-round grazing horses in France indicated that crude protein intake was very low six months per year, and that adult maintenance requirement was met only in April–September [2]. This estimation was based on fecal analyses of crude protein and an observed positive correlation between dietary and fecal crude content. The use of fecal crude protein analysis to monitor pasture crude protein content and intake needs, however, to be further validated.

Year-round grazing systems may have the potential to reduce feed costs, but may also support horses' natural behaviors and contribute to increased biological diversity. Abandonment of natural and semi-natural grasslands and forest encroachment, induced by lack of large herbivores in open landscapes, has caused loss of flora and fauna biodiversity in Sweden [3]. Studies in European countries, including Sweden, indicate that grazing horses can be used instead of cattle and sheep to promote biological diversity [4–9]. Horses remove more vegetation per unit body mass than cattle [10], create mosaic patches of short and tall grass, and leave more broad-leaved plants than cattle [10]. Horses prefer grasses [2], but their intake of forbs and shrubs may increase during periods of intense grazing in winter and spring [11,12], and they may also perform bark-stripping [13]. Use of horses in a year-round grazing system could therefore have great impacts on the landscape and biological diversity [14].

The overall aim of this study was to describe the seasonal and land-to-land variation in pasture quality in a Swedish forest-grassland landscape grazed year-round by horses, and compare it with that in adjacent exclosure areas mown monthly. A second aim was to investigate fecal sampling as a measure of pasture quality. The results are discussed in relation to whether the fodder quality was acceptable to meet energy and protein requirements in horses, and to the possible impact on pasture diversification. The hypotheses tested were that pasture energy and protein content can meet animal requirements but with differences between land areas; that horse grazing alters pasture energy and nutrient composition compared with mowing; and that fecal crude protein (CP) content is correlated with pasture nutrient concentration.

2. Materials and Methods

The study was carried out between May 2014 and September 2016 in Krusenberg, Uppsala, Sweden (59°44′8″ N, 17°38′58″ E). During the 15 years preceding the study, mean daily temperature April–October was 12.4 ± 5.0 °C ($\pm$SD) and mean precipitation was 1.7 ± 4.3 mm/day, while in November–March the values were -0.6 ± 5.4 °C and 1.4 ± 2.7 mm/day, respectively (Swedish Meteorological and Hydrological Institute (SMHI) weather station Uppsala Aut, https://www.smhi.se/klimatdata).

In our study, the summer was defined to start after the first four consecutive days with mean temperature >+5 °C in spring and to end after the first four days with <+5 °C in fall, which defined the start of winter. Based on this definition, the summer season started on 11 April 2015 and on 30 March 2016, while the winter season started on 9 November 2014 and on 10 October 2015. The study was approved by Uppsala animal welfare ethics committee (license number: C28/14). Data on daily temperature and monthly precipitation during the study period were retrieved from the Department of Earth Sciences, Uppsala University, Sweden (www.geo.uu.se).

2.1. Horses and Management

Twelve one-year old Gotlandsruss stallions (mean body weight 185 ± 21 kg at the start) from six different breeders were used in the study. Gotlandsruss is a native Swedish horse breed that has been present on the island of Gotland from at least the seventeenth century [15] and probably the thirteenth century. The horses were divided into three groups of four and allocated to three enclosures at the start of the experiment on 21 May 2014. The horses were kept without supplementary feeding throughout the study. To avoid the grazing preferences of an individual horse or group affecting pasture composition, the groups were rotated between the enclosures on 27 May 2015 and 20 May 2016, i.e., each group grazed each enclosure for one growing season. In January 2016, one individual was excluded from the study due to an injury. Each enclosure contained a 16 m^2 shelter (Figure 1). Water was offered in automatic water troughs, located in the forest, during summer, spring, and fall. During winter, when the temperature was below 0 °C, water was offered once/day in plastic troughs. In all enclosures, water was also available in streams in the forest, even during winter. A salt block with trace minerals (May 2014–August 2014: Ab Hansson & Möhring, Halmstad, Sweden, content (mg/kg): Zinc 300, manganese 200, copper 80, iodine 50, selenium 20, cobalt 12; August 2014–September 2016: Standard, KNC, Netherlands, content (mg/kg): Zinc 810, copper 220, iodine 100, selenium 20) was provided in all enclosures. Horses were dewormed five times during the study period, using Banminth (Pyrantel, Zoetis Finland Oy, Helsinki, Finland), Equimax (Ivermectin and Praziquantel, Sofarimex Indústria Química e Farmacêutica Ltd., Cacém, Portugal), or Cydectin (Moxidectin, Zoetis Manufacturing & Research Spain, Gerona, Spain). Once per month, fresh grab sample (approximately 300 g) of faces were collected immediately after defecation from a minimum of two horses/enclosure (in total a minimum of six samples). These samples were stored at −20 °C until analysis.

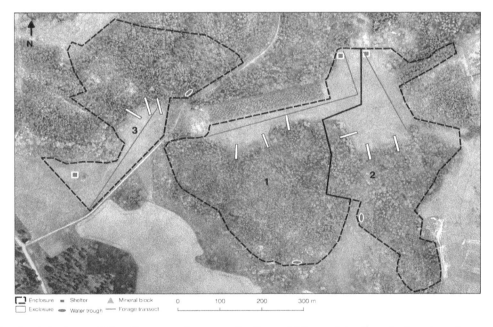

Figure 1. Aerial view of the three enclosures (1, 2, and 3) used in the study, showing position of shelters, water troughs, mineral blocks, exclosures, and pasture transects. Photo taken on 24 May 2016 at 150 m altitude.

2.2. Enclosures and Exclosures

The three enclosures (En1–En3) were 13, 11, and 10 ha in size, respectively, and consisted of approximately 1/3 fields and 2/3 forest (Table 1). Each was surrounded by electric fencing. Dominating vegetation types, according to the classification used by the Swedish land survey (lantmateriet.se), were recorded in plots 15 m^2 placed at 35 m spacing in a grid and located with GPS [16]. The vegetation was then merged into three different vegetation classes: grassland, forest, and semi-forest. Forest was defined as forested areas with >30% crown coverage and semi-forest with 10–30% crown coverage, while grassland had <10% crown coverage. In addition, plant species and species coverage were recorded in a 20 × 20 cm square in the plots (Table 1). The fields had not been grazed by horses for at least 10 years, but En1 and En2 had been grazed by cattle and En3 had been used for production of conserved forage.

In each enclosure, three exclosures each measuring 42.5 m × 5 m were fenced off using electric fencing (Figure 1). All exclosures were placed in the edge zone between forest and open field, with 20 m of the exclosure in the forest and 22.5 m in the open field.

Table 1. Area (ha) of vegetation types within each enclosure and dominant plant species identified in a vegetation inventory performed in all three enclosures in May 2014 (study start).

Vegetation Type, ha	Enclosure 1	Enclosure 2	Enclosure 3
Grassland	2.7	3.3	2.7
Forest	10.7	5.8	6.8
Semi-forest [a]	0	1.3	0.2
Total area	13.4	10.3	9.7
Dominant Plant Species in Grassland, %			
Grasses	60	57	26
Yarrow (*Achillea millefolium*)	10	14	
Dandelion (*Taraxacum* spp.)	7	10	47
Hempnettle (*Galeopsis tetrahit*)		6	
White clover (*Trifolium repens*)			5
Dominant Ground Cover in Forest, %			
Mosses	46	29	55
Grasses	11	26	12
Bilberry (*Vaccinium myrtillus*)	12	7	10
Lingonberry (*Vaccinium vitis-idaea*)	5	4	3

[a] Semi-forest includes forest areas dominated by deciduous forest with a large proportion of grass in the ground cover.

2.3. Pasture Sampling

The pasture in the open field areas was sampled in the second week of each month all year round, except when the ground was covered with snow (0–29 cm, December 2014–March 2015, January 2016, and March 2016). Four types of sample were collected: Forage, volume, graze, and exclosure.

Forage samples were collected by clipping a grab sample of vegetation 5 cm from the ground every 10 m along a transect crossing the open fields. In En2 and En3, samples were collected along that one transect, but due to the shape of En1, the sampling line was L-shaped and longer, and samples were collected every 20 m to retrieve the same amount of samples representing the approximate same size of grassland.

Volume samples were taken along the same transects as the forage samples, but every 50 m (100 m in En1), collecting all vegetation 5 cm from the ground within a 30 cm × 30 cm square (0.09 m^2). The volume samples were placed in plastic bags and weighed later for determination of pasture quantity.

Graze samples were collected in the area where horses were grazing at the time of sampling. Vegetation was cut close to the ground, i.e., the height at which the horses were assumed to graze.

Exclosure samples were taken within a 50 cm × 50 cm quadrat (0.25 m^2) in the open field part of the exclosures, 2 m from the fence (Figure 1). All vegetation above 5 cm from the ground within the quadrat was collected by mowing with a scissor. Grassland production was assessed by weighing the Exclosure samples, determining the dry matter (DM) content, and calculating the amount of DM per hectare.

In February and April 2016, only graze samples were collected, due to small sample size/no sample for the other three sample types. At all sampling sites except those for the graze samples, grass height was measured using a herbometer (Herbometre, AGRO-Systémes, La membrolle sur Chosille,

France), with a 30 cm × 30 cm square plate placed on top of the vegetation. On occasion, horses were observed eating bilberry plants (*Vaccinium myrtillus* L.) and in December 2014, random samples of bilberry, without berries or leaves, were collected at the time of grazing/browsing. All samples were stored at −20 °C until analysis.

2.4. Analyses of Chemical Composition

The exclosure samples from the three exclosures within each enclosure were pooled to one sample before analysis of nutrient content. To determine DM content, pasture and feces samples were dried at 60 °C for 24 h and milled in a 1 mm hammer mill (Kamas, Slagy 200 B, Malmö, Sweden). A 2 g subsample was then dried at 103 °C for 16 h. Ash content was determined by incinerating a 2 g sample at 550 °C for 3 h, after which the residue was cooled and weighed. Organic matter (OM) content was calculated by subtracting the content of ash from the DM content. Digestibility coefficient of organic matter (VOS) and metabolizable energy (ME) content were determined in vitro according to Lindgren 163 [17]. The ME content is, however, based on the ME for ruminants, so was adjusted for horses using the following equation derived by Jansson et al. [18]:

$$ME_{horse} = 1.12(ME_{ruminant}) - 1.1$$

The concentration of neutral detergent fiber (NDF) was determined according to Chai and Uden [19]. Analysis of CP was performed according to Kjeldahl [20], where ammonia nitrogen concentration was determined by direct distillation with a Kjeltec 2460 analyser (Foss, Hilleröd, Denmark) and N content was multiplied by 6.25 to give the CP content. To estimate the amount of digestible CP (dCP) [19], the following equation used was:

$$dCP = 0.939\text{-}31.1/g\ CP\ kg/DM$$

The ratio between digestible CP and ME (RdCPME) was also calculated, since this is an established measure of horse feed quality in Sweden [18]. In samples retrieved in June 2016 (see Table S1), macronutrient concentrations were analyzed by inductively coupled plasma-atomic emission spectroscopy (ICP-AES) using Spectro Flame equipment (SPECTRO Analytical Instruments, Kleve, Germany). Due to small sample size, values for exclosures in En2 and En3 in October 2014 are missing.

2.5. Statistical Analyses

All statistical analyses were performed in Statistical Analysis Systems package 9.4 (SAS Institute Inc., Cary, NC, USA). Differences were considered significant at $p < 0.05$. Values presented are least square means (LSmeans) ± standard error (SE).

To study if climate differed significantly between years during the study period, effects of year on air temperature and precipitation were estimated in a mixed model including an interaction between season and year.

Test of differences in pasture nutrient content between samples retrieved in exclosures and the samples retrieved in the enclosures, as well as possible differences between the different enclosure sample types (Forage, Volume, and Graze), were performed using a mixed model, with enclosure as repeated measurement and an effect of interaction between enclosure and time period (year and month). When the p-value for the interaction effect between enclosure and time period was <0.1, a separate analysis without the interaction effect was run. If no interaction effect is reported, values refer to the latter analysis. As analysis of differences in pasture nutrient content showed no significant difference between forage and volume samples, these were pooled before further analysis.

To test if different land areas, i.e., enclosures, responded differently in terms of nutrient content, as well as pasture quantity on horse grazing and mowing, a mixed model with enclosure as repeated measurement was used. The same model was also used to test if season (summer/winter) and year affected the nutritional content and pasture quantity. An analysis including the effect of interaction between year and season was also performed.

3. Results

Mean air temperature and precipitation did not differ between the three study years ($p > 0.05$). During summer, mean temperature was 13.3 ± 5.8 °C and mean precipitation was 1.8 ± 4.3 mm/day. During winter, the corresponding values were 1.3 ± 5.1 °C and 1.0 ± 2.2 mm/day.

3.1. Variation in Pasture Nutrient Content Between Exclosures and Enclosures, and Between Sample Types

The exclosure samples showed lower DM and higher CP and ME contents than the forage and volume samples from the enclosures, but there was no difference in NDF content (Table 2). The graze samples showed higher nutrient contents than the other three sample types (with the exception of energy content in exclosure samples, which was similar) (Table 2). There was a significant interaction between sample type and time period for ME per kg OM and CP as a percentage of OM ($p < 0.05$), where ME and CP remained at a high concentration in the period May–September in the graze samples, while decreasing from August onwards in the other sample types.

The bilberry shrubs sampled in the forest had the following composition: DM 45%, CP 7% of DM, digestible CP 32 g/kg DM, ME 3.2 MJ/kg DM, and OM 97% of DM.

Table 2. Content of dry matter (DM), metabolizable energy (ME) per kg organic matter (OM), crude protein (CP) as % of OM, and neutral detergent fiber (NDF) as % of OM in three different types of pasture samples collected in three enclosures grazed by horses, and in three exclosures per enclosure, monthly between May 2014 and September 2016, except for December 2014–March 2015 and January-April 2016.

Sample	Enclosures			Exclosures	p
	Forage	Volume	Graze		
ME, MJ/kg OM	10.0 ± 0.2 [a]	10.0 ± 0.2 [a]	10.9 ± 0.2 [b]	10.5 ± 0.2 [b]	0.0002
CP, % of OM	11.9 ± 0.4 [a]	11.6 ± 0.5 [a]	17.0 ± 0.4 [b]	13.7 ± 0.4 [c]	<0.0001
NDF, % of OM	57.2 ± 0.7 [a]	56.7 ± 0.9 [a]	51.0 ± 0.7 [b]	56.8 ± 0.8 [a]	<0.0001
DM, %	34 ± 1 [a]	34 ± 1 [a]	29 ± 1 [b]	27 ± 1 [b]	<0.0001

[a,b,c] Different superscript letters indicate significant differences within rows ($p < 0.05$).

3.2. Variation in Pasture Quantity

Mean grass sward height was lower and DM and ME content/ha were higher in enclosures compared with exclosures, but there were also differences between the enclosures (Table 3). During the summer in 2016, grass sward height decreased in both enclosures and exclosures compared with in 2015, but DM and ME content/ha only decreased in exclosures (Table 4).

Table 3. Mean grass sward height and content of dry matter (DM) and metabolizable energy (ME) in the enclosures grazed by horses and in the exclosures within each enclosure mown monthly from May 2014 to September 2016. Exclosures were mown at the same spots as sward height measurements were made.

	Enclosures				Exclosures				p Means
	En1	En2	En3	Mean	En1	En2	En3	Mean	
Grass sward height, cm	5.5 ± 0.1 [a]	7.7 ± 0.2 [b]	5.5 ± 0.2 [a]	5.3 ± 0.1	4.3 ± 0.4 [a]	4.9 ± 0.4 [a,b]	6.0 ± 0.4 [b]	5.6 ± 0.4	<0.05
DM, kg/ha	957 ± 114 [a]	1393 ± 114 [b]	741 ± 114 [a]	886 ± 73	525 ± 111	487 ± 111	721 ± 111	326 ± 76	<0.0001
ME, MJ/ha	8494 ± 895 [a]	$11,204 \pm 928$ [b]	7351 ± 895 [a]	7470 ± 636	5208 ± 1171	4912 ± 1171	7760 ± 1171	3048 ± 701	<0.0001

[a,b,c] Different superscript letters indicate significant differences ($p < 0.05$) between enclosures (En1, En2, En3).

Table 4. Mean summer season grass sward height and content of dry matter (DM) and metabolizable energy (ME) per year in the enclosures grazed by horses and in the exclosures within each enclosure mown monthly from May 2014 to September 2016. Exclosures were mown at the same spots as sward height measurements were made.

	Enclosures			Exclosures		
	2014	**2015**	**2016**	**2014**	**2015**	**2016**
Grass sward height, cm	8.9 ± 0.21 [a]	5.9 ± 0.2 [b]	5.2 ± 0.2 [c]	-	7.1 ± 0.4 [a]	5.4 ± 0.4 [b]
DM, kg/ha	1560 ± 160 [a]	863 ± 176 [b]	770 ± 176 [b]	714 ± 96 [a]	760 ± 105 [a]	260 ± 105 [b]
ME, MJ/ha	13,858 ± 1293 [a]	8133 ± 1416 [b]	8176 ± 1466 [b]	7201 ± 1044 [a]	7869 ± 1144 [a]	2810 ± 1144 [b]

[a,b,c] Different superscript letters indicate significant differences between years ($p < 0.05$).

3.3. Variation in Pasture Nutrient Content between Enclosures

In exclosure samples, there was no general effect of the different enclosures on any of the nutritional parameters analyzed (Table 5). In the pooled forage + volume samples and in graze samples, the content of ME, CP, and NDF/kg OM showed differences ($p < 0.05$) between enclosures (Table 5).

Table 5. Content of dry matter (DM), metabolizable energy (ME) per kg organic matter (OM), crude protein (CP) as % of OM, and neutral detergent fiber (NDF) content as % of OM in four types of pasture samples collected in three enclosures (En1–En3) grazed all year round by Gotlandsruss and in three exclosures per enclosure. Samples were collected monthly between May 2014 and September 2016, except for December 2014–March 2015 and January–April 2016. LSmeans ± SE, p-values indicate the general effect of enclosure.

Sample Type	En1	En2	En3	p
Exclosure				
ME, MJ/kg OM	10.8 ± 0.2	10.9 ± 0.2	11.2 ± 0.2	0.2925
CP, % of OM	14.9 ± 0.4	14.3 ± 0.4	13.9 ± 0.4	0.1955
NFD, % of OM	55.0 ± 1.1 [a]	54.7 ± 1.1 [a,b]	51.8 ± 1.1 [b]	0.0924
DM, %	27 ± 1	29 ± 1	26 ± 1	0.2722
Enclosure Graze				
ME, MJ/kg OM	10.4 ± 0.2 [a]	10.7 ± 0.2 [a]	11.5 ± 0.2 [b]	0.0029
CP, % of OM	16.9 ± 0.8 [a,b]	16.0 ± 0.8 [a]	18.8 ± 0.8 [b]	0.0427
NFD, % of OM	52.7 ± 1.4 [a]	53.9 ± 1.3 [a]	45.6 ± 1.3 [b]	0.0001
DM, %	30 ± 1 [a,b]	31 ± 1 [a]	27 ± 1 [b]	0.0159
Enclosure Forage + Volume				
ME, MJ/kg OM	10.1 ± 0.1 [a]	9.3 ± 0.1 [b]	10.6 ± 0.1 [c]	<0.0001
CP, % of OM	11.5 ± 0.4 [a]	11.9 ± 0.4 [a,b]	12.5 ± 0.4 [b]	0.1405
NFD, % of OM	54.7 ± 1.1 [a]	59.9 ± 1.1 [b]	54.2 ± 1.1 [a]	0.0007
DM, %	34 ± 1 [a,b]	37 ± 1 [a]	31 ± 1 [b]	0.0083

[a,b,c] Different superscript letters within rows indicate significant differences ($p < 0.05$) between enclosures (En1, En2, En3).

3.4. Variation in Pasture Quality Between Years, Seasons, and Months

As pasture was not sampled in all months throughout the year, only differences within seasons between years are presented (Figure 2). During the study period, summer pasture NDF concentration decreased in the enclosures, while ME and CP concentrations increased (Figure 2). In the exclosures, summer pasture CP remained unchanged between years, while ME concentration increased and NDF concentration showed a decrease.

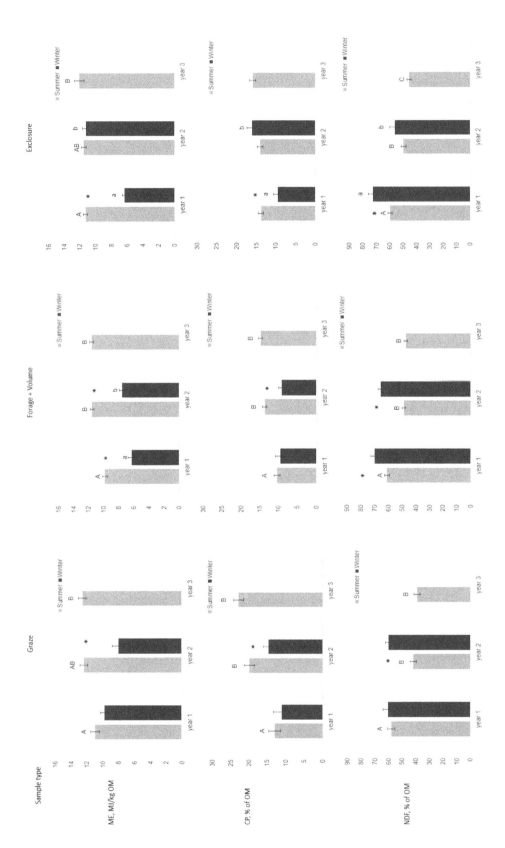

Figure 2. Content of metabolizable energy (ME) per kg organic matter (OM), crude protein (CP) as % of OM, and neutral detergent fiber (NDF) as % of OM in three types of pasture samples (graze, forage + volume, exclosure) collected monthly in three enclosures grazed by 12 Gotlandsruss between May 2014 and September 2016. One year is defined as start of summer season to end of winter season. Each enclosure contained three exclosures. An asterisk indicates significant difference between seasons within year ($p < 0.05$). Different lowercase letters (a, b) indicate differences between years within winter season, while different uppercase letters (A, B) indicate differences between years within summer season ($p < 0.05$).

Mean monthly nutrient composition and RdCPME in forage + volume and graze samples in all three enclosures during the whole study period are shown in Tables S2 and S3. Mean ME content ranged from 4.9 ± 0.5 to 12.0 ± 0.6 MJ per kg DM, mean CP content ranged from 7 ± 1 % to 24 ± 3 %, mean digestible CP per kg DM ranged from 37 ± 12 g to 190 ± 26 g, and mean RdCPME ranged from 6.4 ± 1.1 to 15.9 ± 2.6 (Table S2).

3.5. Fecal Composition and Correlation with Pasture Nutrient Content

The OM and CP content in feces, but not the DM content, were dependent on the individual horse, but overall the OM content in feces was lower in En3 than in En1 and En2 (Table 6). Content of CP, both as % of OM and as % of DM, was lowest in En1 and highest in En3 (Table 6). Concentrations of CP in feces were lower in winter than in summer (Figure 3). Within season, fecal CP concentration increased with year.

Table 6. Content of dry matter (DM), organic matter (OM), and crude protein (CP) as % of DM and as % of OM in feces from Gotlandsruss, divided equally between three enclosures (En1–En3) and grazing all year round.

	En1	En2	En3	p
DM, %	20 ± 0.3 [a]	19 ± 0.3 [b]	21 ± 0.3 [a]	<0.0001
OM, %	79 ± 0.5 [a]	80 ± 0.5 [a]	75 ± 0.5 [b]	<0.0001
CP, % of DM	8.4 ± 0.2 [a]	9.1 ± 0.2 [b]	9.7 ± 0.2 [c]	<0.0001
CP, % of OM	10.8 ± 0.3 [a]	11.5 ± 0.3 [b]	13.1 ± 0.3 [c]	<0.0001

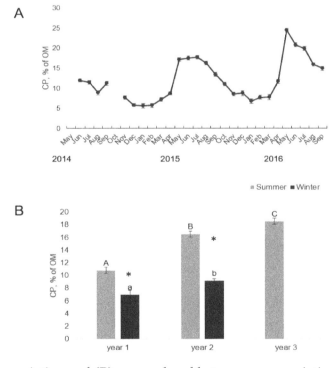

Figure 3. (**A**) Monthly variation and (**B**) seasonal and between-year variation in crude protein (CP) content (LSmeans ± SE) as % of organic matter (OM) in feces from Gotlandsruss grazing without supplementary feeding from May 2014 to September 2016. One year is defined as start of summer season to end of winter season.

The content of DM, OM, and CP in feces showed moderate to strong correlations with the DM, NDF, OM, and CP concentrations in graze samples and in forage + volume samples (Tables 7 and 8).

Table 7. Correlation coefficient (*r*, *p*-value) between nutrient concentrations in graze samples and in fecal samples from 12 Gotlandsruss kept in three enclosures between May 2014 and September 2016 without supplementary feeding. DM = dry matter, OM = organic matter, CP = crude protein, VOS = digestibility coefficient of organic matter, NDF = neutral detergent fiber.

Graze Samples		Fecal Samples			
		DM	OM	CP, % of DM	CP, % of OM
DM	*r*	0.37703	ns	−0.30271	−0.28796
	p	0.0032		0.0198	0.0270
VOS	*r*	−0.58502	−0.32107	0.62955	0.64978
	p	<0.0001	0.0132	<0.0001	<0.0001
NDF	*r*	0.36083	0.53560	−0.69221	−0.75195
	p	0.0050	<0.0001	<0.0001	<0.0001
OM	*r*	−0.41346	0.36627	ns	ns
	p	0.0011	0.0043		
CP, % of DM	*r*	−0.35722	−0.48196	0.60511	0.66090
	p	0.0055	0.0001	<0.0001	<0.0001
CP, % of OM	*r*	ns	−0.55235	0.49911	0.57349
	p		<0.0001	<0.0001	<0.0001
NDF, % of OM	*r*	0.48350	0.46697	−0.75952	−0.80312
	p	0.0001	0.0002	<0.0001	<0.0001

Table 8. Correlation coefficient (*r*, *p*-value) between nutrient concentrations in forage + volume samples and in fecal samples from 12 Gotlandsruss kept in three enclosures between May 2014 and September 2016 without supplementary feeding. DM = dry matter, OM = organic matter, CP = crude protein, VOS = digestibility coefficient of organic matter, NDF = neutral detergent fiber.

Volume + Forage Samples		Fecal Samples			
		DM	OM	CP, % of DM	CP, % of OM
DM, %	*r*	0.58646	ns	−0.45184	−0.44157
	p	<0.0001		0.0003	0.0005
VOS	*r*	−0.62653	−0.44629	0.75445	0.76465
	p	<0.0001	0.0004	<0.0001	<0.0001
NDF, %	*r*	0.57023	0.48945	−0.84249	−0.86310
	p	<0.0001	<0.0001	<0.0001	<0.0001
OM, %	*r*	ns	ns	ns	ns
	p				
CP, % of DM	*r*	−0.38627	−0.43487	0.64359	0.67240
	p	0.0025	0.0006	<0.0001	<0.0001
CP, % of OM	*r*	−0.38872	−0.42891	0.64552	0.67353
	p	0.0023	0.0007	<0.0001	<0.0001
NDF, % of OM	*r*	0.56563	0.50105	−0.84228	−0.86426
	p	<0.0001	<0.0001	<0.0001	<0.0001

4. Discussion

The results obtained in this three-year study showed that horse grazing altered pasture nutrient composition and diversified pasture chemical composition (between enclosures) to a greater extent than mowing. This indicates that horses can manage pasture and are therefore suitable for year-round grazing in Sweden, as a means to increase pasture diversity. To our knowledge, this is the first study at Nordic latitudes to evaluate the effect on pasture chemical composition of year-round grazing by horses without supplementary feeding. Year-round grazing is currently not practiced in the region because of the lack of vegetation growth in winter (i.e., temperatures below 5 °C for more than four days). However, the study area is within a zone suggested to be suitable with respect to climate conditions for rewilding of horses, although local biotic factors, land cover, and soil type may influence the degree of suitability [21]. The study was conducted between 2014 and 2016, under temperature and precipitation conditions typical for the region, and the results are therefore of general relevance for this form of horse management.

4.1. Pasture Quality and Effects of Sample Type

One likely explanation for the diversified nutrient content of grazed pasture is the grazing behavior of horses. Horses perform selective grazing [22,23], and also create mosaic landscape patterns [24], as some areas are frequently grazed close to the ground while others are avoided. In the present study, this was reflected in the higher content of ME and CP and lower content of NDF in the graze samples, compared with the forage and volume samples, indicating an earlier botanical stage. An additional explanation for the altered nutrient content may be the change in botanical composition reported previously for the study area [4]. For example, grazing favored prostrate plant species (low plant height at maturity).

The three enclosures all had different qualities. En1 and En2 showed a different response to En3 with respect to sward height and nutrient content, for example. The reason is unclear, but En3 had previously been cultivated (forage production) and En1 and En2 had been grazed by cattle. En2 was also the enclosure with the lowest grazing pressure, as it had a larger grass area (3.3 ha, compared with 2.7 ha in En1 and En3), and the plant composition differed (Table 1). Mowing of the exclosures was performed without selection, creating similar plant stress between enclosures, which might have evened out initial differences in species and chemical composition. Horses, on the other hand, graze selectively [22,23] and might have favored different species and areas in the three enclosures, as well as putting more stress on the plants by grazing them shorter than mowing (i.e., <5 cm). In addition, the trampling effect of horses might have affected the botanical composition and, therefore, also the nutrient content.

Compared with the exclosure areas, from which grazing horses were excluded, grazed pasture showed both lower (volume and forage samples) and higher (graze samples) concentrations of CP, depending on sample type. The difference in composition between sample types may be due to the forage + volume samples and the exclosure samples including only plant parts >5 cm. These plant parts may be of a later botanical stage [25], as reflected by the higher content of NDF compared with graze samples collected close to the ground. Both the graze and forage + volume samples showed relatively high correlations with fecal NDF and CP content, indicating that they may both be valid sampling options when measuring the nutrient content of pasture consumed by horses.

Observations from Sweden indicate that heavily grazed grasses may have higher CP and energy content in October than grasses subjected to lower grazing pressure [26]. This is likely due to old biomass being replaced by new, nitrogen-rich leaves as grazing increases defoliation [27], and is presumably the reason for the higher nitrogen content in the graze samples in the present study. During summer seasons, ME and CP increased in graze and forage + volume samples, while they were unchanged in exclosure samples and remained high in the graze samples for longer than in the other sample types. These results are in accordance with observations in sheep pastures that frequent grazing during the vegetation period improves nutrient quality compared with mowing [28]. The results confirm that grazing managed at the right intensity can enhance the quality of pasture [29].

4.2. Pasture Quantity

Pasture dry matter production in the study area was within the range reported for other horse and ruminant year-round grazed pastures in Germany [1], and also similar to that reported for natural pastures in Sweden grazed by cattle and sheep [30]. In addition, based on the 10-year mean for forage harvest from cultivated grassland, the years included in the study seem to be representative for the region [31].

The pasture quantity values determined in the exclosures could be regarded as a measure of overall pasture production, while the volume samples collected in the enclosures could be considered a measure of the amount of pasture available to the horses. The exclosures were mown at the same sites as the exclosure samples were taken. In contrast, the volume samples were collected by walking along a straight line, and with this method, sampling sites may have varied slightly between sampling occasions. Comparisons of pasture and energy quantity between enclosures and exclosures should therefore be made with caution. However, the lack of differences in grass sward height between enclosures and exclosures implies that mowing to 5 cm every month resulted in a similar rate of vegetation removal as horse grazing within the enclosures.

Pasture quantity decreased over the study years in both enclosures and exclosures (2015 and 2016). This may therefore be an effect of annual variation, rather than an effect of horse grazing, on pasture production. To evaluate the long-term impact of horse grazing on pasture production, much longer studies are required.

4.3. Ability of Pastures to Meet the Nutritional Requirements of Horses

Despite the cold winters in the region, CP content during winter (7–11% of DM) was within the range reported for other European winter pastures [1]. Assuming a maximum DM intake capacity of 3% of body weight [32–34], a 250 kg adult stallion with a CP requirement of 432 g/day [35] would, in theory, manage with a CP concentration in pasture of 6% of DM, which is just below the lowest value recorded in the present study. However, as found for other European pastures [1], the amount of pasture available and the ME content during winter were insufficient to maintain body condition in horses. In addition, a snow layer of >10 cm covered the ground for 14–31 days/winter in our study, making the pasture more difficult to access. At the beginning of winter, when snow was still absent, the mean energy content of the pasture was estimated to be 6200 MJ ME/ha in the grasslands. Assuming a winter season lasting five months, and that all vegetation sampled here could be consumed by the horses, this would supply each horse with 10 MJ ME/day, which represents 30% of the estimated daily requirements [35]. However, in reality the samples would probably contain plants generally not consumed by the horses and the energy requirement of the horses may have been higher during cold spells.

The insufficient levels of energy and CP in the pasture were reflected in loss of body weight and body condition in the horses during the winter months (unpublished data). On the other hand, during the growing season, pasture provided a surplus of energy great enough for the horses to store body fat, compensating for energy deficiency in winter. En2 was the only enclosure where no horse at any time would have required supplementary feeding to maintain a functional body condition. An energy content in pasture of at least 12,000 MJ ME/ha in November therefore seems sufficient to avoid horses becoming underweight at the given animal density (approximately 1000 kg horse on 3 ha of grassland and 7 ha forest) and summer conditions. However, the lack of need for supplementary feeding in En2 could also be due to this enclosure containing a rather large area (1.3 ha) of semi-forest with some grass in the understory.

Interestingly, during wintertime, horses spent more time in the forest (unpublished data). Therefore, tree materials, some grass, and bilberry plants probably comprised a greater proportion of the horses' diet during winter. The chemical analysis of bilberry plants showed rather low contents of ME and CP. However, it should be noted that the analytical methods used are designed for grasses and legumes, and may be less relevant for shrubs.

4.4. Correlation of Pasture Quality and Fecal Composition

The correlations between fecal and pasture concentrations of CP (as % of DM and OM) were moderate to strong ($r = 0.57$–0.67). These were similar to correlations reported for stabled horses fed a forage-only diet [36]. The fairly high correlation implies that a fecal grab samples could be used to give a rough estimate of CP intake in grazing horses. However, the method may not apply if the horses are growing or mares are lactating, as their nutrient requirements are higher than those of adult horses [35].

The trend for fecal CP concentration to be correlated with season is similar to that reported for bachelor horses in Camargue [2] and feral horses in Canada [37]. However, the seasonal variation in the present study was greater, ranging from 5.6% of DM in January to 17.0% of DM in May. The increment in fecal CP content with year reflected the pasture composition, as CP content in pasture vegetation was also higher in the second and third summer than at the start of the study.

Surprisingly, the correlation with fecal CP concentration was slightly stronger with forage + volume samples than with graze samples. The highest CP values were observed in Graze samples and the CP content was periodically much higher than the requirement. However, this was not reflected in higher CP content in feces compared with the forage + volume samples, probably because more of the easily digestible nitrogen was excreted with urine and not with feces [38]. The intention with collecting the graze samples was to get a more accurate estimate of the nutritional composition of the pasture actually consumed by the horses. However, both the graze samples and the grab sample of the feces were spot-samples taken at random times and the graze samples represented conditions at those times, while fecal samples would consist of digesta ingested hours to days before sampling [38]. This may be another reason why the CP content of forage + volume samples correlated better with CP concentration in feces.

4.5. Practical Implications

Our study provides practical data on the quality and quantity of Swedish pasture grazed year-round by horses. The area of semi-natural pastures in Sweden is decreasing [39]. At the same time, the number of horses in Sweden is increasing, from 85,000 in 1970 [40] to now approaching 355,000, which makes horses more common than dairy cows [31]. There is, thus, great potential for using horses in landscape conservation in Sweden. However, most horses are stabled for most of the year and the main roughage fed is hay or haylage harvested from cultivated leys. The results in the present study indicate that pasture grazed year-round south of latitude 60° N in Sweden can have a sufficient energy and nutrient content to meet the nutritional recommendations of adult horses for at least 10 months per year (Table S3, no data for January and March) and that the amount of pasture may be a limitation. The results presented here could be used as the basis for recommendations on utilization of semi-natural pastures by horses even outside the growing season, as an alternative to feeding hay or haylage, for example. Compared with feeding conserved forage, pasture provides increased opportunities for horses to express their natural behavior and requires less resources than harvesting, conservation, and transportation of hay/haylage. The low content of energy in shrubs such as bilberry means that forest pastures could be suitable to meet the feeding behavior requirement of obese horses, although this would require further evaluation. Increased grazing of semi-natural pastures would also increase biological diversity [4] and help preserve agricultural landscapes. Moreover, our data on the nutrient composition of Swedish semi-natural pastures support the suggestion [21] that they could be suitable for future rewilding of horses at this latitude.

5. Conclusions

Compared with mowing, year-round grazing by horses in Sweden increased pasture nutrient quality and diversity. This indicates that year-round grazing by horses in Sweden could be used as a general tool to increase biodiversity. Pasture sampling method affected the pasture quality results but, overall, CP content was sufficient to meet the horses' requirements year-round, while the energy content and pasture availability may be a limitation during winter. Fecal grab samples proved to give a fairly good estimate of CP intake in grazing horses, but should be complemented with analysis of pasture quality for pregnant, lactating, and young horses with high CP requirements.

Author Contributions: S.R. contributed to the methodology, validation, formal analysis, investigation, data curation, and original draft manuscript preparation. A.S. contributed to the methodology, investigation, review and editing, and visualization. A.J. contributed to the conceptualization, methodology, formal analysis, investigation, resources, reviewing and editing, project administration, and funding acquisition.

Acknowledgments: We would like to thank our colleague Carl-Gustaf Thulin for contributing to the conceptualization and for commenting on the manuscript, and students Oceane Martinét, Elin Mattsson, and Karin Näslund for contributing to data collection. We also thank staff at the feed laboratory, Department of Animal Nutrition and Management, SLU.

References

1. Gilhaus, K.; Hoelzel, N. Seasonal variations of fodder quality and availability as constraints for stocking rates in year-round grazing schemes. *Agric. Ecosyst. Environ.* **2016**, *234*, 5–15. [CrossRef]

2. Duncan, P. Horses and grasses: The nutritional ecology of equids and their impact on the Camargue. *Ecol. Stud.* **1991**, *87*, 1–287.

3. Gärdenfors, U.; Simán, S.; Lundkvist, K. *Rödlistade arter i Sverige 2005*; ArtDatabanken i samarbete med Naturvårdsverket: Uppsala, Sweden, 2005.

4. Garrido, P.; Mårell, A.; Öckinger, E.; Skarin, A.; Jansson, A.; Thulin, C.G. Experimental rewilding enhances grassland functional composition and pollinator habitat use. *J. Appl. Ecol.* **2019**, *56*, 946–955. [CrossRef]

5. Loucougaray, G.; Bonis, A.; Bouzille, J.-B. Effects of grazing by horses and/or cattle on the diversity of coastal grasslands in western France. *Biol. Conserv.* **2004**, *116*, 59–71. [CrossRef]

6. Öckinger, E.; Eriksson, A.K.; Smith, H.G. Effects of grassland abandonment, restoration and management on butterflies and vascular plants. *Biol. Conserv.* **2006**, *133*, 291–300. [CrossRef]

7. Osoro, K.; Ferreira, L.; García, U.; García, R.R.; Martinez, A.; Celaya, R. Grazing systems and the role of horses in heathland areas. In *Forages and Grazing in Horse Nutrition*; Wageningen Academic Publishers: Wageningen, The Netherlands, 2012; pp. 137–146.

8. Putman, R.; Pratt, R.; Ekins, J.; Edwards, P. Food and feeding behaviour of cattle and ponies in the New Forest, Hampshire. *J. Appl. Ecol.* **1987**, *24*, 369–380. [CrossRef]

9. Yunusbaev, U.; Musina, L.; Suyundukov, Y.T. Dynamics of steppe vegetation under the effect of grazing by different farm animals. *Russ. J. Ecol.* **2003**, *34*, 43–47. [CrossRef]

10. Menard, C.; Duncan, P.; Fleurance, G.; Georges, J.Y.; Lila, M. Comparative foraging and nutrition of horses and cattle in European wetlands. *J. Appl. Ecol.* **2002**, *39*, 120–133. [CrossRef]

11. Gudmundsson, O.; Dyrmundsson, O. Horse grazing under cold and wet conditions: A review. *Livest. Prod. Sci.* **1994**, *40*, 57–63. [CrossRef]

12. Pratt-Phillips, S.E.; Stuska, S.; Beveridge, H.L.; Yoder, M. Nutritional quality of forages consumed by feral horses: The horses of Shackleford Banks. *J. Equine Vet. Sci.* **2011**, *31*, 640–644. [CrossRef]

13. Kuiters, A.; Van der Sluijs, L.; Wytema, G. Selective bark-stripping of beech, Fagus sylvatica, by free-ranging horses. *For. Ecol. Manag.* **2006**, *222*, 1–8. [CrossRef]

14. Helmer, W.; Saavedra, D.; Sylvén, M.; Schepers, F. Rewilding Europe: A New Strategy for an Old Continent. *Rewilding European Landscapes*. Available online: https://link.springer.com/content/pdf/10.1007%2F978-3-319-12039-3.pdf (accessed on 20 March 2019).

15. Linné, C.V. *Öländska och Gotländska Resa, Faximil Edition Produced 1940*; AB Malmö.Ljustrycksanstalt: Malmö, Sweden, 1745.

16. Skarin, A. Habitat use by semi-domesticated reindeer, estimated with pellet-group counts. *Rangifer* **2005**, *27*, 121–132. [CrossRef]

17. Lindgren, E. *Vallfodrets näringsvärde bestämt in vivo och med olika laboratoriemetoder*; Dep of Animal Nutrition and Management: Uppsala, Sweden, 1979.

18. Jansson, A.; Lindberg, J.; Rundgren, M.; Müller, C.; Connysson, M.; Kjellberg, L.; Lundberg, M. *Utfodringsrekommendationer för häst*; Swedish University of Agricultural Sciences: Uppsala, Sweden, 2011.

19. Chai, W.H.; Uden, P. An alternative oven method combined with different detergent strengths in the analysis of neutral detergent fibre. *Anim. Feed Sci. Technol.* **1998**, *74*, 281–288. [CrossRef]

20. Kjeldahl, J. A new method for the determination of nitrogen in organic matter. *Z. Anal. Chem.* **1883**, *22*, 366–382. [CrossRef]

21. Naundrup, P.J.; Svenning, J.-C. A geographic assessment of the global scope for rewilding with wild-living horses (Equus ferus). *PLoS ONE* **2015**, *10*, e0132359. [CrossRef] [PubMed]

22. Carson, K.; Wood-Gush, D.G.M. Equine behaviour: II. A review of the literature on feeding, eliminative and resting behaviour. *Appl. Anim. Ethol.* **1983**, *10*, 179–190. [CrossRef]
23. Fleurance, G.; Duncan, P.; Fritz, H.; Cabaret, J.; Gordon, I.J. Importance of nutritional and anti-parasite strategies in the foraging decisions of horses: An experimental test. *Oikos* **2005**, *110*, 602–612. [CrossRef]
24. Carpenter, D.; Hammond, P.M.; Sherlock, E.; Lidgett, A.; Leigh, K.; Eggleton, P. Biodiversity of soil macrofauna in the New Forest: A benchmark study across a national park landscape. *Biodivers. Conserv.* **2012**, *21*, 3385–3410. [CrossRef]
25. McDonald, P. *Animal Nutrition*; Pearson Education: London, UK, 2002.
26. Andersson, A. *Näringsvärde i betesgräs från naturliga betesmarker*; Swedish University of Agricultural Sciences: Uppsala, Sweden, 1999.
27. Holland, E.A.; Detling, J.K. Plant response to herbivory and belowground nitrogen cycling. *Ecology* **1990**, *71*, 1040–1049. [CrossRef]
28. Kleinebecker, T.; Weber, H.; Hoelzel, N. Effects of grazing on seasonal variation of aboveground biomass quality in calcareous grasslands. *Plant Ecol.* **2011**, *212*, 1563–1576. [CrossRef]
29. Bilotta, G.S.; Brazier, R.E.; Haygarth, P.M. The Impacts of Grazing Animals on the Quality of Soils, Vegetation, and Surface Waters in Intensively Managed Grasslands. *Adv. Agrono.* **2007**, *94*, 237–280.
30. Steen, E.; Matzon, C.; Svensson, C. *Avkastning På Naturbeten*; Markväxt, Lantbrukshögskolan: Uppsala, Sweden, 1972.
31. Swedish Board of Agriculture. Jordbruksstatistisk sammanställning 2017. Available online: http://www.jordbruksverket.se/omjordbruksverket/statistik/statistikomr/jordbruksstatistisksammanstallning/jordbruksstatistisksammanstallning2017.4.695b9c5715ce6e19dbbaacb1.html (accessed on 20 March 2019).
32. Cymbaluk, N.F.; Christison, G.I.; Leach, D.H. Energy uptake and utilization by limit-fed and adlibitum-fed growing horses. *J. Anim. Sci.* **1989**, *67*, 403–413. [CrossRef] [PubMed]
33. Cymbaluk, N.F.; Christison, G.I. Effects of diet and climate on growing horses. *J. Anim. Sci.* **1989**, *67*, 48–59. [CrossRef] [PubMed]
34. LaCasha, P.A.; Brady, H.A.; Allen, V.G.; Richardson, C.R.; Pond, K.R. Voluntary intake, digestibility, and subsequent selection of Matua bromegrass, coastal bermudagrass, and alfalfa hays by yearling horses. *J. Anim. Sci.* **1999**, *77*, 2766–2773. [CrossRef] [PubMed]
35. NRC. *National Research Council Committee Nutrient Requirements of Horses*; National Academies Press: Washington, DC, USA, 2007.
36. Ringmark, S.; Jansson, A. Effects of crude protein content in forage-only diets fed to horses. In Proceedings of the 5th Nordic feed Science Conference, Uppsala, Sweden, 10–11 June 2014.
37. Salter, R.E.; Hudson, R.J. Feeding ecology of feral horses in western Alberta. *J. Range Manag.* **1979**, *32*, 221–225. [CrossRef]
38. Connysson, M.; Muhonen, S.; Lindberg, J.E.; Essen-Gustavsson, B.; Nyman, G.; Nostell, K.; Jansson, A. Effects on exercise response, fluid and acid-base balance of protein intake from forage-only diets in Standardbred horses. *Equine Vet. J.* **2006**, *38*, 648–653. [CrossRef] [PubMed]
39. Swedish Board of Agriculture. Jordbruksmarkens användning 2015. Available online: https://www.jordbruksverket.se/webdav/files/SJV/Amnesomraden/Statistik,%20fakta/Arealer/JO10/JO10SM1601/JO10SM1601_ikortadrag.htm (accessed on 10 June 2019).
40. Dyrendahl, S. Från arbetshäst i jordbruk och skogsbruk till sport- och rekreationshäst. *K. Skogs. Lantbr. Akad. Tidskr.* **1988**, *20*, 239–262.

6

Eye Blink Rates and Eyelid Twitches as a Non-Invasive Measure of Stress in the Domestic Horse

Katrina Merkies [1,2,*]**, Chloe Ready** [1]**, Leanne Farkas** [1] **and Abigail Hodder** [1]

[1]   Department of Animal Biosciences, University of Guelph, Guelph, ON N1G 2W1, Canada
[2]   Campbell Centre for the Study of Animal Welfare, University of Guelph, Guelph, ON N1G 2W1, Canada
*   Correspondence: kmerkies@uoguelph.ca

Simple Summary: Eye blink rate has been used as an indicator of stress in humans and, due to its non-invasive nature, could be useful to measure stress in horses. Horses exhibit both full and half blinks as well as eyelid twitches. We exposed 33 horses to stressful situations such as separation from herdmates, denied access to feed and sudden introduction of a novel object, and determined that full and half eye blinks decrease in these situations. Feed restriction was the most stressful for the horse as indicated by increased heart rate, restless behaviour and high head position. The decrease in eye blink rate during feed restriction was paralleled with an increase in eyelid twitches. There was no increase in eyelid twitches or heart rate with the other treatments indicating that the horses did not find these overly stressful, but they did focus their attention more during these situations. Observation of eye blinks and eyelid twitches can provide important information on the stress level of horses with a decrease in eye blinks and an increase in eyelid twitches in stressful environments.

Abstract: Physiological changes provide indices of stress responses, however, behavioural measures may be easier to determine. Spontaneous eye blink rate has potential as a non-invasive indicator of stress. Eyelid movements, along with heart rate (HR) and behaviour, from 33 horses were evaluated over four treatments: (1) control—horse in its normal paddock environment; (2) feed restriction—feed was withheld at regular feeding time; (3) separation—horse was removed from visual contact with their paddock mates; and (4) startle test—a ball was suddenly thrown on the ground in front of the horse. HR data was collected every five s throughout each three min test. Eyelid movements and behaviours were retrospectively determined from video recordings. A generalized linear mixed model (GLIMMIX) procedure with Sidak's multiple comparisons of least squares means demonstrated that both full blinks (16 ± 12^b vs. 15 ± 15^b vs. 13 ± 11^b vs. 26 ± 20^a full blinks/3 min $\pm$ SEM; a,b differ $p < 0.006$) and half blinks (34 ± 15^{ab} vs. 27 ± 14^{bc} vs. 25 ± 13^c vs. 42 ± 22^a half blinks/3 min $\pm$ SEM; a,b,c differ $p < 0.0001$) decreased during feed restriction, separation and the startle test compared to the control, respectively. Eyelid twitches occurred more frequently in feed restriction ($p < 0.0001$) along with an increased HR ($p < 0.0001$). This study demonstrates that spontaneous blink rate decreases while eyelid twitches increase when the horse experiences a stressful situation.

Keywords: spontaneous blink rate; eyelid twitches; stress; horse; behaviour; welfare

1. Introduction

Stress is defined as the response of an organism to environmental stimuli that threatens its internal equilibrium [1]. As a prey species, the domestic horse (*Equus caballus*) has developed adaptive fear and flight responses when faced with external stressors [2]. However, modern husbandry practices routinely subject horses to aversive stimuli such as transportation, social isolation and medical intervention.

Identifying indicators of stress in the horse is fundamental for the welfare of the animal itself and the safety of the handler [3].

Various physiological measures can be used to assess stress responses in animals including heart rate and heart rate variability [4], blood or salivary cortisol [5]. However, these measures have their limitations, including the increase in stress due to the invasive nature of drawing blood, for example [6]. As a result, researchers have explored behavioural indicators to augment physiological data. For example, horses exposed to various stressors demonstrate higher head carriage [7], focused orientation of the ears [8], increased vocalizations [9] and increased mouth movements [10]. Assessing stress responses in animals appears more accurate when using a combination of both behavioural and physiological indicators [11,12]. A novel scale developed to identify stress-related behaviours subjected 32 horses to known stressful husbandry practices including the sound of electric coat clippers, social isolation and grooming procedures [13]. Moderate to high levels of stress showed an increase in oral behaviours, flared nostrils and flattened or pinned ears which correlated with an increase in heart rate (HR) and salivary cortisol [13].

However, evidence of behaviours associated with stress in horses is conflicting. Horses subjected to two stressful handling tasks—walking across a tarpaulin and walking through streamers attached to an overhead pole—displayed an increase in heart rate variability and eye temperature [14]. The time taken or willingness to complete each task was not associated with physiological indicators, showing that the horses did experience stress even when not overtly displaying stress behaviours [14]. Further, horses undergoing a hair clipping procedure, a known aversive management practice, showed compliant behaviour while displaying an increase in HR, salivary cortisol and eye temperature [15]. These studies suggest that a horse's level of compliance and/or ability to tolerate stressors is not indicative of their level of arousal, and influences such as training may overshadow emotional responses [15].

Understanding the response of an animal to external stressors through valid behavioural indicators can be challenging and subjective, however behaviour is an easily observable and non-invasive measurement [16]. Identifying valid indicators of stress is essential to understanding the animal and ultimately improving welfare.

Eye-blink rate has been used as a non-invasive measure of arousal to predict stress levels in humans [17]. Blinking is defined as a quick movement of the eyelid that opens and closes the palpebral fissure and is composed of three different blinks: spontaneous, reflex and voluntary [18]. The levator palpebrae superioris muscle of the upper eyelid is primarily responsible for opening the eyelid, whereas the orbicularis oculi muscle encircling the palpebral fissure works to close the eyelid [19]. Upon close observation, different eyelid movements are noticeable, ranging from full blinks (complete closure of both eyelids with concomitant suppression of vision) to partial blinks (incomplete closure of the eyelids) and eyelid twitches (movement of the upper eyelid through innervation of the levator palpebrae superioris muscles with no movement of the orbicularis oculi muscles) [20]. Partial blinks have been observed in humans focused on computer terminal displays [21] and have been used as a diagnostic for dry eye disease [22]. Partials blinks have also been documented in both dogs [23] and cats [24]. Due to the large size and lateral placement of a horse's eyes, identifying eyelid movements is easily observable. Although little investigation has been done on half blinks in horses, it has been incorporated into the Equine Facial Action Coding System (EquiFACS) as its own action unit [25].

Spontaneous blinks are uniquely different from voluntary and reflexive blinks, as they can represent a range of information processing functions spanning attention and working memory [26]. Humans subjected to stressful stimuli through social and emotional recollection tests exhibited an increase in spontaneous blink rate and a similar trend has been demonstrated in guinea pigs that are in states of emotional arousal following handling [27]. Although an increase in spontaneous blink rate has been observed in humans subjected to stressful and neurologically-demanding stimuli, spontaneous blink rate has also been found to decrease when the subject is most attentive while performing demanding tasks or exposed to stressful visual stimuli [28,29]. This suggests that humans reduce their spontaneous blink rate when perceiving visual stimuli in order to maximize the amount of information entering the nervous system; thus, increased spontaneous blink rate potentially hinders attention in humans and the ability to perceive immediate stimuli [28]. Therefore, the influence of specific stressors such as visual stimuli, emotional anxiety and/or neurological levels of arousal that initiate a fight or flight response must be considered when investigating behavioural responses in horses [17,28].

While spontaneous blink rate has been used as a non-invasive measure of stress in humans, little research has been applied to using spontaneous blink rate as a behavioural indicator in horses, and no research has differentiated between different eyelid movements. This study aims to investigate the use of spontaneous blink rate and eyelid movements as a non-invasive measure of stress in domestic horses (*Equus caballus*) in response to induced, external stressors. Based on the limited and contrasting evidence reviewed, sources of stressors including feed restriction and separation from herdmates were selected to induce social and neurological states of arousal. In comparison, the startle test was chosen to induce a stress response from visual stimuli. We hypothesized that spontaneous blink rates in horses would significantly increase during feed restriction and separation from conspecifics and decrease during the startle test in comparison to the control.

2. Materials and Methods

This project was approved by the University of Guelph's Animal Care Committee (AUP #3143) in compliance with the Canadian Council on Animal Care guidelines for the use of animals in research.

2.1. Subjects and Housing

This study used 33 riding lesson horses (*Equus caballus*) from three different facilities in Eastern Ontario, Canada. The horses had a mean age of 11 ± 6 years, and all horses were in good health with no documented instances of digestive issues or ulcers that might impact the feed restriction treatment. As a variety of breeds were represented, they were categorized into four categories: Thoroughbred and Thoroughbred crosses ($n = 10$), warmbloods (e.g., Hanoverian; $n = 8$), stock (e.g. Quarter Horse; $n = 7$) and ponies (e.g., Welsh; $n = 8$). Before and after each treatment, the horses were housed in their usual stalls and followed their regular regimes, including being turned out with their normal herdmates and maintaining their normal exercise routines. Their diets and feeding schedules remained the same throughout the study, with the exception of the feed restriction treatment when feed was withheld.

2.2. Treatments

Each horse was exposed to each of the four treatments in randomized order.

(i) Control: the horse was observed individually for three minutes in their normal turnout environment, with the exception of the presence of the handler and observer. The horse was surrounded by, or within sight of, their paddock mates. Observations occurred during quiet times at the facility with no expectation of predictable events such as riding or feeding.

(ii) Feed restriction: the horse was tied individually in their stall during their regular afternoon feeding time, and was observed while feed was withheld for three minutes, during which time the horse watched their neighbouring conspecifics being fed.

(iii) Separation: the horse was led individually from their normal environment to an isolated testing arena. Once there, the horse was asked to stand and was observed for three minutes. There was no visual contact with their conspecifics although auditory contact was still possible. Observations occurred during quiet times at the facility with no expectation of predictable events such as riding or feeding.

(iv) Startle test: the horse was led individually from their normal environment to an isolated testing arena where they were unable to see conspecifics although auditory contact was still possible. Once there, the horse was asked to stand while a ball was thrown suddenly on the ground 2 m in front of the horse (Figure 1), and the horse was observed for three minutes. Observations occurred during quiet times at the facility with no expectation of predictable events such as riding or feeding.

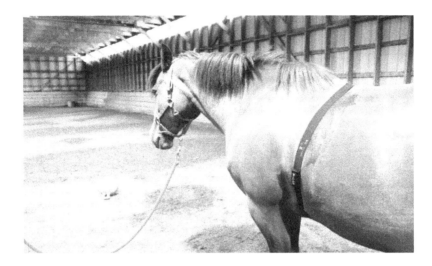

Figure 1. Horse with heart rate (HR) monitor (Polar RS800) during the startle test. The ball was tossed approximately 2 m in front of the horse. The handler maintained a fairly loose lead. The observer (not visible in this photo) was positioned about 3 m from the horse's right eye.

2.3. Data Collection

All data at any one facility were collected over two or three days, and all treatments for any one horse were tested on a single day. At least 10 min prior to each treatment, each horse was outfitted with a heart rate monitor (Polar RS800, Lachine, QC, USA; Figure 1) to allow them to acclimatize to the monitors. Heart rate (HR) was collected every five s throughout each treatment.

During each treatment, a handler held the horse by a lead rope attached to the halter, and the same handler was used throughout the study. The handler held the horse in place on a fairly loose lead (1 m), with just enough contact to maintain the head relatively still without restricting movement. A single observer was used throughout the study and maintained a position about 3 m perpendicular to the right eye of the horse. The observer videotaped all treatments using a Panasonic 2MOS video camera, with focus maintained on the right eye of the horse. One individual coded all the behaviours (Table 1) retrospectively from the videos using Observer XT (Version 12.0, Noldus, Leesburg, VA, USA). All occurrences of eye blinks and eyelid twitches were tallied, while the total duration of ear movement, head movement, mouth movement and restlessness was calculated over each 3 min treatment and reported as a percentage of total time.

2.4. Data Analysis

The data was exported from Observer XT to Microsoft Excel 2011 (Supplementary File SBrate data.xlsx). All statistical analyses were carried out using SAS (Version 9.4, Toronto, ON), and significance was considered to be $p < 0.05$.

As there was no missing data, $n = 33$ for all analyses. Residuals were graphically inspected to determine the fit of the model, and horse HRs were log transformed since they did not achieve a normal distribution (Kolmogorov–Smirnov test, $p < 0.01$). A generalized linear mixed model procedure was used to analyse the effect of treatment on behaviour and HR, with location, age, breed and treatment as the independent variables and horse as the random factor. Sidak's methodology was used to test multiple comparisons of least squared means for each behaviour across treatments.

Table 1. Ethogram of behaviours observed in horses ($n = 33$) during each of the four treatments—control, feed restriction, separation from conspecifics and the startle test. Adapted from [13,25,30,31].

Behaviour	Description
Eye—full blink	The right eye becomes momentarily but completely closed
Eye—half blink	The right upper lid moves toward the lower lid of the eye but does not cover the eye completely
Eyelid—twitch	Fine fibrillar movement of the skin involving the levator palpebrae superioris muscle of the upper eyelid
Ears—forward	The right ear points forward in an attentive manner
Ears—sideways	The right ear is angled to the side
Ears—back	The right ear is turned backward
Head—above withers	The right eye level goes above the height of the withers
Head—even with withers	The right eye is even with the height of the withers
Head—below withers	The right eye level drops below the height of the withers
Oral behaviour	The lips are in motion, either with mouth shut, with the tongue licking or coming out of the mouth, or chewing
Restlessness	Any movement made by the legs, including movement that causes the horse to move out of view of the camera

3. Results

On average, horses performed full blinks 8–9 times/min in the absence of any stressors. This rate decreased to 5 blinks/min in the presence of any external stressors. Conversely, eyelid twitches increased from about 2/min in the control situation to 6/min during feed restriction. Full eye blinks occurred more often during control than during any other treatment ($F(3,95) = 9.88$, $p < 0.0001$; Figure 2). Half blinks occurred most often during control and feed restriction treatments, and least often during separation or startle test ($F(3,95) = 10.65$, $p < 0.0001$; Figure 2). Eyelid twitches were more evident during the feed restriction treatment than during any other treatment ($F(3,95) = 9.46$, $p < 0.0001$; Figure 2).

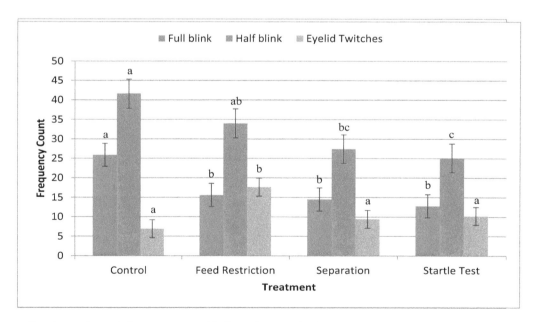

Figure 2. Total number of observations of full eye blinks, half blinks and eyelid twitches (±SD) in horses ($n = 33$) over a 3 min observation period during control, feed restriction, separation from conspecifics or a startle test. a,b,c differ across treatments $p < 0.0001$.

Horse heart rate was higher during feed restriction (44 ± 13 beats per minute (bpm)) and lower during separation (37 ± 7 bpm) and the startle test (37 ± 8 bpm) compared to the control (39 ± 8 bpm) ($F_{(3,92)} = 306.12$, $p < 0.0001$). There was no effect of facility ($p > 0.05$) on the behaviours or HR.

The horses' right ear was forward more often during separation and the startle test ($F_{(3,95)} = 8.29$, $p < 0.0001$; Figure 3), whereas it was more often sideways during feed restriction and the control ($F_{(3,95)} = 22.53$, $p < 0.0001$). There was no difference among treatments for the percentage of time the horses had their ears back ($F_{(3,95)} = 0.82$, $p > 0.49$).

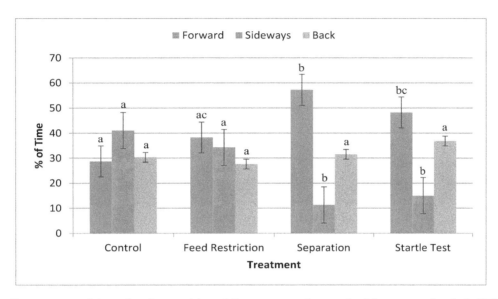

Figure 3. Percentage of time that horses' ($n = 33$) ears were forward, sideways or back ($\pm$SD) over a 3 min observation period during feed restriction, separation from conspecifics, a startle test or control. a,b,c differ across treatments $p < 0.0001$.

Horses held their head raised more frequently during feed restriction ($F_{(3,95)} = 30.02$, $p < 0.0001$; Figure 4) and held their head low more often during the control treatment and startle test ($F_{(3,95)} = 7.15$, $p = 0.0002$).

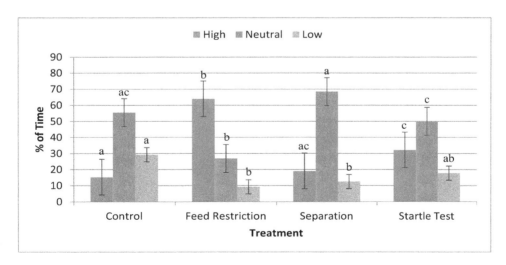

Figure 4. Percentage of time that horses' ($n = 33$) heads were held high, neutral or low ($\pm$SD) over a 3 min observation period during feed restriction, separation from conspecifics, a startle test or control. a,b,c differ across treatments $p < 0.0001$.

Oral behaviours were most evident during the feed restriction, with significantly fewer during separation and the startle test ($F_{(3,95)} = 11.42$, $p < 0.0001$). Horses were more often restless during feed restriction than separation or the startle test ($F_{(3,95)} = 6.78$, $p = 0.0003$).

4. Discussion

The aim of this study was to determine changes in eyelid movement in horses during exposure to stressful stimuli. It was expected that spontaneous blink rate would increase during exposure to mental stressors such as feed restriction and separation from conspecifics and decrease during exposure to visual stimuli via the startle test. However, our results showed a decrease in both full and half blinks in response to each of the test situations.

Feed restriction was clearly the most stressful situation for the horse demonstrated by an increase in HR, restlessness, oral behaviours and percentage of time the head was held high. However, as HR increased only slightly, it may be concluded that this was only a mildly stressful situation. During separation from conspecifics and the startle test, horses had their right ear forward for more of the time, indicating focused attention in front of them that presumably maximized visual information processing. The horses may also have been attentive to sounds emanating outside of the test arena, although such sounds could come from any direction. Heart rate during these two situations was slightly decreased, showing that while attention was focused, physiologically the horses did not appear stressed. It could be that the presence of the human handler and observer may have had an appeasing effect on the horse, as Merkies et al. [7,10] demonstrated a decrease in HR when any human was present rather than the horse being completely alone.

Like full blinks, half blinks were most evident in the control environment, with the lowest frequency in the startle test where visual stimulation is important for the horse to process information about their environment. Similarly, in humans, the number of both full and partial blinks decreased when they were asked to perform a reading test requiring visual concentration [32]. Also shown in guinea pigs, eye blink rate decreased during handling compared to the control, although this study did not verify the stressful effects of handling [27]. Conversely, both full and half blinks characterized in cats increased under conditions of induced fear [24].

A recent study investigating stress responses in horses observed a decrease in mean spontaneous blink rate only during the first minute of a 10 min sham clipping procedure [33]. This decline preceded the onset of a significant increase in spontaneous blink rate for the following nine min, suggesting that the initial reduction may be characteristic of a fight or flight response, allowing the horse to visually fixate on the stimulus before responding accordingly. In our current study, blink rate was calculated as a mean over three min. It could be that a longer time frame may have noted a subsequent increase in spontaneous blink rate similar to Mott et al.'s [33] study. It would have been interesting if Mott et al. [33] had collected data on differing eyelid movements.

Our results showed an increase in eyelid twitches during feed restriction, which the horses found stressful. Conversely, separation from conspecifics and the startle test did not evoke an increase in eyelid twitches, suggesting that eyelid twitches are more prevalent during stressful situations as the horses did not appear to find these two latter situations stressful.

Research investigating horse facial expressions led to the development of the equine Facial Action Coding System (EquiFACS) for use in determining behavioural indicators of a horse's emotional state [25]. In accordance with EquiFACS, the Equine Pain Face and Horse Grimace Scale (HGS) were created after observing changes in equine facial expressions when horses were subjected to painful procedures. These scales identified notable changes in eye expression associated with horses experiencing negative situations [31,34]. An angled eye was found to be indicative of stress by comparing the shape of the eye with the appearance of eye-white during treatments associated with negative arousal [35]. The appearance of "worry wrinkles", defined as a contraction of the levator anguli oculi medialis muscle and the corrugator supercilii muscle, are prominent during situations involving negative arousal such as food competition, in contrast to relaxed eyes during positive stimuli such as grooming [35]. These "worry wrinkles" may be similar to eyelid twitches that appear to increase under stressful situations. As Hintze et al.'s [35] study analysed photographs, what they may have witnessed was an eyelid twitch in a static moment. Furthermore, a more recent study showed that worry wrinkles could be assessed systematically regardless of horse sex, age or breed, with the exception of angle. In this instance, thoroughbred horses displayed less contraction of the eyelid muscles than ponies, with coldbloods displaying the strongest contraction [36].

In horses, an increase in spontaneous blink rate was associated with a more anxious temperament accompanied by more movement, while a decrease in spontaneous blink rate was associated with more docile behaviour [37]. The authors proposed that these differences in temperament may be directly related to striatal dopamine levels, with anxious horses having elevated dopamine while docile horses have lower levels of dopamine. In humans, spontaneous blink rate has been positively associated with striatal dopamine production. For example, it is well known that spontaneous blink rate is decreased in humans experiencing Parkinson's disease, an affliction that causes reduced functioning due to a decrease in dopamine production, whereas patients with schizophrenia (a hyperdopaminergic condition) exhibit increased anxiety and increased blink rate [38]. Additionally, Colzato et al. [39] demonstrated that both high and low spontaneous blink rates caused a decrease in performance in humans undergoing a start–stop task whereas those with average spontaneous blink rates performed best. These results emphasize the effect of tonic dopamine levels on spontaneous blink rate, which is an important baseline to understand changes in blink rates as a result of induced environmental stressors. It may be that a reduced spontaneous blink rate could indicate underlying pathologies such as depression, or alternatively that activity level may not necessarily reflect the level of arousal.

To our knowledge, this is the first report of the significance of eyelid twitches in horses. Blinking, and even more so eyelid twitches, are relatively easy measures to examine in horses, although the observer must be fairly close to the horse in question. Monitoring changes in blink rate, and in particular eyelid twitches, could alert the observer to changes in the level of arousal of the horse. A decrease in spontaneous blink rate concomitant with an increase in eyelid twitches may indicate a stressful situation for the horse, whereas a decrease in spontaneous blink unaccompanied by an increase in eyelid twitches may indicate an environment that is engaging but not stressful to the horse. Further research investigating specific eyelid movements in relation to level of arousal could give us insight into the emotional responses of horses. For example, in humans, facial electromyography has been successfully used to correlate facial muscle activation to positive or negative emotions. Since we cannot ask horses to self-report how they are feeling, physiological measures that differentiate between pleasant and unpleasant experiences may allow us to infer underlying emotions. Further investigation of changes in spontaneous blink rate and eyelid twitches over varying time spans is needed to identify patterns and temporal trends in response to stressful stimuli.

5. Conclusions

Horses exposed to stressful environments decrease their spontaneous eye blink rate and increase the frequency of eyelid twitches. However, if the environment is simply visually stimulating, eyelid twitches do not appear to increase even if eye blink rate decreases. Monitoring spontaneous blink rate is a sensitive metric of neural activity and differentiating eye blinks from eyelid twitches may provide insight on the level of arousal of the horse.

Author Contributions: Conceptualization, K.M.; methodology, K.M. and C.R.; formal analysis, K.M.; investigation, K.M., C.R. and L.F.; resources, K.M.; data curation, K.M. and L.F.; writing—original draft preparation, L.F. and A.H.; writing—review and editing, K.M.; supervision, K.M.; project administration, K.M.; funding acquisition, K.M.

Acknowledgments: The authors would like to thank the three stables who volunteered the use of their school horses for data collection. Also, thanks to Amelia Garnett for coding many of the videos.

References

1. Ramos, A.; Mormede, P. Stress and emotionality: A multidimensional and genetic approach. *Neurosci. Biobehav. Rev.* **1998**, *22*, 33–57. [CrossRef]
2. Budzynska, M. Stress reactivity and coping in horse adaptation to environment. *J. Equine Vet. Sci.* **2014**, *34*, 935–941. [CrossRef]
3. Yarnell, K.; Hall, C.; Royle, C.; Walker, S.L. Domesticated horses differ in their behavioural and physiological responses to isolated and group housing. *Physiol. Behav.* **2015**, *143*, 51–57. [CrossRef] [PubMed]
4. Von Borell, E.; Langbein, J.; Després, G.; Hansen, S.; Leterrier, C.; Marchant-Forde, J.; Marchant-Forde, R.; Minero, M.; Mohr, E.; Prunier, A.; et al. Heart rate variability as a measure of autonomic regulation of cardiac activity for assessing stress and welfare in farm animals—A review. *Physiol. Behav.* **2007**, *92*, 293–316. [CrossRef] [PubMed]
5. Janczarek, I.; Wilk, I.; Stachurska, A.; Krakowski, L.; Liss, M. Cardiac activity and salivary cortisol concentration of leisure horses in response to the presence of an audience in the arena. *J. Vet. Behav.* **2019**, *29*, 31–39. [CrossRef]
6. Palme, R. Non-invasive measurement of glucocorticoids: Advances and problems. *Physiol. Behav.* **2019**, *199*, 229–243. [CrossRef]
7. Merkies, K.; Sievers, A.; Zakrajsek, E.; MacGregor, H.; Bergeron, R.; König von Borstel, U. Preliminary results suggest an influence of psychological and physiological stress in humans on horse heart rate and behaviour. *J. Vet. Behav.* **2014**, *9*, 242–247. [CrossRef]
8. McKinney, C.; Mueller, M.K.; Frank, N. Effects of therapeutic riding on measures of stress in horses. *J. Equine Vet. Sci.* **2015**, *35*, 922–928. [CrossRef]
9. Visser, E.K.; van Reenen, C.G.; Hopster, H.; Schilder, M.B.H.; Knaap, J.H.; Barneveld, A.; Blokhuis, H.J. Quantifying aspects of young horses' temperament: Consistency of behavioural variables. *Appl. Anim. Behav. Sci.* **2001**, *74*, 241–258. [CrossRef]
10. Merkies, K.; McKechnie, M.J.; Zakrajsek, E. Behavioural and physiological responses of therapy horses to mentally traumatized humans. *Appl. Anim. Behav. Sci.* **2018**, *205*, 61–67. [CrossRef]
11. Mason, G.; Mendl, M. Why is there no simple way of measuring animal welfare? *Anim. Welf. Sci.* **1993**, *2*, 301–319.
12. Hall, C.; Huws, N.; White, C.; Taylor, E.; Owen, H.; McGreevy, P. Assessment of ridden horse behavior. *J. Vet. Behav.* **2014**, *8*, 62–73. [CrossRef]
13. Young, T.; Creighton, E.; Smith, T.; Hosie, C. A novel scale of behavioural indicators of stress for use with domestic horses. *Appl. Anim. Behav. Sci.* **2012**, *140*, 33–43. [CrossRef]
14. Squibb, K.; Griffin, K.; Favier, R.; Ijichi, C. Poker Face: Discrepancies in behaviour and affective states in horses during stressful handling procedures. *Appl. Anim. Behav. Sci.* **2018**, *202*, 34–38. [CrossRef]
15. Yarnell, K.; Hall, C.; Billett, E. An assessment of the aversive nature of an animal management procedure (clipping) using behavioral and physiological measures. *Physiol. Behav.* **2013**, *118*, 32–39. [CrossRef]
16. Mendl, M.; Burman, O.H.P.; Paul, E.S. An integrative and functional framework for the study of animal emotion and mood. *Proc. Biol. Sci.* **2010**, *277*, 2895–2904. [CrossRef]
17. Giannakakis, G.; Pediaditis, M.; Manousos, D.; Kazantzaki, E.; Chiarugi, F.; Simos, P.G.; Mariasa, K.; Tsiknakis, M. Stress and anxiety detection using facial cues from videos. *Biomed. Signal Process. Control* **2017**, *31*, 89–101. [CrossRef]
18. Cruz, A.A.V.; Garcia, D.M.; Pinto, C.T.; Cechetti, S.P. Spontaneous eyeblink activity. *Ocul. Surf.* **2011**, *9*, 29–41. [CrossRef]
19. Koo Lin, L. Eyelid anatomy and function. In *Ocular Surface Disease: Cornea, Conjunctiva and Tear Film*; Holland, E.J., Mannis, M.J., Lee, W.B., Eds.; Elsevier: Cambridge, UK, 2013; pp. 11–15. ISBN 978-1-4557-2876-3.
20. Blount, W.P. Studies of the movements of the eyelids of animals: Blinking. *Exp. Physiol.* **1927**, *18*, 111–125. [CrossRef]
21. Portello, J.K.; Rosenfield, M.; Chu, C.A. Blink rate, incomplete blinks and computer vision syndrome. *Optom. Vis. Sci.* **2013**, *90*, 482–487. [CrossRef]
22. Jie, Y.; Sella, R.; Feng, J.; Gomez, M.L.; Afshari, N.A. Evaluation of incomplete blinking as a measurement of dry eye disease. *Ocul. Surf.* **2019**, *5*. [CrossRef]

23. Nakajima, S.; Takamatsu, Y.; Fukuoka, T.; Omori, Y. Spontaneous blink rates of domestic dogs: A preliminary report. *J. Vet. Behav.* **2011**, *6*, 95. [CrossRef]
24. Bennett, V.; Gourkow, N.; Mills, D.S. Facial correlates of emotional behaviour in the domestic cat (Feliscatus). *Behav. Proc.* **2017**, *141*, 342–350. [CrossRef]
25. Wathan, J.; Burrows, A.M.; Waller, B.M.; McComb, K. EquiFACS: The equine facial action coding system. *PLoS ONE* **2015**, *10*, e0131738. [CrossRef]
26. Rubin, M.; Denise, H.; Dipanjana, D.; Melara, R. Inhibitory Control under Threat: The Role of Spontaneous Eye Blinks in Post-Traumatic Stress Disorder. *Brain Sci.* **2017**, *7*, 16. [CrossRef]
27. Trost, K.; Skalicky, M.; Nell, B. Schirmer tear test, phenol red thread tear test, eye blink frequency and corneal sensitivity in the guinea pig. *Vet. Ophthalmol.* **2007**, *10*, 143–146. [CrossRef]
28. McIntire, L.K.; McKinley, R.A.; Goodyear, C.; McIntire, J.P. Detection of vigilance performance using eye blinks. *Appl. Ergon.* **2014**, *45*, 354–362. [CrossRef]
29. Weiner, E.A.; Concepcion, P. Effects of affective stimuli mode on eye-blink rate and anxiety. *Clin. Psych.* **1975**, *31*, 256–259. [CrossRef]
30. Lemasson, A.; Boutin, A.; Boivin, S.; Blois-Heulin, C.; Hausberger, M. Horse (*Equus caballus*) whinnies: A source of social information. *Anim. Cognit.* **2009**, *12*, 693–704. [CrossRef]
31. Gleerup, K.B.; Forkman, B.; Lindegaard, C.; Andersen, P. An equine pain face. *Vet. Anaesth. Analg.* **2015**, *42*, 103–114. [CrossRef]
32. Argilés, M.; Cardona, G.; Pérez-Cabré, E.; Rodríguez, M. Blink rate and incomplete blinks in six different controlled hard-copy and electronic reading conditions. *Investig. Ophthalmol. Vis. Sci.* **2015**, *56*, 6679–6685. [CrossRef]
33. Mott, R.; Hawthorne, S.; McBride, S. Spontaneous blink rate as a measure of equine stress. In Proceedings of the 14th Conference of the International Society for Equitation Science, Rome, Italy, 21–24 September 2018.
34. Dalla Costa, E.; Minero, M.; Lebelt, D.; Stucke, D.; Canali, E.; Leach, M.C. Development of the Horse Grimace Scale (HGS) as a pain assessment tool in horses undergoing routine castration. *PLoS ONE* **2014**, *9*, e92281. [CrossRef]
35. Hintze, S.; Smith, S.; Patt, A.; Bachmann, I.; Wurbel, H. Are eyes a mirror of the soul? What eye wrinkles reveal about a horse's emotional state. *PLoS ONE* **2016**, *11*, e0164017. [CrossRef]
36. Schanz, L.; Krueger, K.; Hintze, S. Sex and age don't matter, but breed type does—Factors influencing eye wrinkle expression in horses. *Front. Vet. Sci.* **2019**, *6*, 154. [CrossRef]
37. Roberts, K.; Hemmings, A.J.; Moore-Colyer, M.; Parker, M.O.; McBride, S.D. Neural modulators of temperament: A multivariate approach to personality trait identification in the horse. *Physiol. Behav.* **2016**, *67*, 125–131. [CrossRef]
38. Jongkees, B.J.; Colzato, L.S. Spontaneous eye blink rate as predictor of dopamine-related cognitive function—A review. *Neurosci. Biobehav. Rev.* **2016**, *71*, 58–82. [CrossRef]
39. Colzato, L.S.; van den Wildenberg, W.P.M.; van Wouwe, N.C.; Pannebakker, M.M.; Hommel, B. Dopamine and inhibitory action control: Evidence from spontaneous eye blink rates. *Exp. Brain Res.* **2009**, *196*, 467–474. [CrossRef]

7

Green Assets of Equines in the European Context of the Ecological Transition of Agriculture

Agata Rzekęć [1], Céline Vial [1,2,*] and Geneviève Bigot [3]

[1] Research Unit MOISA (Marchés, Organisations, Instituts et Stratégies d'acteurs)-French National Research Institute for Agriculture, Food and Environment (INRAE), CIHEAM-IAMM, CIRAD, Montpellier Supagro, Univ Montpellier, 34060 Montpellier, France; agata.rzekec@gmail.com

[2] Pôle Développement, Innovation, Recherche-French Institute for Horse and Horse Riding (Ifce), 61310 Exmes, France

[3] Université Clermont Auvergne, AgroParisTech, French National Research Institute for Agriculture, Food and Environment (INRAE), VetAgro Sup, Research Unit Territoires, 63000 Clermont-Ferrand, France; genevieve.bigot@irstea.fr

* Correspondence: celine.vial@inrae.fr

Simple Summary: Equines have a peculiar place in our society. From livestock to sport, through to landscape managers and leisure partners, equines show a wide range of little-known environmental advantages and assets. Today's wake-up calls about the environment are progressively putting pressure on stakeholders of the agricultural sector, including the equine industry. This study focusses on the main environmental consequences of equine use and possession in Europe based on scientific and technical sources under the lens of five leading sectors where equines show unique impacts as green assets. Now, more than ever before, it is important to highlight the role of equines as a green alternative in political debates and management practices to give them the place equines deserve in the ecological transition of agriculture.

Abstract: Despite the decline of equine populations in the middle of the 20th century, the European horse industry is growing again thanks to economic alternatives found in the diversification of the uses of equines (sports, racing, leisure, etc.). Equines have many environmental advantages, but the fragmentation of the sector and the lack of synthetic knowledge about their environmental impacts do not enable the promotion of these assets and their effective inclusion in management practices and European policies. To highlight the equine environmental impacts, a literature review was carried out to cover the main European stakes. This work led to the identification of five "green assets", fields where equines show unique environmental advantages compared to other agricultural productions. These green assets are linked to the nature of equines (grazing and domestic biodiversity), to their geographical distribution (land use), and to their use by human beings (tourism and work). Today, when searching for sustainable solutions to modern environmental issues, the use of equines is a neglected green alternative. Better knowledge and use of equine green assets could partly respond to more ecological agricultural needs and contribute to the development of this animal industry, which has a place in regional development and in Europe's sustainable transition.

Keywords: equine; horse; environment; green assets; land use; equine grazing; domestic biodiversity; equine and equestrian tourism; equine work; multifunctional review

1. Introduction

In the European Union (EU), after World War II, equine numbers declined drastically because of the motorization of transport (estimates generally agree that horse numbers decreased approximately

90% in Europe by the 1950s [1]). For example, in France, the total number of equines was evaluated to be three million at the beginning of the 20th century but was less than half a million at the end of this century (Figure 1). Before 1950, horses were largely used for agriculture, transportation, and the army. This was particularly the case for heavy (or draft horses) (represented by the light color in Figure 1), but also for saddle horses (represented by the dark color in Figure 1). The European community was built after the Second World War to maintain peace and ensure the autonomy of its inhabitants regarding basic necessities, particularly food products. To this end, a Common Agricultural Policy (CAP) (which still exists today) was created to improve agricultural production, first to improve cereal yields and then to improve animal productions. However, equines were not included in the CAP's plans. In this post-war period, the market situation was geared toward productivism, where the disinterest in equines as a source of power and their absence in development policies led the European equine population to collapse. In France, the decline of heavy horses led to the construction of national programs in 1980 to develop meat production [2,3]. After 1970, saddle horses began to be used for other purposes (sport and leisure), which explains the progressive increase in their numbers (Figure 1). This trend was similar in Sweden. In 1920, there were 700,000 horses; then, their numbers decreased to around 95,500 in 1980 before increasing again [4]. However, today, the equine population is still lower (48%) than it was in 1920 (362,700 heads in 2010) [5]. However, the recent increases in equine livestock have not been observed in Mediterranean countries: 87% of horses were lost in Greece between 1983 and 2000, and 31% and 36% were lost in Spain and Portugal, respectively, between 1987 and 2000 [6]. This population decline could be linked to difficulties in national economies during this period.

Today, Europe has 88.4 million cattle, 150 million pigs, 86.8 million sheep, and 12.7 million goats to ensure the animal protein needs of the European population (Eurostat, 2017), but Europe has only six million equines [7], according to the European Horse Network (a non-profit network of stakeholders acting at the world, European, national, or regional level within the European equine sector).

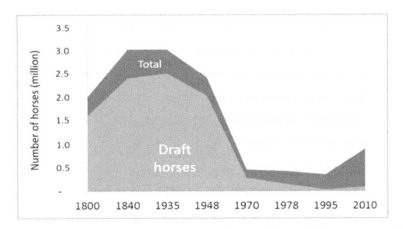

Figure 1. Evolution of the horse population in France from 1800 to 2010 (translated from French [8]).

Since the end of the 20th century, the equine industry has undergone significant evolutions linked to the diversification of equine uses: first, in terms of sports, racing, and leisure; and second, in terms of meat and milk productions, traction, therapy, and even companionship. This gave rise to debates on the status of equines (as a farm animal or pet) between European countries but also inside each country. For example, in the United Kingdom, horses are seen as companion animals, whereas in other European countries—France, Germany, or Sweden for example—equines are considered livestock [9].

These issues are particularly problematic because the European agricultural census could be a powerful tool to quantify the equine population at the European level, but today, this census underestimates the equine population because only equines kept by farmers are counted (EU Regulation (EC) No 1166/2008 19.11.2008), whereas, for example, in France, only half of the equine livestock was kept on farms in 2010 [10]. This problem is similar in other European countries (in Germany, for example) [11]. No other database exists at this scale, so the current official figures are misleading.

To counter this, the European Union required the creation of a central national database for each Member State that would identify all equines. This requirement was presented in the EU Regulation (EC) No 262/2015 3.03.2015, and the creation of these databases remains a work in progress in some member states today [9]. This lack of data concerning equine numbers in Europe complicates descriptions about their importance and impacts.

At present, the diversity of equine uses leads to a large variety of impacts on the environment, especially when activities are exclusive to equines, such as sports or racing. This creates difficulties in listing and evaluating the environmental impacts of the entire equine industry. However, in today's European context, local authorities aim to maintain rural activities and to support agriculture in its sustainable transition. These terms, in the EU context, include all policies that search for the transformation of current societal systems to minimize negative effects on the environment and promote innovative projects [12]. A variety of new uses of equines could meet these challenges. In particular, the European Horse Network has expressed the need for a foundation of scientific resources to build arguments and promote equines in European policies and debates. Consequently, this study describes equines not only as animal producers, but also as ecosystem service providers, especially for land use and biodiversity conservation. This choice consolidates the fact that equines are not only seen as a source of agricultural goods (leisure diversification or meat production, for example), but generate a wide range of other externalities.

The aim of this review is to highlight the most important services provided by equines for the environment at the European level. In order to answer public and professional stakeholders' questions about the inclusion of equines in public policies according to priority, we first met with stakeholders to understand their main issues. Then, a literature review of the available knowledge on the green assets of equines was conducted. Consequently, this paper examines the environmental assets (and limits) of equines that appear to be most important in the context of European policies about agriculture and rural development.

2. Choice and Definition of the Main Green Assets

To achieve this goal, the reflection about equine environmental assets started with fifteen interviews among key stakeholders coming from various institutes at the French and European level: the European Commission, the European Parliament, the French Ministry of Agriculture, the International Federation of Equestrian Tourism, the French Permanent Representation in Brussels, and the European Federation of Working Horses. The aims of the interviews were also to estimate the main issues of the equine industry toward environmental challenges. The interviews were semi-directed and were conducted from May 2019 to July 2019. The choice of respondents was made in order to have a large panel of expertise scales (France or Europe) and professional functions (researchers, institutional stakeholders, and professional representatives). The distribution of these 15 interviews was as equal as possible in function and level of decision (Table 1). The high number of interviews from European institutional stakeholders was undertaken to have a better view of the diversity of the issues at this level. Put another way, it was not easy to find professional representatives concerned with the green assets of equines at the European level.

Table 1. Distribution of interviewees according to the scale of expertise and the function.

Scale of Expertise/Function *	Researchers	Institutional Stakeholders	Professional Representatives
France	3	3	3
Europe	1	4	1

* Respondents who answered questions about the French equine industry are counted in the group "Scale of expertise: France." Those who answered questions with a European point of view are counted in "Scale of expertise: Europe".

These interviews entailed a reflection based on sectors where equines have particular and specific impacts on the environment, here called "green assets".

Globally, the most cited impact is grazing, which was described positively, e.g., in terms of pasture maintenance, complementarity with other livestock, and as a carbon sink, even if some respondents insisted on the destructive nature of equine grazing when mismanaged. Equine work and tourism were seen as green alternatives even if the dynamism of equine work was perceived quite negatively, at least in Western Europe. Another point was the key role of buildings and infrastructures related to equines. These buildings are, in most cases, made from wood, which is a renewable material with positive technical characteristics, e.g., isolation, and favorable effects on the landscape. Manure methanization was also mentioned. Indeed, recycling manure by producing energy is a promising way to improve the environmental image of the equine industry. More generally, according to the respondents, equines have quite a positive impact on the environment since they do not eject as much methane as cattle do, they graze, there is a very rich domestic biodiversity that permits a wide range of uses, even as alternatives for engines in agriculture, for example. Often, no negative effects were spontaneously mentioned by stakeholders when asked about environmental impacts.

This process led to an evaluation of the whole European equine industry that considered the current European context and challenges. Green assets were directly or indirectly linked to European rural policies: the maintenance of open areas through grazing, agritourism, and the maintenance of endangered breeds [7].

Green assets are generally directly linked to:

- The inherent nature of equines (as non-ruminant herbivores whose species presents a high biodiversity of breeds).
- The geographical repartition of equines and their particular land use.
- Their use by humans, which can offer environmental benefits. Even if all human activities with equines benefit from a green image linked to the use of an animal, only some of these uses have positive environmental impacts. Of course, the impacts of equine breeding, whatever the equine's future use, are considered when examining green assets for land use, breed biodiversity, and grazing. We decided to focus on two equine uses that generate specific environmental advantages: equine work and tourism.

This initial information was completed through a literature review of the newest sources possible to build the first state-of-the-art overview of this subject. This review mainly focused on Europe, even though some sources also concerned other continents. In addition to international references, many countries, especially in Europe, developed studies published only in their national language. Consequently, this study was supplemented by French-language references. From March 2019 until November 2019, we used two main databases, Web of Science and Google Scholar to search for articles and reports. To select documents, we searched for the terms "equid*", "equine*," "environment*," "horse*," and all words related to green assets, in both French and English.

This literature review highlighted key arguments for five green assets:

- Equine grazing: Equine grazing incidence is unique because of this animal's morphology and its physiology specificities, especially regarding ruminants (the main herbivores raised on grasslands). In particular, equine grazing is done on different patches made of lawns and high grasses.
- Domestic biodiversity: Human and environmental selection has led to a rich diversity of equine breeds all over the world. Some of these breeds are currently endangered and their conservation is an important issue, which could be introduced in European policies.
- Land use: Equines are present in various areas, especially where other livestock is presently absent. This land use is directly linked to the place of equines in society; as livestock, it is possible to find equines in farms and large areas, but as family pets, equines can be encountered near houses, sometimes on small plots of lands that are not usable for agriculture.
- Tourism: Equines can be used as a means of transport but also as travel companions to discover wild countries and landscapes.

- Equine work: Equines are also used in tourism, cities, and agriculture as a source of energy, whereas other livestock are not, at least in Europe.

3. State of the Current Scientific Knowledge Concerning the Five Equine Green Assets

3.1. How does the Inherent Nature of Equines Impact the Environment?

Equines are non-ruminant herbivores. Equine grazing impacts pastures differently than cattle, sheep, or goat grazing thanks to the particular physiology and morphology of equines, who have a double row of incisors and a high capacity for ingestion linked to the absence of a rumen. They adapt their diet easily according to the available forage. Their behavior also differs from that of cattle in their feed preferences and greater movement when grazing (as there is no rumination rest). These differences induce various impacts on grasslands according to the whether horses graze alone or are associated with ruminants.

3.1.1. Equine Grazing

Grasslands in Europe and Their Maintenance

Meadows are lands covered by grasses and legumes and mainly aim to be a feed source for livestock through grazing or mowing. Meadows are known to be a carbon sink: they can stock 60 to 70 tons of carbon per hectare in temperate areas [13,14]. Moreover, the presence of legumes allows for the fixation of nitrogen from the air to the ground. There are different types of meadows: permanent grasslands (retained over 5 years, with possible reseeding after 5 years); semi-natural grasslands (a particular kind of permanent grassland because they exist for more than five years and are known to be among the most species-rich habitats in Europe [15]); cultivated grasslands (seeded each year (or more often than 5 years), and they are the most commonly used in current breeding systems [15]); and other less productive grasslands present in arid or rugged areas, which are called rangelands (shrub lands, steppes, alpine communities, marshes, tundras, etc.) [16]. In Europe, grasslands cover 21% of the agricultural land in the European Union, while croplands cover 22% and woodlands cover 38% (Eurostat, 2015). However, there is a decrease in permanent grasslands in Europe [15], while pastures are known to have different positive impacts: they are seen as pleasant and aesthetic [16] and provide multifunctional goods [15] that produce agricultural commodities and maintain biodiversity, soil, and water quality, even in suburban areas [17], in addition to meeting the needs of herbivorous productions. Animal grazing presents three general consequences: (1) it maintains a certain level of vegetal biomass [18] with the control of invasive species (through the intake of plants [19] and trampling [19]); (2) it has effects on plant metabolism (defence, resistance, and avoidance) [19]; and (3) it enables the creation of ecological niches [20].

Equine Grazing Specificities

Equines show specificities that are morphological, physiological, and behavioral. First, equines have two rows of incisors that allow them to graze on short grasses [16] lower than cattle are able to and reach young plants that are easily digestible [20]. This leads to a panel of feed preferences. Equines are graminoids feeders [16] and make less use of forbs and legumes than cattle [15]. Equines' preferences change when there is feed shortage. Equines move more easily toward less palatable grasses species than cattle [21], particularly in winter [16]. This may be explained by their physiology. The absence of rumination permits more time for feed intake [22]; less methane emissions compared to cattle (100 kg/CH_4/dairy cow in Western Europe [23], more specifically: 117.9 kg CH_4/dairy cow/year in France [24], and 18.0 kg CH_4/horse/year in Western Europe [23] and 20.7 kg CH_4/horse/year in France [24]); and no limitation of intake capacity due to rumen volume [25]. Consequently, the global intake capacity of equines can be bigger than that of cattle [16] as equines graze longer than cattle [26]. Most equines (except high-performance horses like racehorses) do not need high-quality feed; instead, they are able to live on low-nutrition feed [27]. Moreover, their physiology explains their adaptation to a low starch diet because of the low activity of amylase [15]. They also seem not to be able to digest the secondary metabolites of some plants (such as shrubs) [26]. Because of this, there should be a lower control of shrubs when equines graze them compared to other herbivorous species. However, in extensive Mediterranean conditions with a low stocking rate, shrubs are controlled by local equines [16]. Equines do not appear to be affected by shrubs' defences, such as thorns [26].

As a consequence of these specificities, equines can adapt their intake through a reduction of feed resources during harsh environmental periods [16]. Some primitive equine breeds are better able to mobilize the bodily reserves they gain during summer to survive during the winter if grass is not sufficient [16] than some specialized equine breeds. For example, in Iceland, despite the cold climate, equine grazing occurs even during winter and is supplemented with the hay refused by cattle or sheep [16]. Likewise, equines can be raised in Camargue, where plants are halophilic and resources are scarce [16].

Equines impact pasture differently than other herbivores because their grazing behavior induces a particular heterogeneity of plant cover [28]. Some areas with high plants are not grazed by equines and could be used as latrines [15], but they could be grazed if there were a lack of resources [27]. Nevertheless, there is also the risk of overgrazing and soil erosion in equine pastures [15]. Equines can impact a pasture, for example, by trampling during periods of exercise [29] or if they are concentrated in small areas. Finally, gnawing on trees in semi-natural pastures was reported [20] due to the lack of minerals [15], but this can be reduced with supplementation [30].

Studies on exclusive equine grazing show the positive impacts on flora, including an increase of legumes in France [20], a control of competitive grass [15,20], and control of some shrub species in Camargue [20] (e.g., Vaccinium *myrtillus* by trampling [16,31]), alongside negative impacts, such as the limited control of fast forest regeneration in boreal conditions [15] and an increase of foams in Iceland [20]. In particularly harsh conditions, some lands are maintained by equines to decrease fire risks, for example, by reducing the aerial biomass of gorse in Galicia [16]). Equines are able to preserve and maintain pastoral biodiversity [15] by grazing in areas abandoned by agriculture. This process is identified as a specific threat to habitats and species by the European Union, as the invasion of some plants left non-grazed by livestock leads to landscape closure [25].

Equine grazing also impacts fauna:

- The populations of small herbivores increased due to high-quality vegetative regrowth.
- Insectivorous birds, such as the spoonbill, appeared in pastures grazed by equines in the Netherlands (as with ducks in French wetlands) [20].
- The wolf population in Galicia, Spain, was maintained partly thanks to ponies bred in semi-natural conditions. These one were the preferential feed source of wolves [27], which consequently did not attack other livestock. As a result, farmers felt less disposed to shoot them down. Finally, ponies were an indirect way to conserve the wolf population.

Pastures can be grazed by equines only, by two or more herbivorous species, or mowed. All these types of pasture management methods allow animals to feed and exert their own effects on the environment (biomass, grassland evenness, and effects on butterflies) depending on the area (grazing seems to be a better option in Central Europe and mowing may be more suitable in Southern Europe) [32]. In boreal conditions, continuous equine grazing seems to be less beneficial to biodiversity than alternative grazing regimes (late grazing, years without grazing, and mowing) [15]. In Sweden, year-round equine grazing increased the pasture quality and diversity compared to mowing [28].

The Impact of Cattle and Horse Mixed Grazing on Meadows

The heterogeneity of plant cover may be beneficial for flora and fauna but can lead to pasture destruction when it is mismanaged. Indeed, non-grazed areas can be invaded by shrubs, while grazed areas can be overgrazed. To prevent this situation, mixed grazing with two herbivorous species could be a solution. Mixed grazing is "a practice of stocking two or more species of grazing or browsing animals on the same land unit, not necessarily at the same time but within the same grazing season" [16]. Cattle and equines graze preferentially in similar habitats, such as grasslands [26]. This can lead to a competition for resources but also complementarity because of their different foraging behaviors. A survey carried out on farmers who raised both (dairy and/or beef) cattle and horses in the French central mountains highlighted the equines' ability to exploit the grasses refused by cattle (total surveyed farmers: 25) [33]. In Massif Central (French mountains), when equines were introduced on cattle pastures, their nutritive value increased because of the development of higher nutritive plants (within the following conditions: 350–440 kg live weight per hectare and as many horses as cattle) [16,20]. When bred with cattle, equines could graze in poorer pastures not useable by cattle because of their low nutritive requirements [33]. Moreover, in winter, equines can remain on pastures and continue maintaining them while cattle are held in stables [16]. However, mixing cattle and equine grazing can generate problems, especially when increasing the number of animals in the pasture. For example, when the equine stocking rate is too high, cattle may be disadvantaged because they are unable to eat enough [27,34]. The main consequences of mixed grazing on grassland management are:

- A lower workload: Because their grazing behavior is different from and complementary to cattle, equine replace the use of machines for mowing grass refused by cattle and for crushing wastelands [22,33]. In the absence of equines, farmers would need to use the roller chopper more frequently [22].
- A decrease of the parasitic burden: This may be explained by different host sensitivities and improved nutritional status (intake of various plant species) [16,35].
- A better control of woody species: In Massif Central, woody species were better controlled when pastures were grazed by cattle and horses than by cattle only [20].

In France, equines are often raised with cattle in grassland areas [10]. A study of 51 farms located in the highlands, breeding cattle (for beef or milk production) and heavy horses together, highlighted the common practices and confirmed the preceding results about the advantages of mixed grazing [22]:

- Equine grazed mostly after dairy cattle but grazed simultaneously with suckling cows and heifers.
- Equine grazing helped to remove grasses refused by milked cows.
- Equine were present on small plots, fields far from stables, and on poor pastures.

In this survey, equines comprised on average about 10% of the total livestock in terms of livestock units. Heavy horses were bred for meat production mainly. This breeding process is seen as a complement to the production of cattle for use in grasslands but is not considered a significant source of income. On the other hand, thanks to equine pasture management, the mechanical maintenance of meadows decreased, which can be seen as an indirect contribution to the greater efficiency of forage systems in the highlands.

Beyond grazing specificities, the inherent nature of equines leads to another environmental asset: equine species present high biodiversity among their breeds, which is shaped by the environment and human beings.

3.1.2. Domestic Biodiversity

Several equine breeds are well adapted to poor grasslands and semi-wild breeding systems [18,26,27] because their format and size seem to ease their growth in these specific areas [36].

For a long time, and still today, natural and human selection have affected breeds, including those that live in semi-natural or particularly harsh conditions [25]. Human beings selected the most suitable breeds for different uses (traction, meat production, racing, and sport), leading to a high variety of sizes, formats, and phenotypes of equines sorted into breeds. A breed may be defined as a breeding pool of individuals that share a common phenotype (which is typically purely morphological, i.e., coat color, height, etc.) [25]. Worldwide, in 2011, there were 397 equine breeds according to the Universal Equine

Life Number (the international lifelong identification number for equine, www.ueln.net). Germany (46 breeds), France (37), and the United States (34) were the countries with the greatest numbers of breeds. Of all these breeds, three-quarters were saddle horses (sport, recreation, and ponies), 15% were heavy horses (France had the highest heavy breeds number with 9), and racing breeds represented only 5% of breeds [37]. Today, uses of equines are diversifying, and competition between studbooks is increasing because of the internationalization of the equine industry. Consequently, some equine populations collapse when they are not useful anymore. For example, worldwide, 60 donkey breeds are known, but only 28 have had their morphologies described [38]. Only one donkey breed out of 28 is considered to be not threatened by the Food and Agriculture Organization of the United Nations, as 80% of the donkey population disappeared over 20 years in Europe [ibid.].

Being part of a breed means being identified in a studbook, which is a book of genealogy where breed standards are established. A closed studbook does not allow foals whose parents are not identified in the book; this keeps the breed "pure." However, to improve the traits of interest, studbooks may be opened to some other breeds that will shape the breed. As an example, the pure-bred Arabian and thoroughbred breeds underlie 25% of the genetic variability of 500,000 saddle horses from 55 different breeds, born between 2002 and 2011 in Europe [39]. A consequence of this intensive direct selection is the specialization of breeds toward modern uses. This specialization is a threat to the versatile, multi-skilled, and generalized breeds. Conversely, these breeds can be seen as an emblem of a region that needs to find new approaches to remain competitive. Finding economically sustainable alternative activities in environmental conservation projects (e.g., animal traction, equestrian tourism, and meat and milk production) would be an interesting way to conserve local and endangered breeds.

European aids for endangered breeds help breeders maintain their activity, even if equine production is no longer economically competitive (Regulation (EU) 2016/1012 of the European Parliament and of the Council of 8 June 2016). This underlines a contradiction in common European programs that give funds based on a small number of animals to preserve animal biodiversity and not just their skills [40]. This kind of breed development is not sufficient to ensure economic viability [41]. However, these programs conserve invaluable environmental services and cultural heritage [25,42].

Equines, because of their essential nature, directly impact the environment. Nevertheless, these impacts may differ depending on the location of the animals.

3.2. How Does the Spatial Repartition of Equines Impact Land Use?

Equines require forage areas for grazing and for preserved fodder harvesting. Equine grazing may lead to the maintenance of open areas and a possible improvement in the agronomic quality of grasslands [43]. Breeding farms, riding schools, racecourses, and trails are other kinds of indirect land uses by equines. A French region typology of equine farms highlighted that equines are present in 91% of all cantons (exclusive of Corsica) [44], which means that the equine industry extends to various areas. Equines are also present in suburban areas: 75% of equines are encountered in the most inhabited areas in Sweden, for example [45]. This is also the case in Scotland [46], Germany [47], France [48], and Belgium [17]. Equines are also present in Polish post-agricultural lands and forests [49], Spanish heathlands [27], and British grasslands [25]. Studies across Europe agree that equines, whatever their use, are located in various kinds of lands: (a) suburban areas, (b) rural areas, and (c) sensitive areas, such as mountains [17,44–48].

3.2.1. Suburban Areas

Pastures in suburban areas are seen as an extensive method of farming and a good way to conserve water quality; however, these pastures are disadvantageous for agriculture [30] because of their small available surface, the presence of housing and non-rural neighbors, and land conflicts. Equines are kept in suburban areas thanks to urban demand [47]; equines are not only seen as a source of agricultural income [50] but as a leisure activity [51] or even as a family "member" whose place is near the home [43]. As a link between urbanization and rurality [51], equines are present in transitional areas that have been abandoned by agriculture but have not yet been developed by urbanization [43]. This kind of "sub-agriculture" [43] may be called "soft urbanization" [17] because equines are a spatial and functional link between residential areas and agriculture, and are also of concern in land conflicts with other agricultural productions. In fact, these conflicts for land force equines to reach the edges of urban areas [43]. Nevertheless, equines remaining in these areas mobilize a large array of services, including veterinarians, feed industries, trainers, equipment sellers, and transporters [47].

Some studies quantify the presence of equines in suburban areas, underlying their growing importance. In France, research conducted on 49 municipalities showed that equines use between 1% and 3.5% of suburban areas depending on the region [43]. In Sweden, it was shown that the density of equines can increase up to 6 horses/km^2 near urban areas but that specialized equine farms near suburban areas keep more equines than farms in rural areas [51].

In terms of environmental impacts, the presence of equines is comparable to the introduction of nature in the city, which improves landscapes [51]. Equines can graze on small plots near forests and gardens (for example, in Flanders (Belgium) and the Netherlands [17]). Their presence yields positive changes, such as positive land management, added landscape value, and job creation [43,45]. Equine grazing is perceived as being a positive element of landscape management in Belgian suburban areas [17] and in Sweden, where the terms used by inhabitants to describe the equine presence were globally positive: ecology, landscape managing, and useful [51]. The presence of equines in these areas opens the possibility to include equines in reflections about urban planning [43,51]. Nevertheless, the high density of equines in small areas could pose a threat to the environment [47] through problems, such as overgrazing, droppings concentration, and destruction of landscapes, because of the creation of mismanaged infrastructures (overgrazed paddocks, and horse-riding rings) [45]. Equines are also a source of odor, insects, and lack of safety [43,45], thereby leading to land conflicts. In Belgium, because of the population density of the country, farmers and equine owners fight for land [52]. Indeed, only 10% of land is considered to be rural (depending on the density of inhabitants) in this country, but agriculture is present in 45% of the total national area, and 1/3 of grasslands are grazed by equines, underlying their high presence in suburban grasslands [17]. The presence of equines in these areas does not always imply grazing because equines can be fed with cereals and preserved forage. In some cases, the available area does not allow grazing, and equines are kept in small unproductive plots or stables [34]. For example, in the Berlin suburban region, there is no grazing in specialized and intensive farms [47]. This raises specific issues related to overgrazing and manure management. More studies about grazing in these specific areas are needed [17].

It is important to note that the presence of equines in suburban areas depends on urban sprawl (horse riding schools near cities), but sometimes urban sprawl depends on the presence of equines (owners who want to live closer to their horses) [43]. This kind of agriculture in suburban areas could respond to sustainable issues according to some authors [47], but communication actions should be deployed in order to raise awareness about the risks of overgrazing, droppings concentration, manure management, safety, and land laws.

3.2.2. Rural Areas

Rural areas are the main production areas for animal feed and for other derived products, such as straw. In France, equines use from 1.5–6% of the total rural areas. Equine numbers depend on the type of agriculture: the numbers are lower if there is more professional agriculture (e.g., if the lands are used for food production) and they are higher if non-professional agriculture prevails (e.g., retired persons or multi-active land owners) [43]. The presence of equines may be combined with other local agricultural production, thereby maintaining pastures even in intensive agricultural areas [44]. Farmers can experience benefits from the presence of equines on their land, such as selling equine feed, letting the equines graze on unproductive pastures, or receiving manure.

3.2.3. Sensitive Areas

A sensitive area has a long-term capacity to maintain and enhance natural resources, such as soil and water quality, biodiversity, and the landscape. In such areas, agriculture is more constrained than in the lowlands. Mountains are part of this definition. Other kinds of sensitive areas include natural rangelands in harsh climates and abandoned agriculture areas invaded by shrub, wetlands, or heathlands.

Some sensitive areas produce difficulties when using machines because of the slope or soil depth. Consequently, in these areas, animal husbandry seems to be the most adapted solution to preserve landscapes and maintain economic activity. For example, in French plains, farms breed mostly saddle and race horses with high economic value [10], whereas few heavy horses are associated with cattle farming in the uplands, in an attempt to improve grassland management despite their low economic value [22]. In this study, up to 15% of farmers said that they would abandon some parcels of land if they did not possess equines [22]. In Poland, ponies named Konik Polski use forests and post-agricultural areas, exerting several impacts on them, including a reduction of shrubs and bushes through trampling, an increase in coprophagic insects, an increase in birds (thanks to greater food resource diversity and hiding place availability), and an increase in the interest and awareness of nature and ecology for tourists and local inhabitants [49]. In Galicia, Spain, where transhumance was abandoned, the presence of equines raised in semi-wild conditions helped to restore heathlands [16]. Heathlands and grasslands in the United Kingdom are usually grazed by sheep, but there have also been studies on the reintroduction of ponies to these areas, where the authors recognized the value of equines as conservation grazers [25]. Rewilding areas abandoned by agriculture with large mammalians, such as equines, are a proposed solution to maintain the important functional links between plants and pollinators in grassland ecosystems in Sweden [18]. Some releases of equines may be a threat to ecosystems, as sometimes breeds are not indigenous to the region. Specific attention should be given [25] to wisely choosing breeds that are to be introduced in natural areas, as well as the density of the released equines. In the "Parc des Volcans d'Auvergne" (an environmentally protected area in France), some authors prescribed the use of mixed grazing with equines and cattle to restore the area [53]. Rangelands are also affected by the definition of sensitive areas. Mediterranean zones show the possibility to access a diversity of pastures (salty pastures, marshes, and rice stubbles), which complement each other and permit local equine breeders to lengthen the grazing season [16]. In areas where the soil is fragile, like in low mountainous olive groves, donkeys are usually used to control grass growth because they weigh less than other heavy animals, such as cattle or horses [38].

In highlands, preserving natural biodiversity when maintaining human activity is possible through agriculture, and more specifically, through animal husbandry. Indeed, the maintenance of the open landscape is particularly interesting for its biodiversity, as explained above, to preserve vegetal resources in non-arable lands, but also to manage areas for human activities, such as hiking or skiing. Because of their slopes and peculiar climates, arable lands are scarce, and pasture meadows are favored as a source of food for livestock, thereby spearheading agricultural products, such as meat and milk.

To conclude, equines take part in the problems of land pressure but have complementarities to agriculture and urbanization in terms of their functionality for land use and maintenance [43]. Other environmental advantages of equines come from their use by humans for working or leisure activities.

3.3. Animal Uses Serving Environmental Issues

Equines have been used by humans since their domestication. Activities with equines are often seen as "natural." However, not every use of equines is environmentally virtuous; for example, horses travelling by plane for international competitions or races. According to our findings in the literature review, two uses present interesting environmental assets: equine work and tourism.

3.3.1. Equine Work

In this study, a working equine refers to an equine who is used to work with humans; provides energy that can be substituted by other sources of energy, other machines, or types of transport; and generates earnings [54]. Sport and leisure horses, equines used for therapy, race horses, and equines bred for meat or milk productions are not affected by this definition. There exist four primary non-exclusive types of equine work: agriculture (mostly vineyards and market gardening in organic production systems), forests (logging), human transportation (such as equines drawing a carriage for tourists or schoolchildren), and public service missions (watering, garbage collection, and mounted police).

In the world, there are ten times more animals used as sources of traction energy than motorized tractors. In "developed" countries, 26% of the land area is managed by animal traction (versus 52% in "developing" countries) [54], especially in sensitive areas [55] or in mountains [56], where plot structures (slope and soil quality) make mechanization difficult. However, these numbers have decreased over time in Europe. For example, in Poland, the percentage of horses used in agriculture compared to all sources of energy decreased from 93.8% in 1950 to 1.73% in 2009 [57]. In the context of productivity gains in post-war Europe, it was necessary to work on bigger areas in less time. Machines seem to be more adapted to this aim than animals, as their use increases sowing, treatment, and harvesting speeds, along with work efficiency, as well as decreases the time dedicated to crops. Nevertheless, a full replacement of animals used in traction by machines may be perceived as a heritage loss [56] and a threat to the environment (soil quality, for example).

Equines as a Potential Source of Renewable Energy

Equines consume fodder, which is considered a renewable source of energy because it does not involve fossil energy in the narrowest sense (unlike fossil fuels or biofuels) [58]. In fact, biofuels may be considered, in some cases, to be a renewable source of energy, but they need the same arable lands as crops and are the focus of land conflict debates. Grasslands used to feed equines could be located in non-arable parts of the territory, as noted above. A one-day harvest allows for enough forage to feed an equine for one year in Switzerland [54]. A total of 0.6 ha of alfalfa, 0.5 ha of oats, and 0.5 ha of wheat for straw are enough to feed two horses working on 14 hectares for 140 days a year in Croatia; for the remainder of the year, these horses stay in stables, where they are fed the by-products of crops, or graze on roadside vegetation or in orchards [58]. Moreover, grasslands and areas worked with equines can be fertilized with their manure. In the case of biofuels, nitrogen must be imported or manufactured and is spread on plots where it evaporates into the atmosphere, providing the main source for N_2O emissions [58]. The animals must be fed all year, whereas machines can be used occasionally and refuelled infrequently. Despite this disadvantage, equine work allows farmers to attain better feed and energy autonomy [54], to highlight a traditional vision, to be appreciated by urban inhabitants [58], and to maintain a diversified gene pool through the use of local equine breeds [55]. Finally, equine work is considered by some authors as a form of sustainable agriculture [58].

On Arable Lands

On arable lands, soil compaction is known to be the most severe form of degradation in conventional agriculture [55,58]. There is a difference between the paths made by machines (continuous, because of their tyres, with deep soil compaction) and equines (intermittent, because of their hooves, with superficial soil compaction) [58]. The soil porosity was higher after using donkeys or cattle compared to a motorized machine [55]. A comparison between the use of a donkey and a motorized machine for the ploughing, fertilization, and preparation of rapeseed in the context of the high hills in northern Italy was made thanks to the life-cycle assessment (LCA) approach for which inventory data were taken from the GaBi 4.0 database. All aspects related to the life spans of animals were considered, except the end of life: pregnancy, growing and maintenance (health care, feed, keep, and equipment), and work. For machines, material acquisition, manufacturing, utilization, transport, and disposal were considered. This information was acquired through interviews with animal owners, field measurements, and technical reports from manufacturers. The results showed that, for the same amount of carbon emissions (1 kg eqCO$_2$), a donkey was able to prepare 330.63 m^2 of land, whereas the machine prepared only 18.69 m^2 (three operational stages were considered on a 1000 m^2 functional unit: ploughing, application of the fertilizer, and seedbed preparation (harrowing and opening seed furrows). Manure from donkeys was assumed to be applied as a fertilizer, so environmental impacts from fertilizer production were avoided in the case of animal traction) [56]. If the fossil fuel used for machines had been replaced by biofuels, the relative effects on the environment could have been 9% lower. When comparing classic machines and donkeys, these effects were 97% lower [56]. In Ireland, yields were greater when animal traction was chosen after the long-term use of tractors [58]. It is important to note that equine traction is well adapted to small areas. Finally, there are not enough studies about equine work in mountainous areas, where they could have particular assets [55].

In Forest Areas

Equines are known to be more drivable than machines in forests, or on rugged or narrow fields. Thanks to this skill, there is less damage to residual trees [55] because machines need more space to access fields [55] and create disturbances [56]. Without counting trail development costs, the use of an equine was more profitable up to 50 m [55]. This distance increased up to 200 m when trail development costs were considered [55]. A comparison between the use of a mule and a motorized machine in a one-kilometer distant forest was made thanks to the LCA approach and showed that, for the same amount of carbon emissions (1 kg eqCO$_2$), a mule was able to bring 311.30 kg of wood, whereas the machine brought only 79.64 kg of wood (three stages were considered: (i) felling (individual cutting of trees), limbing (removing branches), and bucking (cutting into logs); (ii) yarding (collection of logs); and (iii) transport from the forest to the farm (1 km). The functional unit was set to 100 kg of wood at the warehouse. In the animal traction scenario, a mule was used in stages (ii) and (iii)) [56]. If the fossil fuel used for the machines was replaced by biofuels, the relative effects on the environment could have been 26% lower. When comparing a classic machine and a mule, these effects were 74% lower [56].

Other Agricultural and Territorial Works

Equines can also be used in:

- Old vineyards, because their drivability permits work in narrow rows and on terraced or steeped fields (Douro River Valley, Portugal; Bordeaux, France; Sibeira Sacra region, Spain [55]).
- Greenhouses, because their drivability allows for precise work [54] and can be highlighted in ecological production.
- Natural areas, where they are less noisy, degrade the soil less, and frighten local fauna less [54], thereby enabling them to work in protected and sensitive areas. It is possible to compare this to the consequences of equestrian tourism on wild fauna, which are perhaps less frightened by equines than by pedestrians or bikers [59]. Mules are still present in some European areas, such as national parks, where it is impossible or forbidden to use motorized tractors [38].
- Cities where they decrease the carbon footprint and are used as "city pacification" agents [60].

3.3.2. Tourism

According to the International Federation of Equestrian Tourism (FITE), the term "equestrian tourism," which emerged in the 1950s [61], concerns all outdoor activities with equines outside of residential areas. Indeed, it is necessary to distinguish between equestrian tourism and equine tourism:

- Equestrian tourism comprises itinerant journeys with a ridden or hitched equine or on foot supported by a pack equine.

- Equine tourism concerns all activities devoted to equines, in their presence or not, that attract tourists, including sport events, cultural events, races, fairs, museums dedicated to these animals, riding courses, etc.

In addition, the two kinds of tourism linked to equines can be local (i.e., tourists move inside their region of origin) or non-local (inter-regional, international, etc.).

France is the third-largest country in terms of rider numbers in Europe (behind the United Kingdom and Germany), but is considered to be the leader in equestrian tourism and the first travel destination, with 60,000 kilometres of equestrian trails in 2011 [61]. Other countries highlighted in the European report on the equine industry in 2001 are Greece and Portugal, where donkeys are still used for tourism [62]. Equestrian tourism has expanded through farm diversification called agritourism, which is affected by the European development policies for 2014–2020 [36]. This kind of tourism can involve farmers who want to promote local breeds to preserve culture and tradition, as is the case of Camargue horses related to specific bull farming in marshes of the Rhone River estuary in the south of France (Figure 2) [61] or Icelandic ponies in Iceland [63], where tourists want to find a link between nature, animals, and local culture.

In southern France, near the Mediterranean Sea, a regional park and nature reserve named "Camargue" houses particular flora and fauna adapted to saline conditions. This reserve is composed of marshes and wetlands of the Rhône Delta, spanning 150,000 hectares. Camargue horses live in semi-feral conditions in this area and are ridden by guardians/traditional breeders who raise Camargue bulls used in bullfighting. This traditional work is seen as a part of cultural heritage and is promoted during cultural events. This equine breed is also used in equestrian tourism to explore the Camargue region.

Figure 2. Presentation of the Camargue region and the use of the local equine breeds in this area.

Impacts of Equestrian Tourism

Equestrian tourism is a form of sustainable leisure [61], though there are very few studies on the direct impacts of using equines in tourism; these studies are mainly American (12 in total, as of 2019) and Australian (six in total, as of 2019) [64]. This is why it is important to contextualize the results presented below, as climate and cultural history are not the same in every region. Europe has always been a host to large mammals, whereas Australia never housed such animals before 1800 [64]. A comparison between hiking, cycling, and horse riding [64] shows that the impacts on the environment were the same between these means of transport but they differed in the degree of impact (e.g., soil compaction and erosion, loss of organic matter biomass, and biodiversity losses). The two most severe impacts of equestrian tourism are: (1) nitrification of rivers and soils because of the overconcentration of phosphor in poor soils, and (2) zoochory through fur and manure, which raises the risk of invasive plants being spread in protected areas [64]. On the other hand, this spread may be beneficial for the flora diversity of poor soils.

Equestrian tourism is also an illustration of soft roaming. Indeed, nature-based recreation activities impact wild fauna (most of them, even if non-motorized, have negative impacts on birds in terms of their metabolisms, behaviors, and habitat disturbances [59]). There are no studies concerning equestrian travel, but it is possible to consider a softer approach for wild fauna; wild animals are, perhaps, less frightened by horses than by pedestrians. In addition, the trails used for horse-riding are a softer way to adapt land to tourism than roads. Moreover, in sensitive and protected areas [61], horseback tourism also creates and maintains trails in a useful state for other users.

The negative impacts of equestrian tourism must also be mentioned. Infrastructures are not always adapted to equestrian tourism, and this shortfall can present destructive impacts on the environment.

Indeed, during a long-term journey, a horse must rest every 20–40 kilometres but this is not always accomplished [65]. In Poland, for example, on the longest national equestrian trail (2100 km in 2012), there are only 36 liveries; this means one stable every 41 km [65]. When searching for a campsite or pasture, horse-riders can destroy sensitive or protected areas [64]. Moreover, from surveys carried out on equestrian tourists, security, comfort, and conviviality were more important when travelling than ecosystem conservation [64].

Equine Activities and Tourism

Because of the wide definitions of equine tourism, it is difficult to list all ecological effects of such tourism. However, some reflections about this issue can be highlighted. In France, for example, the professional organization of the horse industry provides awards to infrastructures that put efforts toward improving their environmental impacts, such as riding schools in Camargue [61] or the numerous stud farms and boarding stables all over the country, in the form of environmental labels (www.label-equures.com, accessed 27.11.2019). This is also the start for reflecting upon the environmental impacts of equestrian events (the transport of horses and persons, the use of natural resources, etc.) in France through an evaluation of the most impacting positions, like wash areas [66]. These efforts provide an early ecological wake-up call for all events, including equine-related ones, and need further research.

4. Discussion

4.1. Links between Green Assets and Specific Issues

The green assets highlighted above are not completely independent but are mostly linked to each other.

This fact can be first illustrated through the example of the Camargue horse, which plays a key role in land planning and tourist development. This equine breed is known to be a representative of local cultural heritage (Figure 3). Raised in the wetlands, Camargue horses are robust and largely participate in the maintenance of this sensitive area through shrub grazing. They are also hardy enough to be used to control and move cattle herds in these vast swamps. The Camargue region attracts tourists because of its particular fauna (such as flamingos, horses, bulls, etc.) and flora (halophilic plants). The Camargue equine breed is also well adapted to be ridden by tourists to discover the natural environment of this area. Finally, Camargue horses can also be found in suburban areas, such as in riding schools or pasture areas near houses.

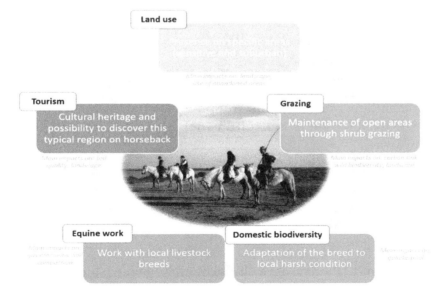

Figure 3. The Camargue horse breed as an image of the link between the five green assets.

Each green asset has its own specificities and issues but some of these issues are common to several green assets. For example, because of the task intensity, animal traction sometimes must be assisted by another source of energy (in cities for example) [60]. To avoid soil compaction, innovative projects have emerged to improve machines, but the use of equines is not always proposed as a solution [58]. Finally, there is a need to improve the equipment for animal traction [55] and to inform stakeholders about this need. Presently, the use of equine work depends on the geographical and market opportunities created thanks to these animals [56]. These opportunities should also be developed for threatened breeds whose population have collapsed since the middle of the 20th century. Regardless of the species, these communities must face common challenges [41], such as how to ensure their competitiveness on the international market while maintaining the appearance of an iconic breed that is illustrative of regional culture and heritage, and how to conciliate the development programs and conservation programs for an already threatened breed. The uniqueness of some breeds is used to bring tourism into the breeds' native region, such as Icelandic ponies in Iceland [63], even if tourism may be a source of conflicts with other users and even sometimes a source of the deterioration of protected areas [61]. These conflicts may also appear in land use. Equine owners are often disadvantaged toward more "agricultural" productions [17]. In suburban areas, equines stand on transitional plots, which are rented or borrowed [43], within the urban network. This presence may lead to disturbances, such as smells and noises. However, in Sweden, a survey made in the suburban areas where equines were encountered showed that the inhabitants were less annoyed by equines than by other disturbances, such as noisy roads or mowers [51]. Urban planning does not consider the specificities of the equine industry. Equine owners want to keep their animals close to their houses and urban centers but are restrained because of land conflicts. If urban planning undertook a multifunctional approach for every activity [17], there would be fewer conflicts of interest [51]. Conversely, intensive equine holdings on small plots, as is often done in suburban areas, can be harmful to the landscape. This is why "horsiculture" is sometimes seen as degrading for the environment [45] as it is directly linked to the specificities of equine grazing that leave areas overgrazed if badly managed. In addition, in most cases, only broodmares and growing horses graze, while stallions or equines in training do not [22]. Moreover, the workload increases when horses and donkeys graze if they are trained daily because, sometimes, it is necessary to pick them up from distant plots [22].

Facing this context, each member state of the European Union has specificities concerning equines. Some countries orient their equine industry more toward sports and racing (e.g., England and Germany) and do not consider equines to be a source for rural development [67], whereas others highlight their equine sector through native breeds and tourism, such as Iceland [68]. Thus, impacts on the environment will be as diverse as the uses of equines and farming in each area. In some countries, equine husbandry is perceived as an intensive process (no grazing, many individuals on small plots, main outlet exported, etc.), for example, in Belgium, the Netherlands, or Luxembourg, because of the lack of available land. In other countries, equines are still used as a source of energy or entertainment, with grazing on large lands and sometimes in semi-wild conditions, such as in Poland [49]. In occidental Europe, land use by equines depends mainly on the type of area (rural, suburban, or sensitive), as discussed in the aforementioned studies. This grouping may be different in other countries, such as Romania, where horses are still used in agriculture [69]. Thus, the numbers of equines may still be anecdotal in suburban areas.

Consequently, it is important to determine how to best take advantage of the green assets of equines.

4.2. How to Better Use Equine Green Assets

4.2.1. For Equine Keepers

This knowledge is partially taught in agricultural training. Indeed, there is a lack of training courses and technical information for small equine owners to teach the specificities of equines' relationship to the environment. Difficulties arise from the atomization of horse owners and the diversity of horse keeping. Moreover, owners mostly raise horses in small areas with low productive value, so the technical improvements of pastures are difficult to apply. The risks of injury or escape have also been is raised by equine breeders, even if equine grazing itself is not to be feared. Advice for management practices needs to be adapted to the specific conditions of each horse keeper according to the type of meadow, the herd size, the physiological and nutritional needs of the equines, and welfare requirements. To improve this situation, communication and teaching about pasture management practices, related risks, and threats are necessary. Raising awareness and improving management practices could improve equine environmental impacts, highlight their green assets, and better use these assets in everyday practice. It is already possible to propose some examples of practical recommendations to equine owners:

- In order to avoid the overconcentration of manure in suburban areas, equine owners may build reliable partnerships with local farmers, who can use manure as fertilizer. Manure can also be recycled and rapidly composted to improve soils in city parks. Another solution is to transport manure to methanation firms for energy production [70].

- Equine grazing has interesting characteristics in pastures and shows complementarities with other herbivores, such as cattle. Thus, associating these animals could be a first step toward improving pasture quality and maintenance.

- When searching to buy an equine for leisure or tourism, looking for a native breed could be a good option if the future owner wants a hardy equine that is well adapted to the local climate. These breeds may value local feed resources better and more cheaply than other breeds. At the same time, this act would help to conserve threatened breeds, facilitating a cultural development of the region and maintaining the genetic diversity of equine species.

- When travelling on horseback, it is important to follow trails that are dedicated to horse riding, to avoid protected areas, and to take care of the paths.

- Equine work represents a diversification opportunity for riding schools. This diversification can be achieved using equines that are already present in the structures for maintenance tasks, either on site or in collaboration with local municipalities (service provision).

Despite these few examples, it is necessary to develop further recommendations for equine keepers in order to establish clear aims, build a reliable argument, and ensure adequate follow-up. These recommendations could be spread by local authorities, teaching centers, and the professional or public institutes responsible for the horse industry. As a contribution to advancing knowledge, this literature review is a first step that will need regular updates to enhance the advisors' arguments.

4.2.2. For Institutional and Political Stakeholders

Equines are often forgotten in political debates as they are seen as both livestock and pets. This duality is exacerbated by the large diversity of stakeholders responsible for the equine industry within Europe, including ministries of agriculture, sports, and tourism; technical institutes; research institutes; national and regional associations; breed associations; and equestrian federations. Moreover, there is a lack of practical information and courses on equine green assets that are usable by stakeholders. Promoting equine grazing and communicating about its benefits on the environment can improve different situations. This could also help integrate equines into political debates and help develop research on this topic. Hence, it is possible to propose some practical recommendations to institutional and political stakeholders:

- Equines are an interesting alternative for the maintenance of small abandoned lands in suburban areas that could be promoted by local authorities.

- In regions where grasslands or rangelands are important, local development policies could include aids in favor of the equine industry, for example, subsides for cattle farmers to also hold equines, or for the creation of numerous platforms to help horse owners meet farmers for feed purchases, pasture grazing, or the use of manure.

- Regional subsidies could support the breeding and keeping of local breeds. These could also be integrated in local tourist events or as a vector for job-creation.

- The trails used for equestrian tourism and camping sites should be framed well to keep tourists from disturbing natural areas. Moreover, every trail should clearly indicate whether it is adapted for equines to ensure that equestrian tourists use the trails safely. Further, linking equestrian tourism stakeholders with stakeholders from the tourism sector or those in charge of protected areas could be an interesting way to develop collaborative actions to support sustainable regional development.

- When discussing new sustainable projects concerning ecological farming or public service missions in cities, equines could be included in the list of suggested alternatives based on the assets presented in this paper if all economic, social, and welfare conditions are fulfilled.

- The new 2020 CAP is in process. At a national level, its measures could better support the equine industry through new agri-environmental measures for equines, such as the use of animal traction, the practice of mixed grazing, the use of local threatened breeds, and the use of equines to maintain vacant suburban plots of land.

Finally, by gathering the available scientific knowledge about equine green assets, this paper offers some common reflections and issues about the place of equines in a sustainable regional development.

5. Conclusions

The equine industry is constantly evolving according to changes in society. One of the next steps is linked to the growing environmental awareness. This issue concerns citizens but also the political spheres, thereby putting pressure on the stakeholders of all economic sectors, including the equine industry. Indeed, in most European countries, environmental issues are not yet considered to be important enough by stakeholders in the equine industry. However, through their green assets, equines can have an active role in ecological transition and debates, both alone and as a complement to other economic productions and services. In the future, it could be interesting to support knowledge exchange in order to progress equine research, thus making this industry more visible and understandable, and to include equines in political debates about the environment and raise awareness about equine uses to avoid radical actions from animal activists. Creating and publishing all kinds of communication media, such as articles, photos, videos, websites, and podcasts, could be a way to reach a larger audience and make equine owners adapt their management practices to better use equine green assets.

From European organizations to society, everyone should be aware of the potential place of equines during the ecological and agronomic transition toward a greener future.

Author Contributions: Conceptualization, A.R.; investigation, A.R.; writing—original draft, A.R.; visualization, A.R.; supervision, C.V.; project administration, C.V.; writing—review and editing, C.V. and G.B.; validation, G.B. All authors have read and agreed to the published version of the manuscript.

Acknowledgments: The work has been carried out with the help of the National Research Institute for Environment and Agriculture (IRSTEA). The authors thank Pascale Heydemann, Amandine Julien, Agnès Orsoni, and Claire Cordilhac from IFCE and Florence Gras from EHN for their input in this study.

References

1.   Evans, R. Introduction to the new equine economy in the 21st century. In *EAAP Scientific Series*; Vial, C., Evans, R., Eds.; Wageningen Academic Publishers: Wageningen, The Netherlands, 2015; Volume 136, pp. 11–18. ISBN 978-90-8686-279-5.

2.   Leteux, S. L'hippophagie en France. La difficile acceptation d'une viande honteuse (archives) [Horse meat in France. Difficult acceptance of a shameful meat]. *Terrains Trav.* **2005**, *9*, 143–158.

3. Rossier, E.; Coleou, J.; Blanc, H. Les effectifs de chevaux en France et dans le monde [Numberof horses in France and in the world]. In *Le Cheval: Reproduction, Sélection, Alimentation, Exploitation [The horse: Breeding, selection, feeding, holding]*; INRA: Paris, France, 1984.

4. Häggblom, M.; Rantamäki-Lahtinen, L.; Vihinen, H. *Equine Sector Comparison between The Netherlands, Sweden and Finland*; Equine Life; MTT Agrifood Research Finland: Jokioinen, Finland, 2008.

5. Swedish University of Agricultural Science; MTT Agrifood Research in Finland; Latvia University of Agriculture. *Current Status of Equine Sector in the Central Baltic Region (Finland, Latvia and Sweden)*; Swedish University of Agricultural Science: Alnarp, Sweden, 2012; p. 47.

6. Vidal, C. Thirty years of agriculture in Europe: holdings with grazing livestock have followed different paths. *Eurostat Stat. Focus* **2002**, *25*, 8.

7. Liljenstolpe, C. *Horses in Europe*; Swedish University of Agricultural Sciences: Uppsala, Sweden, 2009; p. 32.

8. Jez, C.; Coudurier, B.; Cressent, M.; Méa, F.; Perrier-Cornet, P.; Rossier, E. *La Filière Equine Française à L'Horizon 2030 [French Equine Industry in 2030]*; INRA—IFCE: Paris, France, 2012.

9. Engelsen, A.; IFCE; EHN. *What are the Laws for Equines in Europe?* Synthèse; IFCE: Paris, France; Haras Nationaux: Le Pin-au-Haras, France, 2017; ISBN 978-2-915250-56-5.

10. Perrot, C.; Barbin, G.; Bossis, N.; Champion, F.; Morhain, B.; Morin, E. *L'Élevage d'Herbivores au Recensement Agricole 2010. Cheptels, Exploitations, Productions. [Herbivore farming according to the Agricultural Census 2010.Livestock, farms, productions]*; IDELE: Paris, France, 2013; p. 96.

11. Vial, C.; Bigot, G.; Heydemann, P.; Cordilhac, C. Les chiffres clés de la filière équine à l'international: Un essai de collecte d'informations [Key figures of the international equine sector: An attempt of data collection]. *IFCE* **2017**, *6*, 7.

12. Geels, F.; Turnheim, B.; Asquith, M.; Kern, F.; Kivimaa, P. *European Environment Agency Sustainability Transitions: Policy and Practice*; European Environment Agency: København, Denmark, 2019. [CrossRef]

13. Angers, D.A.; Arrouays, D.; Saby, N.P.A.; Walter, C. Estimating and mapping the carbon saturation deficit of French agricultural topsoils: Carbon saturation of French soils. *Soil Use Manag.* **2011**, *27*, 448–452. [CrossRef]

14. Soussana, J.-F.; Loiseau, P.; Vuichard, N.; Ceschia, E.; Balesdent, J.; Chevallier, T.; Arrouays, D. Carbon cycling and sequestration opportunities in temperate grasslands. *Soil Use Manag.* **2004**, *20*, 219–230. [CrossRef]

15. Saastamoinen, M.; Herzon, I.; Särkijärvi, S.; Schreurs, C.; Myllymäki, M. Horse Welfare and Natural Values on Semi-Natural and Extensive Pastures in Finland: Synergies and Trade-Offs. *Land* **2017**, *6*, 69. [CrossRef]

16. Jouven, M.; Vial, C.; Fleurance, G. Horses and rangelands: Perspectives in Europe based on a French case study. *Grass Forage Sci.* **2016**, *71*, 178–194. [CrossRef]

17. Bomans, K.; Dewaelheyns, V.; Gulinck, H. Pasture for horses: An underestimated land use class in an urbanized and multifunctional area. *Int. J. Sustain. Dev. Plan.* **2011**, *6*, 195–211. [CrossRef]

18. Garrido, P.; Mårell, A.; Öckinger, E.; Skarin, A.; Jansson, A.; Thulin, C. Experimental rewilding enhances grassland functional composition and pollinator habitat use. *J. Appl. Ecol.* **2019**, *56*, 946–955. [CrossRef]

19. Lefebvre, T.; Gallet, C. Impacts des grands herbivores sur la végétation des prairies et conséquences sur la décomposition de la litière [Impacts of large herbivores on plants: Consequences for litter decomposition]. *INRA Prod. Anim.* **2018**, *30*, 455–464. [CrossRef]

20. Fleurance, G. Impact du Pâturage Equin sur la Diversité Biologique des Prairies [Impacts of Equine Grazing on Biological Diversity of Pastures]. In Proceedings of the Sommet de l'Elevage, Clermont-Ferrand, France, 2–4 October 2008.

21. Orth, D. Impact sur la végétation ligneuse d'un troupeau mixte de bovins et d'équins en conditions de sous-chargement [Impacts of a mixed herd (cattle/equine) on shrubby vegetation in conditions of low stocking rate]. *Fourrages* **2011**, *207*, 201–209. Available online: https://prodinra.inra.fr/record/179486 (accessed on 8 January 2020).

22. Bigot, G.; Mugnier, S.; Brétière, G.; Gaillard, C.; Ingrand, S. Roles of horses on farm sustainability in different French grassland regions. In *EAAP Scientific Series*; Vial, C., Evans, R., Eds.; Wageningen Academic Publishers: Wageningen, The Netherlands, 2015; Volume 136, pp. 177–186. ISBN 978-90-8686-279-5.

23. U.S. Environmental Protection Agency. 14.4 Enteric Fermentation—Greenhouse Gases. In *Compilation of Air Pollutant Emission Factors*; U.S. Environmental Protection Agency: Washington, DC, USA, 1995; Volume 1.

24. De Cara, S.; Thomas, A. Évaluation des émissions de CH4 par les équins [Evaluation of CH4 emissions by equine]. In *Projections des Missions/Absorptions de Gaz à Effet de Serre Dans Les Secteurs Forêt et Agriculture aux Horizons 2010 et 2020*; INRA: Paris, France, 2008; pp. 146–148.

25. Fraser, M.D.; Stanley, C.R.; Hegarty, M.J. Recognising the potential role of native ponies in conservation management. *Biol. Conserv.* **2019**, *235*, 112–118. [CrossRef]

26. López, C.L.; Celaya, R.; Ferreira, L.M.M.; García, U.; Rodrigues, M.A.M.; Osoro, K. Comparative foraging behaviour and performance between cattle and horses grazing in heathlands with different proportions of improved pasture area. *J. Appl. Anim. Res.* **2019**, *47*, 377–385. [CrossRef]

27. López-Bao, J.V.; Sazatornil, V.; Llaneza, L.; Rodríguez, A. Indirect Effects on Heathland Conservation and Wolf Persistence of Contradictory Policies that Threaten Traditional Free-Ranging Horse Husbandry: Threats to traditional horse husbandry. *Conserv. Lett.* **2013**, *6*, 448–455. [CrossRef]

28. Ringmark, S.; Skarin, A.; Jansson, A. Impact of Year-Round Grazing by Horses on Pasture Nutrient Dynamics and the Correlation with Pasture Nutrient Content and Fecal Nutrient Composition. *Animals* **2019**, *9*, 500. [CrossRef] [PubMed]

29. Bott, R.C.; Greene, E.A.; Koch, K.; Martinson, K.L.; Siciliano, P.D.; Williams, C.; Trottier, N.L.; Burk, A.; Swinker, A. Production and Environmental Implications of Equine Grazing. *J. Equine Vet. Sci.* **2013**, *33*, 1031–1043. [CrossRef]

30. Launay, F.; Genevet, E.; Jouven, M.; Auréjac, R.; IDELE. *Les Parcours: Des Pâtures Intéressantes Pour les Equins Dans les Régions Méditéranéennes [Heathlands: Interesting Pastures for Equine in Mediterranean Regions]*; IDELE, Oier-suamme: Paris, France; Montpellier SupAgro, CE LR: Montpellier, Occitanie, France, 2014; p. 15.

31. Fleurance, G.; Duncan, P.; Farruggia, A.; Dumont, B.; Lecomte, T. Impact du pâturage équin sur la diversité floristique et faunistique des milieux pâturés [Impacts of equine grazing on floral and animal diversity in grazed areas]. *Fourrages* **2011**, *207*, 189–199. Available online: https://afpf-asso.fr/revue/l-utilisation-des-ressources-prairiales-et-du-territoire-par-le-cheval?a=1848 (accessed on 8 January 2020).

32. Tälle, M.; Deák, B.; Poschlod, P.; Valkó, O.; Westerberg, L.; Milberg, P. Grazing vs. mowing: A meta-analysis of biodiversity benefits for grassland management. *Agric. Ecosyst. Environ.* **2016**, *222*, 200–212. [CrossRef]

33. Bigot, G.; Brétière, G.; Micol, D.; Turpin, N. Management of cattle and draught horse to maintain openness of landscapes in French Central Mountains. In Proceedings of the 17th Meeting of the FAO-CIHEAM Mountain Pasture Network, Pastoralism and Ecosystem Conservation, Trivero, Italy, 5–7 June 2013; pp. 72–75.

34. Morhain, B. Systèmes fourragers et d'alimentation du cheval dans différentes régions françaises [Forage and feed system for equine in different French regions]. *Fourrages* **2011**, *207*, 155–163. Available online: https://afpf-asso.fr/revue/l-utilisation-des-ressources-prairiales-et-du-territoire-par-le-cheval?a=1844 (accessed on 8 January 2020).

35. Forteau, L.; Dumont, B.; Sallé, G.; Bigot, G.; Fleurance, G. Horses grazing with cattle have reduced strongyle egg count due to the dilution effect and increased reliance on macrocyclic lactones in mixed farms. *Animal* **2019**, 1–7. [CrossRef]

36. Miraglia, N. Sustainable development and equids in rural areas: An open challenge for the territory cohesion. In *EAAP Scientific Series*; Vial, C., Evans, R., Eds.; Wageningen Academic Publishers: Wageningen, The Netherlands, 2015; Volume 136, pp. 167–176. ISBN 978-90-8686-279-5.

37. Institut Français du Cheval et de L'équitation. Les races d'équidés et leurs utilisations dans le monde [Equine breeds and uses in the world]. In *Panorama Economique de la Filière Equine: Synthèse*; Institut Français du Cheval et de L'équitation: Le Pin au Haras, France, 2011; pp. 34–35, ISBN 978-2-915250-19-0.

38. Camillo, F.; Rota, A.; Biagini, L.; Tesi, M.; Fanelli, D.; Panzani, D. The Current Situation and Trend of Donkey Industry in Europe. *J. Equine Vet. Sci.* **2018**, *65*, 44–49. [CrossRef]

39. Bigot, G.; Vial, C.; Fleurance, G.; Heydemann, P.; Palazon, R. Productions et activités équines en France: Quelles contributions à la durabilité de l'agriculture? [Equine production and activities in France: What contributions to the sustainability of agriculture?]. *INRA Prod. Anim.* **2018**, *31*, 37–50. [CrossRef]

40. Miraglia, N. Equids contribution to sustainable development in rural areas: a new challenge for the third millennium. In *Forages and Grazing in Horse Nutrition*; Saastamoinen, M., Fradinho, M.J., Santos, A.S., Miraglia, N., Eds.; Wageningen Academic Publishers: Wageningen, The Netherlands, 2012; pp. 439–452, ISBN 978-90-8686-755-4.

41. Lauvie, A.; Audiot, A.; Couix, N.; Casabianca, F.; Brives, H.; Verrier, E. Diversity of rare breed management programs: Between conservation and development. *Livest. Sci.* **2011**, *140*, 161–170. [CrossRef]

42. Leinonen, R.-M.; Dalke, K. National Treasure: Nationalistic Representations of the Finnhorse in Trotting Championships. In *Equestrian Cultures in Global and Local Contexts*; Adelman, M., Thompson, K., Eds.; Springer International Publishing: Cham, Germany, 2017; pp. 105–117, ISBN 978-3-319-55885-1.

43. Vial, C. Le développement des activités équestres dans les campagnes françaises: Enjeux et conséquences pour les territoires ruraux et périurbains [Developpment of equine activities in French countryside: Issues and consequences for rural and suburban areas]. In Proceedings of the the the Colloque de Cerisy: Les Chevaux de L'imaginaire Universel aux Enjeux Prospectifs pour les Territoires; Presses Universitaires de Caen, Cerisy-la-Salle, France, 17–22 May 2014;

44. Perret, É.; Turpin, N. Territoires et exploitations équines en France [Equine territories and holdings in France]. *Écon. Rural. Agric. Aliment. Territ.* **2016**, 85–98. [CrossRef]

45. Elgåker, H.; Wilton, B.L. Horse farms as a factor for development and innovation in the urban-rural fringe with examples from Europe and Northern America. In Proceedings of the Nature and Landscape as an Asset to Development in Rural Areas, Copenhagen, Denmark, 8–10 March 2008; University of Copenhagen: Copenhagen, Denmark, 2008; Volume 27–2008, pp. 43–55.

46. Quetier, F.F.; Gordon, I.J. 'Horsiculture': How important a land use change in Scotland? *Scott. Geogr. J.* **2003**, *119*, 153–158. [CrossRef]

47. Zasada, I.; Berges, R.; Hilgendorf, J.; Piorr, A. Horsekeeping and the peri-urban development in the Berlin Metropolitan Region. *J. Land Use Sci.* **2011**, *8*, 199–214. [CrossRef]

48. Vial, C.; Evans, R. (Eds.) *The New Equine Economy in the 21st Century*; EAAP Publication; Wageningen Academic Publishers: Wageningen, The Netherlands, 2015; ISBN 978-90-8686-279-5.

49. Doboszewski, P.; Doktór, D.; Jaworski, Z.; Kalski, R.; Kułakowska, G.; Łojek, J.; Płąchocki, D.; Ryś, A.; Tylkowska, A.; Zbyryt, A.; et al. Konik polski horses as a mean of biodiversity maintenance in post-agricultural and forest areas: An overview of Polish experiences. *Anim. Sci. Pap. Rep.* **2017**, *35*, 333–347.

50. Bailey, A.; Williams, N.; Palmer, M.; Geering, R. The farmer as service provider: The demand for agricultural commodities and equine services. *Agric. Syst.* **2000**, *66*, 191–204. [CrossRef]

51. Elgåker, H.; Pinzke, S.; Lindholm, G.; Nilsson, C. Horse keeping in Urban and Peri-Urban Areas: New Conditions for Physical Planning in Sweden. *Geogr. Tidsskr.-Dan. J. Geogr.* **2010**, *110*, 81–98. [CrossRef]

52. De L'Escaille, T.; (European Land Owners, Bruxelles, Belgium). Personnal communication, 2019.

53. Bigot, G.; Perret, É.; Turpin, N. L'élevage équin, un atout pour la durabilité des territoire ruraux: Cas de la région Auvergne [Equine breeding, an asset for the sustainability of rural areas: case of Auvergne region]. In Proceedings of the Identité, qualité et compétitivité territoriale; Association de Science Régionale de Langue Française, Aoste, Italy, 20–22 September 2010; p. 15.

54. Reynaud, E.; von Niederhaüsern, R.; Ackermann, C. Le cheval de travail en Suisse, enquête 2017 [Working horse in Switzerland, survey 2017]. *Agroscope Transf.* **2018**, 51. Available online: https://www.agroscope.admin.ch/agroscope/fr/home/publications/recherche-publications/agroscope-transfer.html (accessed on 8 January 2020).

55. Almeida, A.; Rodrigues, J. Animal Traction: New Opportunities and New Challenges. In Proceedings of the Farm Machinery and Processes Management in Sustainable Agriculture, IX International Scientific Symposium, Lublin, Poland, 20–24 November 2017; pp. 27–31.

56. Cerutti, A.K.; Calvo, A.; Bruun, S. Comparison of the environmental performance of light mechanization and animal traction using a modular LCA approach. *J. Clean. Prod.* **2014**, *64*, 396–403. [CrossRef]

57. Głębocki, B.; Główny Urząd Statystyczny. Produkcja zwierzęca—Rozwój i zmiany przestrzenne w latach 2002–2010—9.7 konie [Animal production—evolution and land development between 2002–2010—9.7 Horses]. In *Zróżnicowanie Przestrzenne Rolnictwa: Powszechny Spis Rolny 2010: Praca Zbiorowa*; Główny Urząd Statystyczny: Warszawa, Poland, 2014; pp. 402–409. ISBN 978-83-7027-556-3.

58. Gantner, R.; Baban, M.; Glavaš, H.; Ivanović, M.; Schlechter, P.; Šumanovac, L.; Zimmer, D. Indices of sustainability of horse traction in agriculture. In Proceedings of the 3. Međunarodni Znanstveni Simpozij Gospodarstvo Istočne Hrvatske-Vizija i Razvoj/3rd International Scientific Symposium Economy of Eastern Croatia-Vision and Growth, Osijek, Croatia, 22–24 May 2014; Volume 3, pp. 616–626.

59. Steven, R.; Pickering, C.; Guy Castley, J. A review of the impacts of nature based recreation on birds. *J. Environ. Manag.* **2011**, *92*, 2287–2294. [CrossRef] [PubMed]

60. Linot, O. La commission des chevaux territoriaux en France [The commission of territorial horses in France]. In *Les Chevaux: De L'imaginaire Universel aux Enjeux Prospectifs pour les Territoires*; Leroy du Cardonnoy, É., Vial, C., Eds.; Colloques de Cerisy; Presses Universitaires de Caen: Caen, France, 2017; pp. 161–171, ISBN 978-2-84133-864-1.

61. Pickel-Chevalier, S. Can equestrian tourism be a solution for sustainable tourism development in France? *Loisir Soc./Soc. Leis.* **2015**, *38*, 110–134. [CrossRef]

62. EU Equus. *The Horse Industry in the European Union*; Department of Economics Swedish University of Agricultural Sciences: Skara/Solvalla, Sweden, 2001; p. 50.

63. Sigurðardóttir, I.; Helgadóttir, G. Riding High: Quality and Customer Satisfaction in Equestrian Tourism in Iceland. *Scand. J. Hosp. Tour.* **2015**, *15*, 105–121. [CrossRef]

64. Pickering, C.M.; Hill, W.; Newsome, D.; Leung, Y.-F. Comparing hiking, mountain biking and horse riding impacts on vegetation and soils in Australia and the United States of America. *J. Environ. Manag.* **2009**, *91*, 551–562. [CrossRef]

65. Kozak, M.W. Making Trails: Horses and Equestrian Tourism in Poland. In *Equestrian Cultures in Global and Local Contexts*; Adelman, M., Thompson, K., Eds.; Springer International Publishing: Cham, Germany, 2017; pp. 131–152, ISBN 978-3-319-55885-1.

66. Vial, C.; Gouget, J.-J.; Barget, E.; Clipet, F.; Caillarec, C. *Manifestations Equestres et Développement Local [Equestrian Events and Local Development]*, 1st ed.; Synthèse; Institut Français du Cheval et de L'équitation: Arnac-Pompadour, France, 2016; ISBN 978-2-915250-45-9.

67. McCarthy, K.; Cosgrove, A.; Zelmer, A. *A Study into the Business & Skills Requirements of the UK Equine Industry*; Lantra: Warwickshire, UK, 2011; p. 112.

68. Sigurðardóttir, I.; Helgadóttir, G. The new equine economy of Iceland. In *EAAP Scientific Series*; Vial, C., Evans, R., Eds.; Wageningen Academic Publishers: Wageningen, The Netherlands, 2015; Volume 136, pp. 223–236, ISBN 978-90-8686-279-5.

69. World Horse Welfare; Eurogroup for Animals. *Removing the Blinkers: The Health and Welfare of European Equidae in 2015*; World Horse Welfare: London, UK; Eurogroup for Animals: Brussels, Belgium, 2015; p. 122.

70. Bellino, R.; Affeltranger, B.; Battistini, B.; Evanno, S.; Le Pochat, S. Comparative environmental assessment of two systems of agronomic and energetic valorisation of horse manure. In Proceedings of the 2nd LCA Conference, Lille, France, 6–7 November 2012; p. 7.

8

Digestibility and Retention Time of Coastal Bermudagrass (*Cynodon dactylon*) Hay by Horses

Tayler L. Hansen, Elisabeth L. Chizek, Olivia K. Zugay, Jessica M. Miller, Jill M. Bobel, Jessie W. Chouinard, Angie M. Adkin, Leigh Ann Skurupey and Lori K. Warren *

Department of Animal Sciences, University of Florida, Gainesville, FL 32611, USA;
tlhansen@cornell.edu (T.L.H.); creeksidefarm14@gmail.com (E.L.C.); ozugay@ufl.edu (O.K.Z.);
horsewhisperer1230@hotmail.com (J.M.M.); jbrides2@ufl.edu (J.M.B.); jessie23@ufl.edu (J.W.C.);
aadkin@ufl.edu (A.M.A.); leighann.skurupey@ndsu.edu (L.A.S.)
* Correspondence: lkwarren@ufl.edu

Simple Summary: Longer retention of forages with increased fiber concentrations may be a compensatory digestive strategy in horses. We investigated the digestive characteristics of bermudagrass hay, a prominent warm-season grass in the southeast United States that has greater fiber concentrations than other common forages fed to horses. The morphological structure and photosynthetic pathway of warm-season grasses differ from cool-season grasses and legumes which may have important impacts on equine digestion and digesta transit through the gastrointestinal tract. The retention time of Coastal bermudagrass was longer than alfalfa or orchardgrass hay. The digestibility of Coastal bermudagrass decreased with increasing maturity, but the fiber digestibility of alfalfa and orchardgrass was similar to the earliest maturity of Coastal bermudagrass hay. The chemical composition of the plant cell wall influences diet digestibility and is a major difference between warm-season and cool-season forages. The increased retention time of Coastal bermudagrass allows for microbial fermentation to occur longer, adapting to more difficult-to-digest plant cell walls in warm-season forages. The decrease in diet digestibility when horses consume warm-season forages can be reduced by feeding early maturity forage, by harvesting hay at an earlier stage of growth or managing pastures in a vegetative state.

Abstract: Bermudagrass (*Cynodon dactylon*) and other warm-season grasses are known for their increased fiber concentrations and reduced digestibility relative to cool-season grasses and legumes. This study investigated the digestive characteristics and passage kinetics of three maturities of Coastal bermudagrass hay. A 5 × 5 Latin square design experiment was used to compare the digestion of five hays: alfalfa (*Medicago sativa*, ALF), orchardgrass (*Dactylis glomerata*, ORCH), and Coastal bermudagrass harvested at 4 (CB 4), 6 (CB 6), and 8 weeks of regrowth (CB 8). Horses were fed cobalt-ethylenediaminetetraacetic acid (Co-EDTA) and ytterbium (Yb) labeled neutral detergent fiber (NDF) before an 84-h total fecal collection to determine digesta retention time. Dry matter digestibility was greatest for ALF (62.1%) and least for CB 6 (36.0%) and CB 8 diets (36.8%, SEM = 2.1; $p < 0.05$). Mean retention time was longer ($p < 0.05$) for Coastal bermudagrass (particulate 31.3 h, liquid 25.3 h) compared with ORCH and ALF (28.0 h, SEM = 0.88 h; 20.7 h, SEM = 0.70 h). Further evaluation of digesta passage kinetics through mathematical modeling indicated ALF had distinct parameters compared to the other diets. Differences in digestive variables between forage types are likely a consequence of fiber physiochemical properties, warranting further investigation on forage fiber and digestive health.

Keywords: alfalfa; equine; fiber; forage maturity; mathematical modeling; mean retention time; orchardgrass; rate of passage; warm-season grass

1. Introduction

Bermudagrass (*Cynodon dactylon*) is one of the most prominent forages in the southeast United States; however, some horse owners and equine professionals assume that bermudagrass, particularly the Coastal variety, is a lower quality hay due to increased fiber concentrations. Furthermore, feeding Coastal bermudagrass hay in this region has been implicated as a cause of ileocecal impaction in horses [1]. The increased fiber concentrations of Coastal bermudagrass and fine, soft texture have been hypothesized to contribute to impaction [2], but greater fiber concentration is a common characteristic among warm-season grasses. Bermudagrass and other grasses common to subtropical and tropical climates (e.g., bahiagrass, millet, sorghum) possess a series of anatomical and biochemical modifications for C4 photosynthesis that distinguish them from C3 plants. The Kranz anatomy of C4 plants features tightly bundled mesophyll cells that form a ring around bundle-sheath cells. The proximity of mesophyll and bundle-sheath cells allows for carbon concentrating mechanisms in photosynthesis, reducing photorespiration in C4 plants. Plants using C4 carbon fixation are more efficient than C3 carbon fixation in areas of drought, high temperatures, and low nutrient inputs [3]. However, C4 plants tend to a have lower nutritive value via greater fiber concentrations that can lead to decreased animal performance [4].

Greater forage fiber concentrations have long been associated with decreased diet digestibility [5]. Forage digestibility by horses decreases by half a percentage unit for every one percentage unit increase in NDF concentration [6]. Using equine fecal inoculum, Lowman et al. [7] reported that time to reach total gas production took longer for oat (*Avena sativa*) straw and wheat (*Triticum aestivum*) straw compared with alfalfa (*Medicago sativa*) hay and grass haylage. Furthermore, the specific type of dietary fiber (insoluble vs. soluble) alters in vitro digestibility measurements [8]. Not only fiber concentration, but the specific composition of hemicellulose, cellulose, and lignin in the plant cell may alter digestion by horses.

The degradation of forage fiber in the equine gastrointestinal tract may be influenced by digesta rate of passage (ROP); however, a consistent relationship between fiber concentration and digesta mean retention time (MRT) has not been shown in horses. Low nutritional value forages have a longer retention time than high-quality legumes [9], but no difference in MRT was observed when horses were fed similar forage species differing in fiber concentration [10,11]. The influence of fiber concentration on digesta MRT may be confounded by factors such as the level of intake and feed particle size [12]. Furthermore, low-fermentable dietary fibers alter ROP through changes in digesta viscosity in the small intestine [13]. Such changes may not be detectable in total tract mean retention time (TTMRT) calculations.

Several mathematical models have been used to describe digesta passage in ruminants that improve understanding of passage kinetics by estimating retention time in the rumen from fecal marker excretion [14–16]. These models have been applied to equine fecal marker excretion with the hopes of increasing the understanding of digesta ROP in horses [10,11,17–20]. The models described by Dhanoa et al. [14] and Pond et al. [15] have been used most frequently to describe digesta passage in horses. The Dhanoa et al. [14] model is a mechanistic model based on first order kinetics. Digesta flows through an unspecified number of compartments with decreasing compartment retention times [14]. In contrast, the stochastic model described by Pond et al. [15] increases the passage rate of an age-dependent compartment to account for an increased probability of digesta leaving a compartment based on previous residence time in the compartment. These models have not been compared with the same data, due in part to the model equations failing to converge with experimental data collected from horses. With more advanced computer applications, a thorough investigation of model fit can be conducted while also exploring the effect of dietary characteristics on passage parameters in the horse.

We hypothesized that the greater hemicellulose concentration of Coastal bermudagrass would alter digestive characteristics. The objective of this study was to compare the digestibility and MRT of Coastal bermudagrass to alfalfa and orchardgrass (*Dactylis glomerata*) hays, which are other common forages fed to horses. Mean retention time was measured using liquid and particulate phase external markers, and fecal marker excretion was modeled using previously developed equations for marker excretion by ruminants [14,15]. We hypothesized that the use of mathematical modeling would provide a greater understanding of ROP variables than TTMRT alone. Differences in total tract MRT of Coastal bermudagrass compared with alfalfa and orchardgrass hay indicate fiber chemical composition alters digesta movement in the gastrointestinal tract of the horse. Longer digesta retention of Coastal bermudagrass may be an important compensation strategy to maximize the available nutrients from slowly degraded fibers in warm-season grasses.

2. Materials and Methods

All animal protocols were approved by the University of Florida Institutional Animal Care and Use Committee (201509618) under the FASS Guide for the Care and Use of Agricultural Animals in Research and Teaching [21]. This study took place from 1 July 2015 to 9 September 2015, in Gainesville, FL, USA. The mean temperature was 26.3 °C and relative humidity was 88.5% during the study period.

Five mature Quarter Horse geldings (8 ± 3 years, 552 ± 14 kg, BCS 6.0 ± 0.4 [22], mean ± SEM) housed at the University of Florida's Horse Teaching Unit in Gainesville, FL were used in this study. Before the start of the study, horses were fed Coastal bermudagrass hay or kept in warm-season grass pastures. Horses received routine vaccinations and anthelmintic treatment before entering the study. Farrier care was maintained during the study according to standard operating procedures of the Unit. During the study, horses were individually housed in 3.7 m × 3.7 m stalls bedded with wood shaving and provided access to 7.4 m × 18.3 m outdoor, grass-free paddocks with sand footing for 3 h each day for voluntary exercise.

Five hays (Table 1) were used to evaluate 5 forage-based diets (Table 2). Hay was fed at 1.6% body weight (BW) (dry matter (DM) basis). Alfalfa (ALF) and orchardgrass (ORCH) hays were purchased from a commercial hay dealer (Larson Farms; Ocala, FL). Coastal bermudagrass hays were harvested in Alachua, FL at 4 weeks (CB 4), 6 weeks (CB 6), and 8 weeks (CB 8) of regrowth under similar management conditions. The CB 4 and CB 6 were second cuttings, whereas the CB 8 was a first cutting. Based on producer harvesting schedules and study timeline, 8 weeks of regrowth as a second cutting was not feasible for this study. The orchardgrass hay had a high electrolyte concentration, therefore, sodium chloride and potassium chloride were added to each diet to better balance electrolyte intake between diets. Horses were fed a vitamin/mineral pellet (0.1 to 0.125% BW, DM basis) during the evening meal to meet micronutrient requirements [23].

Diets were evaluated in a 5 × 5 Latin square design experiment. A standard 5 × 5 Latin square was randomly selected from Fisher and Yates [24]. Horses were randomly assigned to different rows and each period was considered a column. Each period lasted 14 days and consisted of a 10.5-day restricted intake phase when the ration was split into two equal-sized meals fed at 0730 and 1930 h (Table 2). On day 7, an 84-h total fecal collection that began during the evening meal was conducted to determine diet digestibility and retention time. As part of a companion study [25], horses had ad libitum access to hay for the remaining 3.5 days before the start of the next period.

Table 1. Nutrient composition of feedstuffs.

Nutrient [a]	Alfalfa	Orchardgrass	Coastal 4 Weeks	Coastal 6 Weeks	Coastal 8 Weeks	Vit/Min Suppl 1 [b]	Vit/Min Suppl 2 [c]
DM, %	88.4	90.9	90.0	91.8	91.6	89.4	90.5
DE [d], Mcal/kg	2.50	2.09	1.95	1.90	1.85	2.76	3.31
CP, %	23.2	11.5	18.5	12.7	12.6	15.3	37.2
NDF, %	37.7	57.2	67.5	70.9	73.3	43.3	16.9
ADF, %	29.5	42.0	32.7	34.7	35.1	25.9	8.4
ADL, %	8.3	2.8	4.6	5.0	6.0	n.m.	n.m.
Starch, %	1.3	0.2	1.6	1.6	2.5	n.m.	n.m.
ESC, %	5.9	9.6	4.3	4.4	4.4	n.m.	n.m.
WSC, %	6.3	12.5	3.6	4.3	4.8	n.m.	n.m.

Table 1. *Cont.*

Nutrient [a]	Alfalfa	Orchardgrass	Coastal 4 Weeks	Coastal 6 Weeks	Coastal 8 Weeks	Vit/Min Suppl 1 [b]	Vit/Min Suppl 2 [c]
Ca, %	1.58	0.32	0.57	0.39	0.38	1.31	3.12
P, %	0.24	0.23	0.30	0.27	0.18	1.78	1.19
Na, %	0.067	0.44	0.067	0.023	0.12	0.23	0.40
K, %	2.33	2.12	1.58	1.74	0.82	1.03	1.60
Cl, %	0.93	1.56	0.45	0.33	0.17	0.60	0.75
uNDFom, % [e]	20.9	11.6	21.4	25.8	38.6	n.m.	n.m.

[a] Nutrient composition of forages analyzed by NIRS at Dairy One Inc. (Ithaca, NY, USA). [b] Gro-n-Win alfa (Buckeye Nutrition, Dalton, OH, USA) analyzed by wet chemistry at Dairy One Inc. (Ithaca, NY, USA). [c] Equalizer (Seminole Feed, Ocala, FL, USA) analyzed by wet chemistry at Dairy One Inc. (Ithaca, NY, USA). [d] Digestible energy calculated according to Pagan [26]. [e] Undigestible NDF (ash-free) determined after 240-h in vitro incubation by Dairyland Laboratories (Arcadia, WI, USA). All values are on a 100% DM basis except DM. n.m. not measured.

Table 2. Diet composition and nutrient intake of experimental diets [1].

Item	ALF	ORCH	CB 4	CB 6	CB 8
Ingredient, % DMI					
Alfalfa	93.7				
Orchardgrass		92.8			
Coastal Bermuda, 4 weeks			91.8		
Coastal Bermuda, 6 weeks				91.8	
Coastal Bermuda, 8 weeks					91.4
Vit/Min Suppl 1 [a]	5.9				
Vit/Min Suppl 2 [b]		7.2	7.2	7.2	7.1
Sodium Chloride	0.4		0.3	0.4	0.2
Potassium Chloride			0.7	0.6	1.3
Daily Intake					
DM, % BW	1.71	1.73	1.74	1.74	1.75
DE [c], Mcal/kg BW	0.043	0.038	0.035	0.35	0.034
CP, g/kg BW	3.87	2.31	3.43	2.50	2.48
NDF, g/kg BW	6.47	9.36	11.01	11.56	11.94
ADF, g/kg BW	4.98	6.83	5.34	5.65	5.72
Ca, mg/kg BW	265.9	90.23	130.2	101.4	99.82
P, mg/kg BW	56.2	51.69	62.89	58.09	43.69
K, mg/kg BW	383.1	359.2	332.8	348.4	261.2
Na, mg/kg BW	39.3	75.03	37.92	30.88	35.6
Cl, mg/kg BW	197.8	359.2	174.5	145.9	159.4

[1] Abbreviations. ALF, alfalfa; ORCH, orchardgrass; CB 4, Coastal bermudagrass 4-weeks regrowth; CB 6, Coastal bermudagrass 6-weeks regrowth; CB 8, Coastal bermudagrass 8-weeks regrowth. [a] Gro-n-Win alfa (Buckeye Nutrition, Dalton, OH, USA). [b] Equalizer (Seminole Feeds, Ocala, FL, USA). [c] Digestible energy calculated according to Pagan [26].

External markers were prepared and used to determine digesta MRT for each gelding. A lithium salt of Co-EDTA was prepared according to the methods of Udén et al. [27] as a marker for the liquid phase of digesta. For the particulate marker, Yb-acetate was bound to neutral detergent fiber residue according to Ringler and Lawrence [28]. Bermudagrass hay was chopped by a hammer mill until it passed through a 1.27-cm screen and then boiled in neutral detergent solution for 1 h (60 g of bermudagrass hay per liter of neutral detergent solution). Neutral detergent fiber residue was labeled at a concentration of 100 g of NDF residue/L of 0.007 M Yb solution (prepared by dissolving 2.96 g of Yb (III) acetate tetrahydrate in 1 L of distilled water) [28]. The prepared Co-EDTA was 13.7% Co (DM basis) and Yb-labeled NDF residue was 7304 mg Yb/kg DM.

On day 7 of each period, horses were fed 1.5 mg of each marker per kilogram BW with the evening meal of vitamin/mineral pellets. Marker intake was monitored and spilled feed was immediately returned to the feed bucket to ensure complete marker consumption. On average, horses consumed the markers in 14.7 min (range 9 to 30 min).

Immediately before and during fecal collections, stalls were stripped of bedding and swept clean. All voided feces were collected directly from the floor of rubber-matted stalls. In order to minimize contamination of feces with hay, dirt, and other debris and to prevent the horse from stepping in the feces, stalls were checked for fresh excreta every 15 min. Horses were removed from their stalls in 2 to 4-h intervals and temporarily placed in a stall bedded with pine shavings to allow horses to comfortably urinate. If a horse urinated in their primary stall, urine was removed with a wet-dry vacuum. Horses were hand-walked for two 15-min periods (06:00 and 20:00) each day during fecal collections.

Feces were compiled in 2-h intervals for the first 60-h following marker dosing and then in 4-h intervals from 60 to 84 h post marker dosing. Excreted feces were weighed and homogenized after each time interval with 10% of the feces retained for a 24-h composite sample and a 200-g subsample saved for marker concentration determination. Feces collected the first 12 h post marker dosing were only retained for marker concentration analysis, and feces collected from 12 h to 84 h post marker dosing were used for both marker concentration analysis and 24-h composite samples. During fecal collections, orts were collected prior to the next feeding. Orts were time-matched to 24-h fecal composites to determine diet digestibility. Fecal samples were stored at −20 °C until analysis.

Frozen fecal samples were thawed at 4 °C for 48 h. Fecal samples, representative feed samples from each total fecal collection, and orts were dried in a 60 °C forced air oven until achieving a constant weight. Samples were ground to pass a 1-mm screen using a Wiley Mill prior to laboratory analysis.

Twenty-four-hour fecal composite samples, representative feed samples, and orts were used to determine DM, organic matter (OM), NDF, and ADF digestibility (DMD, OMD, NFD, ADFD, respectively). Samples were dried in triplicate at 60 °C until a constant weight and then ashed at 600 °C for 8 h to calculate OM concentration. Fiber concentrations were sequentially determined using an ANKOM 200 Fiber Analyzer [29]. Heat-stable α-amylase was used in the NDF analysis of all samples. Digestibility was determined as ((Nutrient Intake − Nutrient Output)/Nutrient Intake × 100).

Marker concentrations were determined on fecal samples composited in 2- and 4-h intervals following marker dosing. Fecal samples were dried in triplicate in a 60 °C forced-air oven until a constant weight to determine DM concentration. A 0.500 g subsample was weighed and placed into a Teflon digestion vessel with 8 mL of 15.8 N nitric acid. Samples were sealed and digested for 15 min at 180 °C using a microwave-assisted acid digestion procedure (Anton-Paar, Ashland, VA, USA). Samples were allowed to cool and diluted to 25 mL. Samples were centrifuged at 1050× g for 15 min and the supernatant collected for determination of marker concentrations using inductively coupled plasma spectrometry (Perkin-Elmer, Inc., Shelton, CT, USA) [30,31]. The minimum element detection limit was 0.1 mg/L. Marker recovery was calculated as (Marker Excreted/Marker Dosed × 100).

Total tract MRT was calculated arithmetically according to Blaxter et al. [32] and Thielemans et al. [33]. Total tract MRT calculated according to Blaxter et al. [32] is

$$\text{MRT} = \frac{\sum m_i t_i}{\sum m_i} \tag{1}$$

where m_i = the amount of marker in the ith sample (g) and t_i = time from dosage of the marker to the middle of the ith sampling interval (h). The equation described by Thielemans et al. [33] uses the concentration of the marker in the sample and MRT is calculated as

$$\text{MRT} = \frac{\sum t_i C_i \Delta t_i}{\sum C_i \Delta t_i} \tag{2}$$

where t_i = time from dosage of the marker to the middle of the ith sampling interval (h), C_i = concentration of marker in the ith sample (mg/kg DM), and Δt_i = time interval between the middle of the ith and ith − 1 sample (h).

Fecal marker excretion data were fit with compartment models described by Dhanoa et al. [14] and Pond et al. [15]. The multicompartment model derived by Dhanoa et al. [14] is a mechanistic model based on first order kinetics where marker concentration (mg/kg DM) of the feces can be modeled as

$$\text{Marker Concentration} = Ae^{-k_1 t}e^{-(N-2)e^{-\Delta t}} \tag{3}$$

where A is a scaling parameter, k_1 = rate constant for the first compartment (h^{-1}), t = time from marker dosage (h), $\Delta = k_2 - k_1$ where k_2 is the rate constant for the second compartment (h^{-1}, assuming $k_2 > k_1$), and N = the number of exponentially distributed compartments. The rate constants do not change over time; therefore, the compartments are considered age-independent (the rate digesta leaves a compartment is not influenced by past residence time). The exponentially distributed compartments described by Dhanoa et al. [14] can represent multiple sub-compartments within a larger mixing compartment. The two-compartment model featuring a γ-distribution described by Pond et al. [15] was also fit to fecal marker excretion data (mg/kg DM) as

$$\text{Marker Concentration} = C_2\left[\delta^n e^{-k_2(t)} - e^{-\lambda_1 t}\sum_{i=1}^{n}\frac{\delta^i (\lambda_1 t)^{n-i}}{(n-i)!}\right] \tag{4}$$

where C_2 = the initial concentration in the second compartment if the marker dose had been introduced into the compartment and instantaneously mixed, n = order of the γ-distribution in the first compartment, k_2 = rate parameter for exponentially distributed residence times (h^{-1}), t = time after dosing of marker (h), λ_1 = rate parameter for γ-distributed residence times (h^{-1}), and $\delta = \lambda_1/(\lambda_1 - k_2)$. Time delay was incorporated into the Pond et al. [15] equation by substituting t for t-TT, where t is the time from marker dosing (h) and TT is transit time. Six orders of γ-distribution were analyzed (n = 1, 2, 3, 4, 5, 6) to test the G1G1, G2G1, G3G1, G4G1, G5G1, and G6G1 model described by Pond et al. [15]. If marker residence time is exponentially distributed (n = 1) in a compartment, the compartment is age-independent, indicating that the rate the marker leaves is not dependent on past residence time. However, if the ROP of a marker in a compartment changes over time, the compartment is considered age-dependent. Marker concentration in an age-dependent compartment can be modeled with a γ-distribution of order 2 or greater. Increasing the order of the γ-distribution alters the shape of the curve such that the emergence of marker from the compartment is slowed [15]. Curves were fit using nonlinear least squares methods in MATLAB (Version R2015a, Mathworks, Natick, MA, USA) with model parameter start values randomly assigned (Computer Code S1). Bounds for rate parameters were set between 0 and 1.

Model parameters were used to determine total tract mean retention time (TTMRT) for each fitted equation to fecal marker excretion. For the Dhanoa et al. [14] model, TTMRT (h) was calculated as

$$\text{MRT} = \frac{1}{k_1} + \frac{1}{k_2} + \sum_{i=3}^{N-1}\frac{1}{k_2 + (i-2)(k_2 - k_1)}, \quad k_2 > k_1 \tag{5}$$

where k_1 and k_2 are rate parameters (h^{-1}) of the first and second compartments and N is the number of exponentially distributed compartments. The term $\sum_{i=3}^{N-1}\frac{1}{k_2+(i-2)(k_2-k_1)}$ is said to represent the transit time (TT) of digesta markers or the time from dosing the marker to the first appearance of marker in the collected sample. Total tract MRT (h) for the Pond et al. [15] model was calculated as

$$\text{MRT} = \frac{n}{\lambda_1} + \frac{1}{k_2} + \text{TT} \tag{6}$$

where λ_1 is the age-dependent compartment rate constant (h^{-1}), k_2 is the age-independent compartment rate constant (h^{-1}), n is the order of the γ-distribution, and TT is the transit time (h). The age-dependent compartment MRT (CMRT$_1$) was determined by n/λ_1 and the age-independent compartment MRT

(CMRT$_2$) was determined by $1/k_2$. When n = 1, λ_1 is replaced by k_1, and CMRT$_1$ is an age-independent compartment.

Unless otherwise noted, data are presented as means ± SEM. Data were checked for normality using the Kolmogorov–Smirnov test and the Shapiro–Wilk test. Data were analyzed as a Latin square design using a mixed model ANOVA in SAS (v 3.8 SAS Studio, Cary, NC, USA). Fixed effects included dietary treatment and period, and the random effect was horse. The influence of feeding Coastal bermudagrass (CB 4, CB 6, and CB 8) compared with other hays (alfalfa and orchardgrass) on digestive variables was determined using contrasts. Statistically significant means were separated by Scheffe's method. Model derived TTMRT was compared to arithmetic calculations using both two one-sided tests of equivalence and regression analysis. For equivalence testing, the acceptable difference was 10%. Statistical trends were defined as $p < 0.1$ and differences at $p < 0.05$.

3. Results

3.1. Diet Digestibility

One factor that affects digestibility measurements is feed refusal. Orts were collected during 26 of the 75 daily measurements of intake, most frequently when horses were fed the CB 8 diet. The mean weight of orts was 0.07 kg (DM basis). Ash concentration of hay orts ranged from 24.6% to 66.8%, indicating contamination with sand from the environment. Thus, hay ort weight was corrected by multiplying ort weight by the ratio of ort ash concentration to forage ash concentration. Orts were analyzed for nutrient composition and subtracted from nutrient intake to correct for any feed not consumed by the horses.

Differences in fecal excretion were related to variations in diet digestibility. Horses fed Coastal bermudagrass hay diets defecated 1.4 times more frequently ($p < 0.05$) than when fed alfalfa hay (Table 3). Horses fed CB 6 and CB 8 excreted more feces ($p < 0.05$) than horses consuming ALF, ORCH, or CB 4. Dry matter and OM digestibilities were greatest ($p < 0.05$) for ALF, whereas a reduction in DMD and OMD was observed when horses were fed CB 6 and CB 8. There was a 32.0% reduction ($p < 0.05$) in NDFD and a 47.1% decrease ($p < 0.05$) in ADFD digestibility for the CB 6 and CB 8 diets compared with the other diets.

Table 3. Fecal excretion and diet digestibility of five experimental diets [1] ($n = 5$).

Variable	ALF	ORCH	CB 4	CB 6	CB 8	SEM	Diet [2] p-Value	Contrast [3] p-Value
Defecation Frequency, times/d	10.0 c	11.5 b,c	14.1 a,b	15.3 a	14.0 a,b	0.5	<0.001	<0.001
Fecal Excretion, kg DM/d	3.55 d	4.41 c	5.04 b	5.84 a	5.87 a	0.20	<0.001	<0.001
Fecal DM, %	19.7	20.9	20.0	20.5	22.5	0.42	0.074	0.255
Urination Frequency, times/d	10.6	10.6	8.7	8.3	10.7	0.58	0.324	0.161
Digestibility, %								
DM	62.1 a	51.2 b	47.2 b	36.0 c	36.8 c	2.1	<0.001	<0.001
OM	63.1 a	52.3 b	46.8 c	37.3 d	37.6 d	2.1	<0.001	<0.001
NDF	43.1 a	42.4 a	46.2 a	31.1 b	31.8 b	1.7	<0.001	<0.001
ADF	40.2 a	39.8 a	39.8 a	23.9 b	24.3 b	1.9	<0.001	<0.001

[1] Abbreviations. ALF, alfalfa; ORCH, orchardgrass; CB 4, Coastal bermudagrass 4-weeks regrowth; CB 6, Coastal bermudagrass 6-weeks regrowth; CB 8, Coastal bermudagrass 8-weeks regrowth; SEM, standard error of the mean. [2] Main effect of diet. [3] Contrast between Coastal bermudagrass (CB 4, CB 6, CB 8) and other diets (ALF, ORCH). a,b,c,d Means with unlike superscripts differ ($p < 0.05$).

3.2. Fecal Marker Excretion

3.2.1. Marker Excretion and Recovery

Mean fecal marker excretion is presented in Figure 1. External marker concentrations were detected in feces between 5 to 13 h after feeding horses external markers. Element concentrations were below instrument detection limits by 60 h post marker dosing, thus, fecal samples were only analyzed for marker concentrations to 72 h post marker dosing. A pulsatile pattern was observed in some individual fecal marker excretion curves (Supplementary Figures S1–S5).

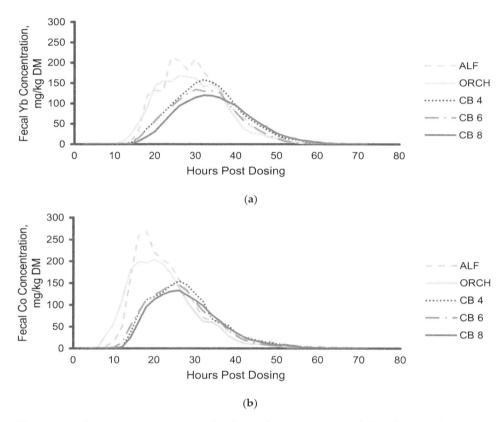

(a)

(b)

Figure 1. Two period moving average of fecal marker excretion of (**a**) Yb and (**b**) Co after dosing external markers (SEM = 12.4 and 13.2 mg/kg DM, respectively). Abbreviations. ALF, alfalfa; ORCH, orchardgrass; CB 4, Coastal bermudagrass 4-weeks regrowth; CB 6, Coastal bermudagrass 6-weeks regrowth; CB 8, Coastal bermudagrass 8-weeks regrowth.

Marker recovery ranged from 73.3 to 97.6% for Yb and 73.9 to 115% for Co. Particulate marker recovery did not differ by diet (Table 4). Liquid marker recovery tended to differ among diets ($p = 0.075$) with mean Co recovery greatest in the ORCH diet and lowest in the CB 4 diet. Particulate and liquid marker recovery did not differ within a horse for each period.

Table 4. Particulate (Yb) and liquid (Co) marker recovery [1] ($n = 5$).

Variable	ALF	ORCH	CB 4	CB 6	CB 8	SEM	Diet [2] p-Value	Contrast [3] p-Value
Particulate, %	80.6	86.3	85.5	82.5	85.0	1.44	0.549	0.948
Liquid, %	85.5	95.1	79.4	80.3	82.6	2.17	0.075	0.025

[1] Abbreviations. ALF, alfalfa; ORCH, orchardgrass; CB 4, Coastal bermudagrass 4-weeks regrowth; CB 6, Coastal bermudagrass 6-weeks regrowth; CB 8, Coastal bermudagrass 8-weeks regrowth; SEM, standard error of the mean. [2] Main effect of diet. [3] Contrast between Coastal bermudagrass (CB 4, CB 6, CB 8) and other diets (ALF, ORCH).

3.2.2. Modeling Fecal Marker Excretion

Seventy six percent of model equations fit fecal excretion data for each horse within a period using initial parameter ranges and start values defined in the program code. When the model did not converge using the code, model parameter ranges were adjusted using curve fitting software in MATLAB to obtain an acceptable fit (as indicated by the R^2 value being non-negative).

Mean model result from all data is depicted for all equations in Figure 2. Mean model fit, parameter values, and retention time from fecal marker excretion of each observation are summarized in Table 5. Model parameters were nonzero ($p < 0.05$) for 47% of the fitted equations. The scaling parameter was less than 0 for the G1G1 model. As the order of the γ-distribution increased for the equations described by Pond et al. [15], TT, $CMRT_2$, and TTMRT decreased, whereas $CMRT_1$ increased. The root mean square error (RMSE) ranged from 2.932 to 42.23 and 3.369 to 29.85 for particulate and liquid fecal marker excretion, respectively (Table 5). The model described by Dhanoa et al. [14] had the lowest RMSE and Akaike's information criterion (AIC) for the particulate and liquid phases of digesta. Among the six two-compartment γ-distributed equations described by Pond et al. [15], the G5G1 model best fit particulate marker excretion and the G4G1 equation best fit liquid marker excretion based on AIC values (Table 5). Because the AIC values increased once the order 5 and order 4 γ-gamma distributions were fit to the particulate and liquid marker excretion, fitting marker excretion to the two-compartment model was terminated at the order 6 γ-gamma distribution.

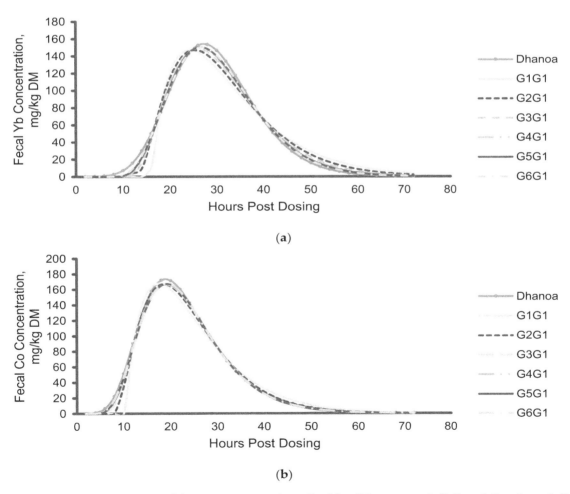

Figure 2. Model result derived from equations described by Dhanoa et al. [14] and Pond et al. [15] applied to all experimental data of (**a**) Yb and (**b**) Co after dosing external markers. Abbreviations. G1G1, first order two-compartment model; G2G1, second order two-compartment model; G3G1, third order two-compartment model; G4G1, fourth order two-compartment model; G5G1, fifth order two-compartment model; G6G1, sixth order two-compartment model according to Pond et al. [15].

Table 5. Model goodness of fit, parameters, and retention time of equations used to fit particulate (Yb) and liquid (Co) fecal marker excretion (diets combined, $N = 25$; means ± SE).

Variable	Dhanoa [a]	G1G1 [b]	G2G1 [b]	G3G1 [b]	G4G1 [b]	G5G1 [b]	G6G1 [b]
Particulate							
RMSE	11.70	20.48	16.30	14.83	14.02	12.43	12.91
AIC	128.8	163.1	148.1	142.3	138.7	134.3	137.0
Models with Nonzero Rate Parameters, %	4	0	20	72	76	84	92
A	$2.68 \times 10^{11} \pm 2.65 \times 10^{11}$						
C, mg Yb/kg digesta		-476 ± 26.6	636 ± 39.6	761 ± 49.6	856 ± 57.6	901 ± 56.3	916 ± 63.0
k_1, h^{-1}	0.265 ± 0.017	0.133 ± 0.007	0.187 ± 0.006	0.223 ± 0.008	0.253 ± 0.008	0.271 ± 0.009	0.277 ± 0.012
k_2, h^{-1}	0.355 ± 0.011	0.142 ± 0.008	0.192 ± 0.007	0.235 ± 0.011	0.279 ± 0.014	0.321 ± 0.018	0.370 ± 0.024
λ_1, h^{-1}							
N	48 ± 12.9						
TT, h	25.8 ± 1.15	17.5 ± 0.71	15.6 ± 0.64	13.4 ± 0.69	11.9 ± 0.66	10.6 ± 0.62	9.50 ± 0.61
$CMRT_1$, h	4.20 ± 0.29	8.01 ± 0.40	10.7 ± 0.34	13.3 ± 0.47	15.0 ± 0.56	16.4 ± 0.63	17.4 ± 0.77
$CMRT_2$, h	2.89 ± 0.096	7.48 ± 0.33	5.46 ± 0.15	4.61 ± 0.17	4.07 ± 0.17	3.82 ± 0.18	3.87 ± 0.26
TTMRT, h	32.9 ± 1.00	33.0 ± 0.99	31.7 ± 0.93	31.3 ± 0.92	31.0 ± 0.90	30.8 ± 0.90	30.8 ± 0.88
Liquid							
RMSE	13.88	17.51	15.57	14.95	14.04	12.43	14.01
AIC	141.1	154.6	147.2	145.0	142.6	143.0	143.5
Models with Nonzero Rate Parameters, %	16	0	52	76	80	84	84
A	$8.00 \times 10^{8} \pm 4.33 \times 10^{8}$						
C, mg Co/kg digesta		-470 ± 26.1	527 ± 21.4	582 ± 33.7	630 ± 42.6	642 ± 47.6	639 ± 66.3
k_1, h^{-1}	0.201 ± 0.020	0.202 ± 0.031	0.166 ± 0.008	0.184 ± 0.011	0.202 ± 0.015	0.206 ± 0.017	0.213 ± 0.020
k_2, h^{-1}	0.443 ± 0.045	0.144 ± 0.006	0.333 ± 0.055	0.430 ± 0.072	0.517 ± 0.086	0.593 ± 0.097	0.655 ± 0.105
λ_1, h^{-1}							
N	$8.34 \times 10^{4} \pm 83191$						
TT, h	17.3 ± 1.27	11.8 ± 0.48	10.1 ± 0.41	8.49 ± 0.38	7.47 ± 0.39	6.32 ± 0.43	5.41 ± 0.46
$CMRT_1$, h	6.18 ± 0.53	6.20 ± 0.40	8.12 ± 0.64	9.81 ± 0.85	10.9 ± 0.97	12.0 ± 1.10	12.8 ± 1.17
$CMRT_2$, h	2.57 ± 0.15	7.23 ± 0.29	6.39 ± 0.34	6.09 ± 0.44	5.85 ± 0.51	5.86 ± 0.52	5.87 ± 0.55
TTMRT, h	26.0 ± 0.92	25.2 ± 0.79	24.6 ± 0.71	24.4 ± 0.68	24.3 ± 0.65	24.1 ± 0.63	24.1 ± 0.63

[a] Dhanoa et al. [14] [b] Pond et al. [15]. Abbreviations. G1G1, first order two-compartment model; G2G_, second order two-compartment model; G3G1, third order two-compartment model; G4G1, fourth order two-compartment model; G5G1, fifth order two-compartment model; G6G1, sixth order two-compartment model according to Pond et al. [15]; RMSE, root mean square error; AIC, Akaike's information criterion; TT, transit time; CMRT, compartment mean retention time; TTMRT, total tract mean retention time.

3.3. Digesta Mean Retention Time

3.3.1. Mean Retention Time Calculated from Model Parameters

The best fitting models to describe marker excretion were used to compare digesta ROP between diets. The Dhanoa et al. [14] and Pond et al. [15] G5G1 models were used to analyze particulate phase ROP. The Dhanoa et al. [14] and Pond et al. [15] G4G1 models were used to analyze liquid digesta ROP.

Particulate TTMRT differed by diet ($p = 0.020$ and $p = 0.022$, respectively; Table 6) when Yb marker excretion was fit to both the Dhanoa et al. [14] and Pond et al. [15] G5G1 model. Rates of passage, k_1 and k_2, did not differ among diets when fecal excretion was fit to equations described by Dhanoa et al. [14]; however, $CMRT_1$ tended to be longer ($p = 0.084$) when horses were fed ALF compared with CB 6. Transit time calculated according to Dhanoa et al. [14] tended to be longer ($p = 0.074$) in CB 8 than ALF. Modeling particulate fecal marker excretion using the G5G1 model [15], λ_1 tended to differ by diet ($p = 0.069$) with λ_1 trending towards being quicker ($p = 0.102$) in ALF than CB 8. The other model parameters k_2 and TT did not differ by diet. Age-dependent compartment mean retention time ($CMRT_1$) was shorter ($p = 0.014$) in ALF than CB 8 and tended to be shorter ($p = 0.065$) than CB 4. There was no difference in the age-independent compartment mean retention time ($CMRT_2$).

Table 6. Model parameters and compartment retention times for particulate (Yb) and liquid (Co) marker fecal excretion [1] ($n = 5$).

Item	ALF	ORCH	CB 4	CB 6	CB 8	SEM	Diet [2] p-Value	Contrast [3] p-Value
Particulate Dhanoa A								
A	2.4×10^9	3.0×10^9	1.3×10^{12}	3.1×10^9	3.5×10^9	4.3×10^8		
k_1, h^{-1}	0.210	0.287	0.248	0.294	0.259	0.0536	0.223	0.239
k_2, h^{-1}	0.357	0.380	0.336	0.372	0.328	0.0450	0.338	0.312
N	107	34	29	36	33	83191	0.179	0.054
TT, h	22.0 [y]	23.5 [x,y]	27.9 [x,y]	26.9 [x,y]	28.7 [x]	1.27	0.035	0.004
$CMRT_1$, h	5.59 [x]	3.77 [x,y]	3.82 [x,y]	3.63 [y]	4.17 [x,y]	0.532	0.049	0.051
$CMRT_2$, h	2.92	2.68	3.03	2.70	3.12	0.148	0.303	0.479
TTMRT, h	30.5 [y]	30.0 [y]	34.7 [x,y]	33.2 [x,y]	35.9 [x]	0.921	0.020	0.003
G5G1 [B]								
C, mg Yb/kg digesta	1134	1007	906	780	681	56.3		
λ_1, h^{-1}	0.396	0.342	0.277	0.319	0.272	0.018	0.069	0.014
k_2, h^{-1}	0.249	0.279	0.280	0.274	0.272	0.009	0.751	0.489
TT, h	11.4	9.02	10.3	11.1	11.4	0.616	0.362	0.948
$CMRT_1$, h	13.6 [b,y]	15.4 [a,b]	18.2 [a,b,x]	16.1 [a,b]	18.5 [a]	0.634	0.008	0.002
$CMRT_2$, h	4.47	3.64	3.59	3.71	3.71	0.182	0.508	0.218
TTMRT, h	29.5 [a,b]	28.1 [b]	32.1 [a,b]	30.9 [a,b]	33.6 [a]	0.900	0.022	0.007
Liquid Dhanoa [A]								
A	1919	3.0×10^9	1.5×10^9	1.3×10^9	9.8×10^7	2.7×10^{11}		
k_1, h^{-1}	0.106 [x]	0.234	0.145	0.111 [y]	0.120	0.0165	0.042	0.023
k_2, h^{-1}	0.708 [x]	0.440	0.373	0.376	0.320 [y]	0.0108	0.042	0.016
N	4.2×10^5	50	49	33	31.2	13	0.661	0.075
TT, h	11.3 [b]	13.6 [a,b]	19.9 [a]	20.7 [a]	20.7 [a]	1.14	0.005	<0.001
$CMRT_1$, h	9.52 [a]	6.14 [b]	5.57 [b]	4.0 [b]	5.58 [b]	0.291	<0.001	<0.001
$CMRT_2$, h	1.75 [b]	2.48 [a,b]	2.75 [a,b]	2.70 [a,b]	3.17 [a]	0.096	0.009	0.005
TTMRT, h	22.6 [c,z]	22.3 [b,c,y,z]	28.2 [a,b,x,y]	27.5 [a,b,c,x]	29.5 [a]	1.00	0.002	<0.001

Table 6. *Cont.*

Item	ALF	ORCH	CB 4	CB 6	CB 8	SEM	Diet [2] p-Value	Contrast [3] p-Value
G4G1 [B]								
C, mg Yb/kg digesta	465	801	614	704	565	43		
λ_1, h^{-1}	1.095 [a,x]	0.491 [a,b,y]	0.359 [b]	0.329 [b]	0.310 [b]	0.0856	0.002	0.001
k_2, h^{-1}	0.107 [b,y]	0.216 [a,b,x]	0.203 [a,b]	0.247 [a]	0.236 [a,b,x]	0.0152	0.010	0.012
TT, h	8.42 [x]	5.45 [y]	8.15	7.34	8.00	0.391	0.042	0.194
CMRT$_1$, h	4.71 [b]	10.3 [a]	12.6 [a]	13.4 [a]	13.6 [a]	0.973	<0.001	<0.001
CMRT$_2$, h	9.44 [a]	5.32 [b]	5.61 [b]	4.20 [b]	4.69 [b]	0.513	0.001	0.002
TTMRT, h	22.6 [b,c]	21.0 [c]	26.4 [a]	25.0 [a,b]	26.3 [a]	0.650	<0.001	<0.001

[1] Abbreviations. ALF, alfalfa; ORCH, orchardgrass; CB 4, Coastal bermudagrass 4-weeks regrowth; CB 6, Coastal bermudagrass 6-weeks regrowth; CB 8, Coastal bermudagrass 8-weeks regrowth; SEM, standard error of the mean; RMSE, root mean square error; AIC, Akaike's information criterion; TT, transit time; CMRT, compartment mean retention time; TTMRT, total tract mean retention time. [2] Main effect of diet. [3] Contrast between Coastall bermudagrass (CB 4, CB 6, CB 8) and other diets (ALF, ORCH). [A] Dhanoa et al. [14]. [B] Pond et al. [15]. [a,b,c] Means with unlike superscripts differ ($p < 0.05$). [x,y,z] Means with unlike superscripts tend to differ ($p < 0.1$).

3.3.2. Arithmetically Calculated Mean Retention Time

Arithmetically calculated particulate digesta MRT (Table 7) using the equation described by Blaxter et al. [32] was longer ($p = 0.019$) than when calculated according to Thielemans et al. [33], but the mean difference was only 0.2 h (30.0 ± 0.88 vs. 29.8 ± 0.88 h). Th method of calculation did not affect liquid MRT (23.4 ± 0.70 vs. 23.5 ± 0.67 h). For both equations, particulate MRT was longer ($p < 0.001$) than liquid MRT.

Particulate MRT differed by diet ($p < 0.001$) and was longer ($p < 0.05$) when horses were fed CB 8 compared with ORCH (Table 7). Liquid MRT differed by diet ($p < 0.001$). Horses fed ORCH had a shorter ($p < 0.01$) liquid MRT than when fed the Coastal bermudagrass diets. Liquid MRT differed ($p < 0.011$) between horses fed ALF compared with CB 4 and CB 8 and tended to be shorter ($p < 0.079$) than horses fed the CB 6 diet.

Table 7. Mean retention time (h) of particulate digesta measured with Yb-NDF and Co-EDTA external markers [1] ($n = 5$).

Equation	ALF	ORCH	CB 4	CB 6	CB 8	SEM	Diet [2] p-Value	Contrast [3] p-Value
Blaxter et al. [32]								
Particulate	29.2 [a,b]	26.9 [b]	31.1 [a,b]	30.0 [a,b]	32.7 [a]	0.88	0.010	0.006
Liquid	21.3 [b,c,y]	20.1 [c]	25.7 [a]	24.3 [a,b,x]	25.9 [a]	0.70	<0.001	<0.001
Thielemans et al. [33]								
Particulate	28.9 [a,b]	27.0 [b]	30.4 [a,b]	29.6 [a,b]	32.6 [a]	0.88	0.010	0.008
Liquid	21.2 [b,c,y]	20.6 [c]	25.4 [a]	24.1 [a,b,x]	25.9 [a]	0.67	<0.001	<0.001

[1] Abbreviations. ALF, alfalfa; ORCH, orchardgrass; CB 4, Coastal bermudagrass 4-weeks regrowth; CB 6, Coastal bermudagrass 6-weeks regrowth; CB 8, Coastal bermudagrass 8-weeks regrowth; SEM, standard error of the mean. [2] Main effect of diet. [3] Contrast between Coastal bermudagrass (CB 4, CB 6, CB 8) and other diets (ALF, ORCH). [a,b,c] Means with unlike superscripts differ ($p < 0.05$). [x,y] Means with unlike superscripts tend to differ ($p < 0.10$).

3.3.3. Comparing Model-Derived and Arithmetically Calculated Mean Retention Time

The mean retention times calculated from Dhanoa et al. [14] and Pond et al. [15] model parameters were compared with arithmetically calculated MRT using two one-sided test of equivalence (TOST) with a 10% difference and regression analysis. Model-derived TTMRT was similar to ($p < 0.05$) MRT calculated according to the equation described by Blaxter et al. [32] or Thielemans et al. [33] for the G2G1, G3G1, G4G1, G5G1 models for the particulate phase of digesta. For the liquid phase of digesta, models described by Pond et al. [15] were similar to ($p < 0.05$) arithmetic MRT, but TTMRT calculated according to Dhanoa et al. [14] was over 2 h shorter than arithmetic MRT.

Arithmetically calculated MRT and model-derived TTMRT are plotted in Figure 3. For the particulate phase of digesta (Figure 2a), the equation relating TTMRT to MRT according to Blaxter et al. [32] was MRT = 0.7929x + 3.903 (RMSE = 1.980; r^2 = 0.8071; p < 0.001) for TTMRT calculated according to Dhanoa et al. [14] and MRT = 0.9737x − 0.05589 (RMSE = 0.6256; r^2 = 0.9807; p < 0.001) when TTMRT according to the G5G1 model described by Pond et al. [15]. When MRT was calculated according to Thielemans et al. [33], MRT = 0.7935x + 3.680 (RMSE = 1.960; r^2 = 0.8105; p < 0.001) for TTMRT calculated according to Dhanoa et al. [14] and MRT = 0.9743x − 0.2772 (RMSE = 0.5592; r^2 = 0.9846; p < 0.001) when TTMRT according to the G5G1 model described by Pond et al. [15]. For the liquid phase of digesta (Figure 2b), the equation relating TTMRT to MRT according to Blaxter et al. [32] was MRT = 0.6329x + 6.956 (RMSE = 2.017; r^2 = 0.6852; p < 0.001) for TTMRT calculated according to Dhanoa et al. [14] and the y-intercept differed (p = 0.007) from zero. Mean retention time calculated according to Blaxter et al. [14] was related to TTMRT calculated according to the G4G1 model described by Pond et al. [15] as MRT = 1.064x − 2.408 (RMSE = 0.6661; r^2 = 0.9657; p < 0.001) and the y-intercept differed (p = 0.028) from zero. When TTMRT was calculated according to Thielemans et al. [33], MRT = 0.5326x + 9.984 (RMSE = 2.565; r^2 = 0.4879; p < 0.001) for TTMRT calculated according to Dhanoa et al. [14] and MRT = 0.8692x + 2.748 (RMSE = 2.129; r^2 = 0.6474; p < 0.001) when TTMRT according to the G4G1 model described by Pond et al. [15]. For the regression equation relating TTMRT from the G4G1 model to MRT calculated according to Thielemans et al. [33], the y-intercept differed from zero (p = 0.003).

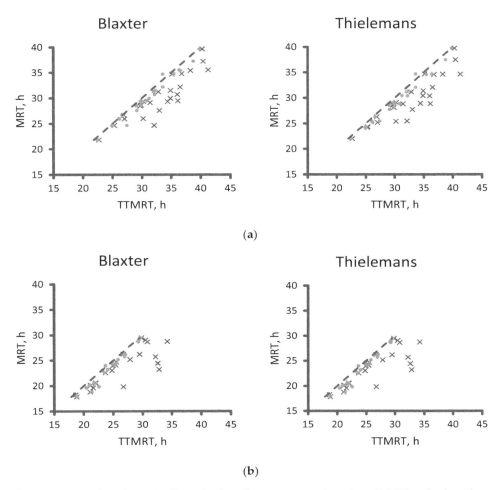

Figure 3. Comparison of arithmetically calculated mean retention time (MRT) calculated according to Blaxter et al. [32] and Thielemans et al. [33] compared with total tract mean retention time (TTMRT) calculated according to Dhanoa et al. [14] (×) and the best fitting two-compartment model described by Pond et al. [15] (•) for the particulate (**a**) and liquid (**b**) phase of digesta. The 1 to 1 line is denoted by the dashed line.

4. Discussion

Increasing the retention time of Coastal bermudagrass may be an important digestive strategy in horses to adapt to the more difficult-to-digest fiber particles of warm-season grasses. Greater retention time of highly fibrous, warm-season forages allows for a lengthened exposure of digesta to microbial degradation. Forage fiber composition may also influence digesta passage rate within the gastrointestinal tract, as observed when mathematically modeling fecal marker excretion when horses were fed alfalfa hay, which has an increased concentration of pectin compared to grasses.

Forage type affected DMD, with the greatest digestibility observed when horses were fed alfalfa. Several other studies have reported greater digestibility of legume hays than cool-season or warm-season grass hays [34–40]. The DMD of alfalfa hay in this study falls within the 54–66% range of alfalfa hay DMD reported in the literature [34–40]. Chemically, legume forages have greater protein and pectin concentrations and decreased insoluble fibers, allowing for a faster rate of digestion [41]. Legumes also contain a greater proportion of more easily digested mesophyll cells compared to grasses. The accumulation of lignin in alfalfa cell walls occurs primarily in alfalfa stems, whereas lignin accumulates in both grass stems and leaves during maturity [42,43]. Thus, there is a greater extent of cell wall digestion of alfalfa leaves compared to grasses [44].

Longer intervals of growth before harvest increased Coastal bermudagrass fiber and lignin concentrations in hay and resulted in reduced dry matter, organic matter, and fiber digestibilities. The internal girder structure of C4 forages firmly links the epidermis to vascular bundles, reducing the rate of digestion [45,46]. Akin et al. [47] reported that even typically highly digestible plant mesophyll cells were only partially degraded with increasing plant maturity in bermudagrass samples. The reported digestibility of Coastal bermudagrass hay ranges from 41–53% [34,35,48–50]. Although the DMD of CB 4 fell within the reported range, the CB 6 and CB 8 had reduced digestibility compared with reported values. Because fiber concentration is negatively correlated with digestibility [6], the comparatively low digestibility of CB 6 and CB 8 to other published data is surprising as the Coastal bermudagrass used in those studies often had a greater detergent fiber concentration than the hays utilized in the current study [34,48–50]. Although the Coastal bermudagrass harvested at four weeks regrowth had greater NDF and hemicellulose concentrations than orchardgrass and alfalfa hays, fiber digestibility was similar. However, a reduction in NDFD and ADFD was observed in the CB 6 and CB 8 diets. Phenolic compounds and hemicellulose composition changes with increasing plant cell maturity [47,51,52] likely leading to the decreased digestibility of Coastal bermudagrass harvested at longer intervals of growth. The results of the current study indicate that differences beyond plant fiber concentration, such as hemicellulose composition and lignin concentration also affect the digestibility of Coastal bermudagrass hay.

One strategy to adapt to the lower digestibility of high fiber C4 grasses such as Coastal bermudagrass is for the digesta retention time to increase. Other high fiber forages, such as oat straw, have also been reported to have increased MRT in equines [9]. Longer exposure to microbial fermentation in the equine hindgut can increase cell wall digestibility and could compensate for a slower rate of degradation of fibers in warm-season grasses [46,53,54]. This selective retention of Coastal bermudagrass hay in the gastrointestinal tract (GIT) may be due to differences in the particle size of digesta, rate of intake, or changes in GIT motility. How fiber affects gastrointestinal transit is not clearly understood in horses, people, or other mammals. One theory is that the fiber can trap water in the gastrointestinal tract, altering the way bacteria and solutes interact in the GIT [55]. In addition to luminal contents, gut motility is also driven by a range of neurohormones. Peptide YY (PYY) and glucagon-like peptide-1 (GLP-1) are important neurohormones regulating colonic motility [56], and the secretion of PYY and GLP-1 increases with the addition of fiber to diets [57,58]. Gut motility, neurotransmitters, and digesta characteristics warrant future investigation for potential mechanisms of regulating digesta transit in horses.

Differences in ROP were identified between forage types that were not apparent in TTMRT, indicating that mathematically modeling fecal marker excretion can advance the study of digesta passage in horses. Alfalfa ROP parameters derived from Dhanoa et al. [14] and Pond et al. [15] differed from other diets, indicating differences in digesta passage kinetics for a legume hay compared to grasses. For the particulate phase of digesta, λ_1 was quicker and $CMRT_1$ shorter in ALF than CB 8, even though no total tract differences were determined when TTMRT was calculated according to the G5G1 model. When particulate $CMRT_1$ was calculated according to Dhanoa et al. [14], $CMRT_1$ was longer in ALF compared with CB 6, a result also not reflected in TTMRT. When the liquid phase of digesta was modeled according to the G4G1 model, the age-dependent compartment ($CMRT_1$) was shorter than the age-independent compartment when horses were fed alfalfa. In contrast, all the grass forages had longer age-dependent CMRT than age-independent CMRT. Alfalfa and other legumes have higher pectin (a soluble fiber) concentrations compared to grasses. Soluble fiber has been shown to increase retention time in the small intestine and delay gastric emptying in other species [59,60]. The effect of fiber type on digesta ROP in the horse may be better elucidated using mathematical models than total tract MRT.

In the current study, compartment models described by both Dhanoa et al. [14] and Pond et al. [15] adequately fit equine marker excretion using non-linear least squares methods by modifying model parameter start values and bounds of rate parameters. Both one- and two-compartment models have been used in the recent literature, with the best fitting model depending on the study. Equations described by both Dhanoa et al. [14] and Pond et al. [15] have failed to produce a solution of acceptable model fit due to lack of convergence between experimental data and the model [10,19,61]. However, computing power and collaboration with computer scientists and modelers can greatly reduce the likelihood of models failing to converge with experimental data. Future studies incorporating mathematical modeling into digesta ROP studies will help to identify the best ways to describe digesta passage in the equine GIT.

Although using the equation described by Dhanoa et al. [14] resulted in an improved fit compared with the Pond et al. [15] models based on AIC values, TTMRT from the best fitting Pond et al. [15] models were more similar to arithmetically calculated MRT. The mathematical basis of these two models differ. The equation described by Dhanoa et al. [14] represents an unspecified number of exponentially distributed compartments, whereas the Pond et al. [15] equations used in the current study represented two distinct compartments plus transit time. The first two compartments in the Dhanoa et al. [14] model ($CMRT_1$ and $CMRT_2$) have the longest retention time, and the remaining compartments are summed to determine transit time. Transit time is often defined as the amount of time from marker dosing to the first appearance of marker in the feces, but Ellis et al. [62] described transit time in the Dhanoa et al. [14] model as the sum of time in the remaining system of mixing compartments. The difference in transit time definition may be part of the discrepancy between transit time calculated from the different model equations. Additionally, the Dhanoa et al. [14] model may represent a set of exponentially distributed compartments that are within a larger mixing structure. The lack of agreement between arithmetic MRT and TTMRT calculated by the Dhanoa et al. [14] model may be explained by these differences.

The use of mathematical models could be further enhanced if theoretical compartments could be correlated to anatomical sections of the GIT. Moore-Colyer et al. [10] hypothesized that the two compartments of the Pond et al. [15] model represented the large colon for the age-dependent compartment and the cecum for the age-independent compartment because retention time in the colon is longer than the retention time in the cecum [63]. Transit time was hypothesized to represent residence time in the remaining structures of the GIT (i.e., stomach and small intestine). In a previous study using similar methodology, modeling fecal marker excretion when horses were fed mainly forage diets resulted in a CMRT similar to MRT in the cecum and colon [64,65]. However, Murray et al. [19] rejected the hypothesis that the age-dependent compartment was the colon because they observed a longer retention time in the age-independent compartment. The longer retention time of the age-independent compartment by Murray et al. [19] may be due to the diet (alfalfa and sugar beet pulp, which are high in soluble fiber), as observed in this study. Thus, diet may have a greater influence on compartment mean retention time than connections with anatomical compartments. Overall, mathematical models show promise to describe passage kinetics in horses, but the physiological relevance of compartment retention times remains unclear.

5. Conclusions

In conclusion, the horse appears to adopt a digestive strategy to decrease the rate of passage of digesta when fed warm-season grass forages. Increasing the retention time allows for fiber particles with greater hemicellulose and lignin concentrations to be exposed to microbial fermentation longer. Because warm-season forages have slower rates of degradation, this change in retention time allows the horse to maximize potential nutrients obtained from the diet. Using mathematical models further characterized differences in digesta ROP between forages which were not apparent when evaluating total tract MRT alone. Minor discrepancies between models and arithmetically calculated MRT were observed and should be resolved for future use. Nonetheless, mathematical modeling should be incorporated into future equine nutrition research to expand knowledge on digesta passage and equine science in general.

Author Contributions: Conceptualization, L.K.W., T.L.H., and E.L.C.; methodology, L.K.W., T.L.H., and E.L.C.; validation, L.K.W., and T.L.H.; formal analysis, T.L.H.; investigation, T.L.H., E.L.C., O.K.Z., J.M.M., J.M.B., L.A.S., J.W.C., A.M.A. and L.K.W.; resources, L.K.W.; writing—original draft preparation, T.L.H.; writing—review and editing, L.K.W., T.L.H.

Acknowledgments: Seminole Feed Inc. (Ocala, FL, USA) provided a vitamin, mineral, protein pellet to feed in this study. The authors are grateful for the staff and students at the University of Florida/Institute of Food and Agricultural Sciences Horse Teaching Unit to complete this study. The authors would like to thank K. Bissinger, K. Bush, S. Chewing, M. Di-Lernia, B. Eubanks, T. Fraguela, J. Gerbert, K. Irvine, A. Mesa, N. Oliver, H. Miller, V. Robbins, D. Van Camp for animal care, sample collection, and laboratory assistance. J. Bauman, B. d'Angelo, and N. Wilkinson were an essential team for marker analysis on fecal samples. We additionally would like to acknowledge the thoughtful discussion provided by J. Brendemuhl, A.L. Fowler, T.J. Hackmann, and L. Sollenberger about this data.

References

1.   Little, D.; Blikslager, A.T. Factors associated with development of ileal impaction in horses with surgical colic: 78 cases (1986–2000). *Equine Vet. J.* **2002**, *34*, 464–468. [CrossRef]
2.   Blikslager, A.T. Colic prevention to avoid colic surgery: A surgeon's perspective. *J. Equine Vet. Sci.* **2019**, *76*, 1–5. [CrossRef]
3.   Moore, K.J.; Boote, K.J.; Sanderson, M.A. Physiology and developmental morphology. In *Warm-Season (C4) Grasses*; Moser, L.E., Burson, B.L., Sollenberger, L.E., Eds.; ASA/CSSA/SSSA: Madison, WI, USA, 2014; pp. 179–216.
4.   Sollenberger, L.E.; Vanzant, E.S. Interrelationships among forage nutritive value and quantity and individual animal performance. *Crop Sci.* **2011**, *51*, 420–432. [CrossRef]
5.   Rohweder, D.A.; Barnes, R.F.; Jorgensen, N. Proposed hay grading standards based on laboratory analyses for evaluating quality. *J. Anim. Sci.* **1978**, *47*, 747–759. [CrossRef]
6.   Hansen, T.L.; Lawrence, L.M. Composition factors predicting forage digestibility by horses. *J. Equine Vet. Sci.* **2017**, *58*, 97–102. [CrossRef]
7.   Lowman, R.S.; Theodorou, M.K.; Hyslop, J.J.; Dhanoa, M.S.; Cuddeford, D. Evaulation of an in vitro batch culture technique for estimating the in vivo digestibility and digestible energy content of equine feeds using equine faeces as the source of microbial inoculum. *Anim. Feed Sci. Technol.* **1999**, *80*, 11–27. [CrossRef]
8.   Sunvold, G.D.; Hussein, H.S.; Fahey, G.C., Jr.; Merchen, N.R.; Reinhart, G.A. In vitro fermentation of cellulose, beet pulp, citrus pulp, and citrus pectin using fecal inoculum from cats, dogs, horses, humans, and pigs and ruminal fluid from cattle. *J. Anim. Sci.* **1995**, *73*, 3639–3648. [CrossRef]

9. Pearson, R.A.; Archibald, R.F.; Muirhead, R.H. The effect of forage quality and level of feeding on digestibility and gastrointestinal transit time of oat straw and alfalfa given to ponies and donkeys. *Br. J. Nutr.* **2001**, *85*, 599–606. [CrossRef]

10. Moore-Colyer, M.J.S.; Morrow, H.J.; Longland, A.C. Mathematical modelling of digesta passage rate, mean retention time and in vivo apparent digestibility of two different lengths of hay and big-bale grass silage in ponies. *Br. J. Nutr.* **2003**, *90*, 109–118. [CrossRef]

11. Miyaji, M.; Ueda, K.; Hata, H.; Kondo, S. Effects of quality and physical form of hay on mean retention time of digesta and total tract digestibility in horses. *Anim. Feed Sci. Technol.* **2011**, *165*, 61–67. [CrossRef]

12. Van Weyenberg, S.; Sales, J.; Janssens, G.P.J. Passage rate of digesta through the equine gastrointestinal tract: A review. *Livest. Sci.* **2006**, *99*, 3–12. [CrossRef]

13. Hooda, S.; Metzler-Zebeli, B.U.; Vasanthan, T.; Zijlstra, R.T. Effects of viscosity and fermentability of dietary fibre on nutrient digestibility and digesta characteristics in ileal-cannulated grower pigs. *Br. J. Nutr.* **2011**, *106*, 664–674. [CrossRef]

14. Dhanoa, M.S.; Siddons, R.C.; France, J.; Gale, D.L. A multicompartmental model to describe marker excretion patterns in ruminant faeces. *Br. J. Nutr.* **1985**, *53*, 663–671. [CrossRef]

15. Pond, K.R.; Ellis, W.C.; Matis, J.H.; Ferreiro, H.M.; Sutton, J.D. Compartment models for estimating attributes of digesta flow in cattle. *Br. J. Nutr.* **1988**, *60*, 571–595. [CrossRef]

16. Grovum, W.L.; Williams, V.J. Rate of passage of digesta in sheep. 4. Passage of marker through the alimentary tract and the biological relevance of rate-constants derived from the changes in concentration of marker in faeces. *Br. J. Nutr.* **1973**, *30*, 313–329. [CrossRef]

17. Austbø, D.; Volden, H. Influence of passage model and caecal cannulation on estimated passage kinetics of roughage and concentrate in the gastrointestinal tract of horses. *Livest. Sci.* **2006**, *100*, 33–43. [CrossRef]

18. Miyaji, M.; Ueda, K.; Hata, H.; Kondo, S. Effect of grass hay intake on fiber digestion and digesta retention time in the hindgut of horses. *J. Anim. Sci.* **2014**, *92*, 1574–1581. [CrossRef]

19. Murray, J.A.M.D.; Sanderson, R.; Longland, A.C.; Moore-Colyer, M.J.S.; Hastie, P.M.; Dunnett, C. Assessment of mathematical models to describe the rate of passage of enzyme-treated or sugar beet pulp-substituted lucerne silage in equids. *Anim. Feed Sci. Technol.* **2009**, *154*, 228–240. [CrossRef]

20. Rosenfeld, I.; Austbo, D.; Volden, H. Models for estimating digesta passage kinetics in the gastrointestinal tract of the horse. *J. Anim. Sci.* **2006**, *84*, 3321–3328. [CrossRef]

21. Federation of Animal Science Societies (FASS). *Guide for the Care and Use of Agricultural Animals in Research and Teaching*, 3rd ed.; FASS: Champaign, IL, USA, 2010.

22. Henneke, D.R.; Potter, G.D.; Kreider, J.L.; Yeates, B.F. Relationship between condition score, physical measurements and body fat percentage in mares. *Equine Vet. J.* **1983**, *15*, 371–372. [CrossRef]

23. NRC. *Nutrient Requirements of Horses: Sixth Revised Edition*; The National Academies Press: Washington, DC, USA, 2007; p. 360.

24. Fisher, R.A.; Yates, F. *Statistical Tables for Biological, Agricultural and Medical Research*; Oliver & Boyd Ltd.: London, UK, 1963; p. 146.

25. Chizek, E.L. Comparison of Feed Intake Behavior between Warm- and Cool-Season Forages Offered to Horses. Master's Thesis, University of Florida, Gainesville, FL, USA, 2016.

26. Pagan, J. Nutrient digestibility in horses. In *Advances in Equine Nutrition*; Kentucky Equine Research, Inc.: Versailles, KY, USA, 1998; pp. 77–83.

27. Udén, P.; Colucci, P.E.; Van Soest, P.J. Investigation of chromium, cerium and cobalt as markers in digesta rate of passage studies. *J. Sci. Food Agric.* **1980**, *31*, 625–632. [CrossRef]

28. Ringler, J.E.; Lawrence, L.M. Development of a method to label forages used in passage rate studies in the horse. *J. Equine Vet. Sci.* **2009**, *29*, 389–390. [CrossRef]

29. ANKOM Technology. Analytical Methods Fiber Analyzer A200. Available online: https://www.ankom.com/analytical-methods-support/fiber-analyzer-a200 (accessed on 20 January 2015).

30. EPA. *Method 3052: Microwave Assisted Acid Digestion of Siliceous and Organically Based Matrices*; EPA: Washington, DC, USA, 1996.

31. EPA. *Method 200.7: Determination of Metals and Trace Elements in Water and Wastes by Inductively Coupled Plasma-Atomic Emmision Spectometry*; EPA: Washington, DC, USA, 1994.

32. Blaxter, K.L.; Graham, N.M.; Wainman, F.W. Some observations on the digestibility of food by sheep, and on related problems. *Br. J. Nutr.* **1956** *10*, 69–91. [CrossRef]

33. Thielemans, M.-F.; Francois, E.; Bodart, C.; Thewis, A. Gastrointestinal transit in the pig: Measurement using radioactive lanthanides and comparison with sheep. *Ann. Biol. Anim. Biochim. Biophys.* **1978**, *18*, 237–247. [CrossRef]

34. Sturgeon, L.S.; Baker, L.A.; Pipkin, J.L.; Haliburton, J.C.; Chirase, N.K. The digestibility and mineral availability of Matua, Bermuda grass, and alfalfa hay in mature horses. *J. Equine Vet. Sci.* **2000**, *20*, 45–48. [CrossRef]

35. Eckert, J.V.; Myer, R.O.; Warren, L.K.; Brendemuhl, J.H. Digestibility and nutrient retention of perennial peanut and bermudagrass hays for mature horses. *J. Anim. Sci.* **2010**, *88*, 2055–2061. [CrossRef]

36. Earing, J.E.; Cassill, B.D.; Hayes, S.H.; Vanzant, E.S.; Lawrence, L.M. Comparison of in vitro digestibility estimates using the DaisyII incubator with in vivo digestibility estimates in horses. *J. Anim. Sci.* **2010**, *88*, 3954–3963. [CrossRef]

37. Cymbaluk, N.; Christensen, D. Nutrient utilization of pelleted and unpelleted forages by ponies. *Can. J. Anim. Sci.* **1986**, *66*, 237–244. [CrossRef]

38. Cymbaluk, N.F. Comparison of forage digestion by cattle and horses. *Can. J. Anim. Sci.* **1990**, *70*, 601–610. [CrossRef]

39. Potts, L.; Hinkson, J.; Graham, B.; Löest, C.; Turner, J. Nitrogen retention and nutrient digestibility in geldings fed grass hay, alfalfa hay, or alfalfa cubes. *J. Equine Vet. Sci.* **2010**, *30*, 330–333. [CrossRef]

40. Crozier, J.A.; Allen, V.G.; Jack, N.E.; Fontenot, J.P.; Cochran, M.A. Digestibility, apparent mineral absorption, and voluntary intake by horses fed alfalfa, tall fescue, and caucasian bluestem. *J. Anim. Sci.* **1997**, *75*, 1651–1658. [CrossRef] [PubMed]

41. Smith, L.W.; Goering, H.K.; Gordon, C.H. Relationships of forage compositions with rates of cell wall digestion and indigestibility of cell walls. *J. Dairy Sci.* **1972**, *55*, 1140–1147. [CrossRef]

42. Albrecht, K.A.; Wedin, W.F.; Buxton, D.R. Cell-wall composition and digestibility of alfalfa stems and leaves. *Crop Sci.* **1987**, *27*, 735–741. [CrossRef]

43. Griffin, J.L.; Jung, G.A. Leaf and stem forage quality of big bluestem and switchgrass. *Agron. J.* **1983**, *75*, 723–726. [CrossRef]

44. Bourquin, L.D.; Fahey, G.C., Jr. Ruminal digestion and glycosyl linkage patterns of cell wall components from leaf and stem fractions of alfalfa, orchardgrass, and wheat straw. *J. Anim. Sci.* **1994**, *72*, 1362–1374. [CrossRef] [PubMed]

45. Wilson, J.R.; Akin, D.E.; McLeod, M.N.; Minson, D.J. Particle size reduction of the leaves of a tropical and a temperate grass by cattle. II. Relation of anatomical structure to the process of leaf breakdown through chewing and digestion. *Grass Forage Sci.* **1989**, *44*, 65–75. [CrossRef]

46. Hastert, A.A.; Owensby, C.E.; Harbers, L.H. Rumen microbial degradation of Indiangrass and big bluestem leaf blades. *J. Anim. Sci.* **1983**, *57*, 1626–1636. [CrossRef]

47. Akin, D.E.; Robinson, E.L.; Barton, F.E.; Himmelsbach, D.S. Changes with maturity in anatomy, histochemistry, chemistry, and tissue digestibility of bermudagrass plant parts. *J. Agric. Food Chem.* **1977**, *25*, 179–186. [CrossRef]

48. Lieb, S.; Ott, E.A.; French, E.C. Digestible nutrients and voluntary intake of rhizomes peanut, alfalfa, bermudagrass and bahiagrass by equine. In Proceedings of the Thirteenth Equine Nutrition and Physiology Symposium, Gainesville, FL, USA, 21–23 January 1993; pp. 98–99.

49. Lieb, S.; Mislevy, P. Comparative intake and nutrient digestibility of three grass forages: Florakirk and Tifton 85 bermudagrasses and Florona stargrass to Coastal bermudagrass fed to horses. In Proceedings of the Seventeenth Equine Nutrition and Physiology Symposium, Lexington, KY, USA, 31 May–2 June 2001; pp. 390–391.

50. LaCasha, P.A.; Brady, H.A.; Allen, V.G.; Richardson, C.R.; Pond, K.R. Voluntary intake, digestibility, and subsequent selection of Matua bromegrass, coastal bermudagrass, and alfalfa hays by yearling horses. *J. Anim. Sci.* **1999**, *77*, 2766–2773. [CrossRef]

51. Akin, D.E.; Hartley, R.D. UV Absorption microspectrophotometry and digestibility of cell types of bermudagrass internodes at different stages of maturity. *J. Sci. Food Agric.* **1992**, *59*, 437–447. [CrossRef]

52. De Ruiter, J.M.; Burns, J.C.; Timothy, D.H. Hemicellulosic cell wall carbohydrate monomer composition in Panicum amarum, P. amarulum and P virgatum accessions. *J. Sci. Food Agric.* **1992**, *60*, 297–307. [CrossRef]

53. Koller, B.L.; Hintz, H.F.; Robertson, J.B.; Van Soest, P.J. Comparative cell wall and dry matter digestion in the

cecum of the pony and the rumen of the cow using in vitro and nylon bag techniques. *J. Anim. Sci.* **1978**, *47*, 209–215. [CrossRef]

54. Coblentz, W.K.; Fritz, J.O.; Fick, W.H.; Cochran, R.C.; Shirley, J.E. In situ dry matter, nitrogen, and fiber degradation of alfalfa, red clover, and eastern gamagrass at four maturities. *J. Dairy Sci.* **1998**, *81*, 150–161. [CrossRef]

55. Eastwood, M.A.; Kay, R.M. An hypothesis for the action of dietary fiber along the gastrointestinal tract. *Am. J. Clin. Nutr.* **1979**, *32*, 364–367. [CrossRef]

56. Wen, J.; Phillips, S.F.; Sarr, M.G.; Kost, L.J.; Holst, J.J. PYY and GLP-1 contribute to feedback inhibition from the canine ileum and colon. *Am. J. Physiol.* **1995**, *269*, G945–G952. [CrossRef]

57. Reimer, R.A.; McBurney, M.I. Dietary fiber modulates intestinal proglucagon messenger ribonucleic acid and postprandial secretion of glucagon-like peptide-1 and insulin in rats. *Endocrinology* **1996**, *137*, 3948–3956. [CrossRef]

58. Cani, P.D.; Lecourt, E.; Dewulf, E.M.; Sohet, F.M.; Pachikian, B.D.; Naslain, D.; De Backer, F.; Neyrinck, A.M.; Delzenne, N.M. Gut microbiota fermentation of prebiotics increases satietogenic and incretin gut peptide production with consequences for appetite sensation and glucose response after a meal. *Am. J. Clin. Nutr.* **2009**, *90*, 1236–1243. [CrossRef]

59. Jenkins, D.J.; Wolever, T.M.; Leeds, A.R.; Gassull, M.A.; Haisman, P.; Dilawari, J.; Goff, D.V.; Metz, G.L.; Alberti, K.G. Dietary fibres, fibre analogues, and glucose tolerance: Importance of viscosity. *Br. Med. J.* **1978**, *1*, 1392–1394. [CrossRef]

60. Schwartz, S.E.; Levine, R.A.; Singh, A.; Scheidecker, J.R.; Track, N.S. Sustained pectin ingestion delays gastric emptying. *Gastroenterology* **1982**, *83*, 812–817.

61. Jensen, R.B.; Austbo, D.; Bach Knudsen, K.E.; Tauson, A.H. The effect of dietary carbohydrate composition on apparent total tract digestibility, feed mean retention time, nitrogen and water balance in horses. *Animal* **2014**, *8*, 1788–1796. [CrossRef]

62. Ellis, W.C.; Matis, J.H.; Hill, T.M.; Murphy, M.R. Methodology for Estimating Digestion and Passage Kinetics of Forages. In *Forage Quality, Evaluation, and Utilization*; Fahey, G.C., Ed.; American Society of Agronomy, Crop Science Society of America, Soil Science Society of America: Madison, WI, USA, 1994; pp. 682–756.

63. Argenzio, R.A.; Lowe, J.E.; Pickard, D.W.; Stevens, C.E. Digesta passage and water exchange in the equine large intestine. *Am. J. Physiol.* **1974**, *226*, 1035–1042. [CrossRef] [PubMed]

64. Miyaji, M.; Ueda, K.; Nakatsuji, H.; Tomioka, T.; Kobayashi, Y.; Hata, H.; Kondo, S. Mean retention time of digesta in the different segments of the equine hindgut. *Anim. Sci. J.* **2008**, *79*, 89–96. [CrossRef]

65. Hansen, T.L. Modeling Digestibility and Rate of Passage in Horses. Master's Thesis, University of Kentucky, Lexington, KY, USA, 2014.

9

Characterization of Feeding, Sport Management and Routine Care of the Chilean Corralero Horse during Rodeo Season

Joaquín Bull [1], Fernando Bas [1], Macarena Silva-Guzmán [2], Hope Helen Wentzel [3], Juan Pablo Keim [4] and Mónica Gandarillas [1,4,*]

[1] Departamento de Ciencias Animales, Facultad de Agronomía e Ingeniería Forestal, Pontificia Universidad Católica de Chile, Avda. Vicuña Mackenna 4860, Santiago 7820436, Chile; jabull@uc.cl (J.B.); fbas@uc.cl (F.B.)
[2] Private statistical consultant, Guardia Vieja 441, Santiago 7510318, Chile; maca.silva.guzman@gmail.com
[3] Escuela de Graduados, Facultad de Ciencias Agrarias, Universidad Austral de Chile, Valdivia 5110566, Chile; hope.wentzel@alumnos.uach.cl
[4] Instituto de Producción Animal, Facultad de Ciencias Agrarias, Universidad Austral de Chile, Independencia 641, Valdivia 5110566, Chile; juan.keim@uach.cl
* Correspondence: monica.gandarillas@uach.cl

Simple Summary: The Chilean corralero horse holds great cultural importance due to its use in Chilean rodeo, the national sport. However, information regarding this breed is sparse, especially husbandry, feeding, and training recommendations, which could present challenges for their proper care. A survey of horse farms in several regions from central to southern Chile was conducted in order to document current management of the Chilean corralero horses which participated in the 2014–2015 Chilean Rodeo Federation season. In the survey, horse owners and trainers were asked about horse gender and size, daily routine, exercise and competition regimen, and feeding practices. All horses in the study were kept in stalls for at least 12 h daily and spent the rest of the day either tied or loose in pens or paddocks. Horses were in moderate- to high-intensity exercise programs, with workouts six days/week and two rodeos per month. Feeding practices varied greatly among farms but most horses received forage (alfalfa or grass hay) and an energy feed (oats, corn, or concentrate), while protein and lipid supplements were less common. The goal of this characterization of current management of the Chilean corralero horse is to contribute to information available about this breed to improve husbandry practices.

Abstract: The aim of this study was to characterize the routine care, training, feeding, and nutritional management of Chilean corralero horses that participated in the rodeos of the Chilean Rodeo Federation. Forty-nine horse farms between the Metropolitan (33°26′16″ south (S) 70°39′01″ west (W)) and Los Lagos Regions (41°28′18″ S 72°56′12″ W), were visited and a survey was conducted on the management and feeding of the Chilean horse. Of the horses which participated in at least one official rodeo in the 2014–2015 season, 275 horses were included in the study. The survey consisted of five questions about general data on the property and the respondent, four questions on the animal characteristics, five questions about where the horses were kept during the day, seven questions to characterize the amount of exercise done by the horse, and 18 questions about feeding practices; additionally, the amount of feed offered was weighed. All horses in this study were in training and kept in their stall for at least 12 h and remained tied or loose for the rest of the day. The intensity of daily exercise of the rodeo Chilean horse could be classified as moderate to heavy and consisted of being worked six days/week and participating in two rodeos/month. Ninety-eight percent of respondents had watering devices in the stables. The diet of the Chilean corralero horse during the training season is based on forages, mainly alfalfa hay, plus oats as an additional energy source. Protein supplements such as oil seed by-products are used less frequently. A wide variation was observed in the diets and quantities of feed offered, which suggests that the feeding management of these individuals is not formulated according to their requirements.

Keywords: Chilean corralero horse; rodeo; feeding practices

---

1. Introduction

Horses served humans for centuries as a source of food, as well as for military, agricultural labor, and sport purposes [1]. The International Federation for Equestrian Sports (FEI) recognizes eight disciplines for competition (dressage, jumping, vaulting, endurance, reining, combined driving, eventing, and para-equestrian). Nevertheless, there are several other widely known equestrian sport competitions worldwide, such as polo and racing.

In Chile, "rodeo" was declared the national sport in 1962 (Decree 269 of the National Council of Sports), when it became part of the Olympic Committee of Chile [2]. This sport is one of the culturally rich activities that take place in the Chilean countryside. It symbolizes traditional cattle-working and is part of the traditional folklife with its own customs [2,3]. After football (soccer), rodeo is the second most popular sport in Chile due to its massive following and because it became an important cultural symbol throughout national history.

The breeding aim for the Chilean corralero horse is to produce a horse that is suitable and functional for saddle and stock work, as well as practicing rodeo [4].

The rodeo is performed in an oval-shaped area and consists of a pair of riders (each rider is called a "huaso") and their horses ("collera") running half laps around the arena while working a steer and attempting to pin it against a large 12-m cushion ("quincha"). Horses run a total of approximately 400 m at 6.95 m/s in each round of the competition [5]. The rodeo is highly regulated by the Chilean Rodeo Federation, and there are 311 officially registered arenas throughout the national territory [6], with 350 competitions per season, which begins in August each year and ends with the national championship in April of the following year.

Despite the national importance of rodeo, there is little information available and easily accessible about the Chilean corralero horse, specifically related to equine numbers, competing animals, feeding type, and routine management. Moreover, unlike other equestrian sports like show-jumping [7], racing [8], and three-day eventing [9], there is no known characterization of feeding management and schedule or dietary ingredients. Since competition horses may develop nutritional and digestive problems such as gastric ulcers, decreased appetite, and weight loss [10], it is important to establish feeding practices of sport horses undergoing hard work. Considering that the rodeo sport season lasts eight months every year, with some horses competing every weekend or every other weekend, a properly designed training routine and sound feeding management is crucial. Moreover, since the Chilean corralero horse is considered a small horse (height at the withers 99.86–136.4 cm) and *huasos* weigh an average of 80.78 + 10.02 kg, some of these horses may be overburdened [5], thus requiring additional energy.

The lack of information on Chilean corralero horses also makes it difficult for breeders, trainers, riders, veterinarians, and owners to develop management and feeding standards. Therefore, this work seeks to contribute with relevant information on diet composition and feed ingredients utilized daily in the diet of Chilean corralero horses actively participating in the national sport. Hence, the objective of this study was to characterize the feeding and nutritional management, routine care, and training of Chilean corralero horses that participate in the rodeos of the Chilean Rodeo Federation.

2. Materials and Methods

2.1. Animals

The target population for this study was healthy Chilean corralero horses that participated in the 2014–2015 season of the Chilean national rodeo competition and that were in training at the time of the visit. The data to identify individuals that participated during that season were provided by the Chilean Rodeo Federation.

The horses' owners were contacted by telephone or e-mail to request their participation in the study, and an interview was scheduled if they were willing to participate. As a result, a personal visit was carried out between October and December of 2015 to 49 farms (275 horses) located between the Metropolitan and the Los Lagos regions (33°26′16″ south (S), 70°39′1″ west (W) to 41°28′18″ S, 72°56′12″ W). Farms were located in the following regions: Metropolitan ($n = 5$), Valparaiso ($n = 3$), O'Higgins ($n = 6$) Maule ($n = 5$), Bío-Bío ($n = 10$), Araucanía ($n = 8$), Los Ríos ($n = 5$), and Los Lagos ($n = 7$).

2.2. Survey Design

The interview consisted of a questionnaire with five questions regarding general farm and respondent information (farm name, location, farm administrator, position, date of the visit), three questions to characterize the animal (registration number, sex, date of birth), five questions about housing conditions and activity (hours spent within the stall, outdoor pen, paddock, and tied), and four questions characterizing the type and amount of daily exercise of the horse (minutes spent walking, trotting, cantering, and in the "bumping activity", on a weekly and daily basis). "Bumping" (topeo) is when the horse performs lateral canter half-passes while maintaining contact at a 45° angle between the horse's chest and the steer's body, using a lateral movement to push slightly on the steer and maintain him at a 45° angle to the horse's body.

The questionnaire also included 18 questions about nutritional management with regard to feeding (feed ingredient type, feeding frequency, and schedule), water availability and source (automatic drinker, water bucket, etc.), access to pasture, hay type, grains, protein supplements, other commercial concentrate or byproduct, minerals, oils, if feedstuffs were moisturized, nutritional advisor, and other information (de-wormer type and frequency, incidence of colic and laminitis per year).

The daily amount of each ingredient offered was weighed with a digital scale. The girth and scapula–ischial length (SIL) were measured using a measuring tape to estimate live body weight, using the formula of Carroll and Huntington [11], where

$$\text{weight (kg)} = (\text{girth (cm}^2) \times \text{SIL (cm)})/11{,}877. \tag{1}$$

The girth corresponds to the thoracic perimeter (cm) and was measured by wrapping a 3-m flexible measuring tape around the girth, just behind the withers, of each horse. The scapula–ischial length (SIL) corresponds to the length from scapula (point of the shoulder) to ischial tuberosity (point of the buttock) in centimeters [12].

The unique registration number and date of birth of the horses were obtained from the owners or the animal's registration certificate.

2.3. Statistical Analyses

Data obtained from the surveys were consolidated in an Excel spreadsheet and reported as a descriptive analysis (average ± standard deviation, frequency, average, minimum, maximum). One-way ANOVA between horses was conducted to determine significant differences among treatments, and a Tukey honestly significant difference (HSD) test was performed whenever there were significant differences. Significance was declared with a p-value lower than 0.05. Pearson correlations were performed between feedstuffs (hay, corn, oats) utilized and incidence of colic and laminitis. The Pearson correlation was performed among different exercises types (walk, trot, canter). Data were processed using Statistica V 7.0.

3. Results

Measurements were collected from 275 horses located on 49 farms, and general characterizations about gender, age, and physical characteristics are presented in Table 1. The evaluated farms had geldings, mares, and stallions, which respectively accounted for 30.6%, 32.7%, and 36.7% of the entire sampled population. The average age per horse was 9.9 ± 2.6 years old. Geldings in this survey were older than mares and stallions ($p < 0.05$).

Table 1. Characterization of horses included in the survey, concerning gender, age, and estimated weight calculated from the girth and scapula–ischial length (SIL).

			Parameter			
Gender	N	%	Age (Years) (Mean ± SD)	Estimated Weight (kg) (Mean ± SD)	Girth (cm) (Mean ± SD)	SIL (cm) (Mean ± SD)
Geldings	84	30.6	10.9 ± 2.4 [a]	379.3 ± 27.9 [b]	166.4 ± 4.3 [b]	150.6 ± 5.1 [b]
Mares	90	32.7	9.7 ± 2.6 [b]	394.2 ± 23.8 [a]	168.3 ± 3.7 [a]	153.1 ± 5.7 [a]
Stallions	101	36.7	9.4 ± 2.6 [b]	377.8 ± 25.8 [b]	165.5 ± 4.4 [b]	151.7 ± 4.9 [ab]
All	275	100.0	9.9 ± 2.6	383.7 ± 26.8	166.7 ± 4.3	151.8 ± 5.3

[a,b] Values within a column with different superscript letters differ significantly at $p \leq 0.05$.

The estimated weight of mares was greater than males (geldings and stallions) ($p < 0.05$), but there was no significant difference between gelding and stallions. When all horses were considered, the average estimated weight was 383.7 ± 26.8 kg. The averages of girth and SIL per horse were 166.7 ± 4.3 cm and 151.8 ± 5.3 cm, respectively, with mares having a larger girth compared to stallions and geldings ($p < 0.05$) and a larger SIL compared to geldings ($p < 0.05$).

Routine care did not follow a common pattern among farms; however, all horses were kept in individual stalls during the night and for part of the day. Eighty-six horses (31.3%) spent the entire day in stalls. Of the rest, 163 horses spend part of the day tied, 24 in pens, and two in paddocks (Table 2).

Table 2. Routine management of Chilean corralero horses.

Hours/Day	Number and Percentage of Horses			
	Stalls	Tied	Pens	Paddock
0–5	0	17 (6.2%) [‡]	0	0
6–10	0	2 (0.7%) [†]	0	0
11–15	172 (62.5%)	144 (52.4%) [†]	24 (8.7%) [†]	2 (0.7%) [†]
15–19	17 (6.2%)	0	0	0
20–24	86 (31.3%)	0	0	0
Total	275 (100.0%)	163 (59.3%)	24 (8.7%)	2 (0.7%)

[†] Animals that spent less than 15 h in a stall; [‡] animals that spent within 15 to 19 h in a stall.

The sport and exercise management routine of the Chilean corralero horses is presented in Table 3. Each farm studied had a unique routine for each of their horses. On average, horses trotted for 8.6 ± 6.6 min/day and cantered for 22.3 ± 6.1 min/day. All farms utilized bumping as an exercise, usually for durations of between 10 and 20 min and two times/week. Lunging was not a common practice among trainers; only 6.5% of horse were lunged for more than 15 min. A Pearson correlation ($r = -0.44$) was detected between trot and canter exercises, indicating that the time spent on one of the two exercises is inversely correlated to the other. There was no correlation detected between exercise routine and number of rodeos annually. The average number of rodeos during the season was 8.6 ± 4.7 per horse; 20 (41.0%), 16 (33.0%), and five (10.0%) farms participated in two, three, and four rodeos per month, respectively. Eight farms participated in one or fewer than one rodeo per month during the 2014–2015 season.

Table 3. Exercise routine (minutes/workout) of Chilean corralero horses.

	Trot	Canter	Lunge	Bumping [†]	Tournaments/Season
N	275	275	275	275	
Mean	8.6	22.3	2.5	16.3	
SD	6.6	6.1	8.7	6.4	
Time spent per activity (minutes/workout)					
0–15	78.5	6.9	93.5	37.8	
15–30	20.4	61.8	2.5	54.2	
30–45	1.1	31.3	3.3	8.0	
>45	0	0.0	0.7	0.0	
Activity (days per week): trot/canter/lunge					
Mean		5.6		2.4	8.6
SD		0.6		0.7	4.7

[†] *Bumping* refers to the exercise in which the horse performs lateral canter half-passes while maintaining contact at a 45° angle between the horse's chest and the steer's body.

Feeding and nutritional management of the Chilean corralero horse during the rodeo season is summarized in Table 4. Results showed that 81.1% of the horses were fed twice daily (morning and noon/night), and 18.9% received three meals (morning, noon and afternoon/evening). The horses that had two rations per day ate during the morning and night (212 horses from 37 farms) or noon and night (11 horses from two farms). None of the horses surveyed had access to pasture.

Table 4. Feedstuffs by type and quantity that were offered to every horse on a daily basis, as fed.

	Horses N (%)	Farms N	Daily feed allowance	SD	Max	Min
Forages (kg/d)						
Alfalfa hay	122 (44.4)	22	8.8	2.3	19.1	4.9
Grass hay	67 (24.3)	13	9.2	1.7	14.3	5.6
Alfalfa cubes and/or pellets	64 (23.3)	8	8.6	0.9	11.3	7.1
Mix	22 (8.0)	6	10.7	2.6	15.7	7.0
	275	49				
Cereal grains (kg/d)						
Oats	57 (77.0)	14	2.6	1.6	6.4	0.6
Corn	17 (33.0)	4	1.3	1.3	3.6	0.2
	74	18				
Wheat by-products (kg/d)						
Wheat bran	93 (60.4)	19	1.9	1.2	5.3	0.5
Wheat middlings	61 (39.6)	11	2.4	1.3	6.0	0.7
	154	30				
Oilseed by-products (kg/d)						
Soybean meal	54 (90.0)	10	0.8	0.4	0.9	0.1
Canola meal	6 (10.0)	1	0.9	0.0	0.9	0.9
	60	11				
Comercial concentrate (kg/d)						
Locally produced	75 (83.3)	17	2.3	2.1	9.0	0.6
Imported	15 (16.6)	4	2.0	1.3	3.8	0.5
	90	21				
Oils (ml/d)						
None	226	39	0			
Corn oil	2	1	200 [†]			

Table 4. *Cont.*

	Horses N (%)	Farms N	Daily feed allowance	SD	Max	Min
Sunflower oil	9	1	45			
Olive oil	2	1	50			
Linseed oil	26	4	60			
Soybean oil	5	1	40			
Fish oil	5	1	30			
Minerals blocks						
No addition	94	17 (34.7%)				
Addition	181	32 (65.3%)				
Suplements (any kind)						
No addition	53	12 (24.5%)				
Addition	222	37 (75.5%)				

[†] Only at one farm during a limited period.

A total of 265 horses were fed hay. Each studied farm used more than one hay type to feed their horses. The average amount of hay offered per horse was 9.0 ± 1.9 kg/day as fed. Alfalfa hay was the most commonly used forage, followed by a pasture hay and alfalfa cubes/pellets. Twenty-seven percent of horses received some type of cereal grains. Oats were the most commonly used grain, followed by corn which was offered fine, ground, or rolled.

Wheat milling by-products were also part of the feed used, where 61.2% of the farms used middlings or bran. The delivered amount of wheat bran ranged from 0.5 kg/day to 5.3 kg/day, whereas wheat middlings were offered at a range of 0.7 to 6.0 kg/day.

A total of 38 farms (77.6% of farms surveyed) did not use oilseed by-products (soybean and canola meal) to feed any of their horses. Eleven farms utilized oilseed by-products in their ration, and only one fed all their horses with oilseed by-products.

Commercial concentrates (based on mixtures of different energetic ingredients such as oats, corn, soybean meal, molasses, wheat middling, and vegetable oils) were used in the Chilean corralero horse diets, with 42.9% (21 of the 49) of the owners including 2.2 ± 2.0 kg daily. Mineral premix and vitamins were used in 181 horses from 32 of the studied farms, whereas 93 horses from 16 farms were not fed supplements. Finally, only 41 horses from 10 farms were fed different oils or fats, with linseed oil being the most popular.

One hundred and ninety-seven horses from 27 farms were dewormed between two and six times per year.

Of 275 horses, 149 horses (54.2%) did not suffer colic during the previous year, whereas 47 (17.1%) and nine (3.3%) horses suffered colic one or two times during the year before, respectively. There was a negative correlation ($r = -0.72$) between the numbers of colic episodes that the horses suffered per year and the amount of hay (other than alfalfa hay) offered, whereas the number of colic episodes that the horses suffered per year was positively correlated with corn grain ($r = 0.6$) and wheat middlings ($r = 0.7$).

Around 63% did not suffer laminitis disease the previous year, and 12% suffered laminitis during that period. The remaining interviewees did not know the answer. A negative correlation ($r = -0.72$) between the incidence of laminitis and the amount of hay was observed.

4. Discussion

Unfortunately, there is little scientific literature describing the characteristics of the Chilean corralero horse. In 1950, Denhardt [13] described the horse as a muscular, strongly built animal, with a broad chest and a good distance between its shoulders. Years later, a review of morphological characteristics of this breed during the rodeo sport season was reported by García et al. [14]. The girth of the Chilean corralero horse usually ranges between 162 and 182 cm for males and 164 and 184 cm for females [15]. In this study, the girth data fell within these ranges. Mares had a larger girth

(168.3 ± 3.7 cm) than both stallions (165.5 ± 4.4 cm) and geldings (166.4 ± 4.3 cm), which was similar to the values obtained by García et al. [14], who reported 170.3 ± 7.1 cm for females and 168.8 ± 5.3 and 169.6 ± 6.4 for stallions and geldings, respectively. However, there is no established reference parameter for the scapula–ischial length (SIL) of this breed. In this study, the average SIL for mares was 153.1 ± 5.7 cm, while it was 150.6 ± 5.1 cm for geldings and 151.7 ± 4.9 cm for stallions. In the case of García et al. [14], mares had an SIL of 147.9 ± 6.3 cm, geldings had an SIL of 148.2 ± 7.9 cm, and stallions were shorter at a length of 145.6 ± 6.4 cm.

In this study, the wither height was not measured but information obtained from the study of Muñoz et al. [5] showed that the standard height of the withers of the breed established at 138–148 cm [15]. The morphological characteristics mentioned above indicate that Chilean corralero horse is considered a small horse. To estimate the average liveweight, the formula proposed by Carroll and Huntington [11] was used, and the estimated weight of the surveyed horses was 383.66 + 26.8 kg. Mares had the greatest estimated average weight of 394.16 ± 23.8, followed by geldings and stallions, with average weights of 379.34 ± 27.9 and 377.8 ± 25.8, respectively. The standard of the Chilean corralero horse establishes that mares are 2 cm longer in SIL than males due to their reproductive anatomy [16]. Several other studies about morphometric measurements of the Chilean horse are available for further information [5].

Stabling is a common practice in sport horse management and is part of the evolution of the horse throughout history, as it changed from being utilized primarily for agricultural and military purposes to use in sports and leisure [17]. This trend to confine the animal allows for a better control of the daily routine, feed intake, reproduction, and health management, among other benefits. To our best knowledge, there is limited information about Chilean corralero horse breeding and management, even during the reproductive or training phases. This research was conducted focusing on the sport phase of those horses exclusively focused on rodeo training. In the rodeo, as in any other equine sport discipline, horses are kept within single stalls which limit locomotion, making horse care easier and more economical [18]. However, the restricted motion and the lack of freedom to express characteristic behaviors may lead to stress and vices [19]. Upon the results of this study, it is clear from the routine handling management questions of Chilean corralero horses that 100% of the horses spent some time of the day within the stall (Table 2), usually more than 12 h a day. Another large percentage (59.3%) of the horses spent between a few hours to up to half of their day tied. Only a small percentage of the horses were left free in pens (8.7%) or paddocks (0.7%). These extensive confinement periods correspond to abnormal behavior and vices in the Chilean corralero horse and, according to Muñoz et al. [20], 10% of the horses studied exhibited these behaviors. Prolonged stabling can result in several undesirable behaviors such as "cribbing" (biting hard, often wooden surfaces, frequently swallowing air in the act) which can generate small fractures (less than 5 mm) in the horses' teeth [21].

Daily exercise routines of the Chilean corralero horse were quite variable among farms. It seems that every trainer and/or owner managed their horses without following a common pattern within the discipline. Overall, Chilean corralero horses exercised six days per week, and the daily exercise consisted of walking, trotting, cantering, and bumping. Total daily exercise lasted 49.6 ± 27.8 min. The NRC [22] categorizes exercise as light, moderate, heavy, and very heavy for horses, depending on the mean heart rate, description, and types of event. In this case, the Chilean corralero horse training can best be categorized as heavy exercise, since, on average, the horse worked 4–5 h/week, during which time the activity was composed of 30% walk, 50–65% trot or canter, and 5% canter, jumping, or other skill. In this case, since there are no data on mean heart rate, bumping may be considered similar to jumping or cantering [22].

Horses are non-ruminant herbivores with an enormous capacity to digest and obtain energy from fibrous ingredients [23]. However, they have a small stomach (8% of the total gastrointestinal tract volume) and, thus, must eat small quantities many times throughout the day. In free-ranging conditions, horses spend between 16 and 20 h per day grazing and, thus, in nature, can consume small amounts of feed throughout the day [24]. Nevertheless, for practical reasons, owners of confined horses may not be able to simulate this feeding delivery [25]. Feeding frequency affects horse health and, thus, performance. In this study, a major portion of the animals surveyed (81.1%) were fed twice a day, whilst the remaining 18.9% were fed three times per day, which is within the recommendations for feeding horses to avoid colic [26]. Furthermore, the number of meals provided for the horses in this study was the same as the number of meals reported in a survey study of show-jumping horses [7], horses in the United Kingdom (UK) with high energy requirements [17], but differed from eventing horses [9] and racehorses [27], which received between one and five meals per day.

With regard to the type and amount of feed offered, this study found a wide variation in the diets and quantities of food offered, which suggests that the feeding management of these individuals is not formulated according to their requirements but rather according to the managers' personal criteria. All horses surveyed had access to some kind of forage in concordance with recommendations by the NRC [22]. An average of 9.0 kg/horse (2.3% live bodyweight, "bwt") of either alfalfa hay and/or pasture hay was offered daily; however, the exact amount varied greatly, between 5.6 and 19.1 kg/horse (as fed). More recently published recommendations for horses with high energy requirements indicate that hay intake should be 20 g dry matter (DM)/kg bwt/day [25]. Considering a 400 kg bwt for the Chilean corralero horse, the minimum forage intake should be 8 kg of DM daily. In contrast to high fiber feeding, 59% of the sampled horses received grains (oats or corn) and/or a commercial concentrate (locally produced or imported). These energetic ingredients were offered from 2.6, 1.3, 2.3, and 2.0 kg per day as fed for oats, corn, local, and imported commercial concentrate, respectively (Table 4). The average grain or concentrate offered was 2.1 kg/horse daily, which represents 0.5% of the horse liveweight. This value is in accordance with the recommendations of Owens [28] and the NRC [22]. Wheat by-products (wheat middlings and wheat bran) were also offered to horses as part of the daily diet as a replacement or partial replacement of energetic feedstuffs. The average offered was 2.1 kg/horse; nevertheless, the NRC [22] lists both by-products as concentrates (3.2 and 3.4 Mcal DE/kg, respectively) even although both are considerably higher in fiber than corn (ADF 15.5 and 12.1% respectively, compared to 3.4% ADF in corn) but similar to oat ADF content (13.5%). As the partial or total replacement of grain with wheat by-products did not follow a common pattern among farms, it is suggested that every owner follows their own criteria when establishing feeding practices for the Chilean corralero horse. Finally, when compared to the 400 kg bwt horses from the NRC tables [22], Chilean corralero horses are fed 2.8% of their bwt, which is slightly greater than the recommendation for horses in heavy exercise. It is worth noting that NRC recommendations do not account for breed-type behavior. Chilean corralero horses have a nervous and alert temperament which may increase the average energy requirements compared with a regular 400-kg horse [29]. Rosselot et al. [29] found that Chilean corralero horses tend to demonstrate proactive avoidance behaviors when presented with a challenge in a handling test; thus, this increased activity in response to environmental stressors could also impact nutrient requirements. Another factor that must be considered is the total rider and saddle weight that the horse supports on its back during competition and daily training. Muñoz et al. [5] conducted a study to determine the weight supported by Chilean corralero horses during a rodeo competition. The estimated back load maximum capacity of these Chilean corralero horses was 115.0 ± 6.1 kg, and the riders weighed on average 80.8 kg with a saddle weight of 11.6 kg; therefore, while the horse is not overloaded, the weight of the rider and saddle corresponds to 25% of the horse's weight and, thus, has a lesser impact on its relative energy expenditure (25.7 Mcal/d DE) as compared to other sports with a lower rider/horse weight ratio such as show jumping and endurance, where the sum of weights for riders and saddle represents approximately 15% of the horses' weight and the energy requirements.

In terms of nutrition, the use of fats and oils was not common, with just 17.8% of horses fed with some oil source. The amount varied greatly but averaged 30–60 mL/day. In general, corn oil was preferred for horse feeding since it was proven that, among vegetable and animal oils, it is the most used and accepted [30]. In contrast, mineral supplementation and the use of other additives/supplements are common practice within the industry. Sixty-five percent of horses were supplemented with mineral blocks, and 75.5% were given some supplement (liver protectors, vitamins premix, creatine sources, and joint supplements, among others).

In this study, colic events were negatively correlated to hay consumption and positively correlated to grain and wheat middling intake. Colic is a complex multifactorial condition and it has some association with feeds and feeding [31]. Grain overloading and, thus, high-starch diets along with low-fiber diets may predispose horses to colic [32]. In this study, grains were included in the diet with a high variation: 2.6 ± 1.6 kg/day of oats on average, ranging from 0.6–6.4 kg/day and 1.3 ± 1.3 kg/day of corn ranging from 0.2 to 3.6 kg/day. As mentioned before, the recommended amount of grain offered to a horse daily is 500 g/100 kg of liveweight. Thus, the Chilean corralero horse (400 kg liveweight) is expected to consume 2.0 kg of grains to supply the energy requirements (20–25 Mcal/d DE) according to NRC guidelines [22]. This recommendation comes from the limited ability of the enzymes in the horses' small intestine to digest starch from cereal grains. When grain, which is high in starch, is overfed, part of the carbohydrates mentioned above are incompletely digested and escape absorption in the small intestine, traveling to the large intestine and promoting a disruption in the microbial population, leading to abnormal pH conditions which compromise intestine health [32]. These alterations may increase the risk of colic, laminitis, and other pathologies.

A characterization of the animal and its routine care was carried out through this study. This information complements all physics hypsometry studies that were done. To reach a better understanding of this breed's nutritional requirements, more data should be taken in situ, such as time dedicated to working in each activity, as well as heart rate, blood metabolites, and other physiological variables.

Chilean legislation for animal welfare only establishes that owners should provide adequate feeding and healthcare to their animals, regardless the animal species. Therefore, the results of this study may provide information which could be considered when creating or enforcing legislation regarding horse care and training.

5. Conclusions

The average Chilean corralero horse spends approximately 12 h daily within a stall, and spends its remaining hours tied, loose in a pen, and/or training. The horse's workload can be classified as moderate to heavy intensity according to the NRC [22], and includes six workouts weekly (walk/trot/canter) with two bumping workouts per week and two competitions per month.

During the training season, the diet of the Chilean corralero horse is based on forages, mainly alfalfa hay, plus oats as an additional energy source. Protein supplements such as oilseed by-products are used less frequently.

Additional studies are required to provide further information on nutritional requirements, feeding management, and exercise in order to improve the raising and management of this breed.

Author Contributions: Conceptualization, J.P.K. and M.G.; data curation, J.B., M.S.-G., and J.P.K.; funding acquisition, F.B. and M.G.; investigation, J.B., J.P.K., and M.G.; methodology, J.P.K. and M.G.; project administration, J.B., J.P.K., and M.G.; resources, F.B.; software, M.S.-G.; supervision, M.G.; validation, M.S.-G.; visualization, M.S.-G. and H.H.W.; writing—original draft, J.B. and M.G.; writing—review and editing, F.B., H.H.W., and J.P.K.

References

1. Siegel, M. *UC Davis School of Veterinary Medicine Book of Horses: A Complete Medical Reference Guide for Horses and Foals*; Harper Collins Publishers: New York, NY, USA, 1996; pp. 3–17.
2. Pérez, R.; García, M.; Cabezas, I.; Guzmán, R.; Merino, V.; Valenzuela, S.; Gonzalez, C. Actividad física y cambios cardiovasculares y bioquímicos del caballo chileno a la competencia de rodeo. *Arch. Med. Vet.* **1997**, *29*, 221–234. [CrossRef]

3. Tadich, T.A.; Araya, O.; Solar, F.; Ansoleaga, N.; Nicol, C.J. Description of the responses of some blood constituents to rodeo exercise in Chilean Creole horses. *J. Equine Vet. Sci.* **2013**, *33*, 174–181. [CrossRef]

4. Hendricks, B.L. *International Encyclopedia of Horse Breeds*; University of Oklahoma Press: Norman, OK, USA, 2007; pp. 121–122.

5. Muñoz, L.; Ortiz, D.; Ortiz, R.; Cabezas, I.; Briones, M. Determination of the back load in horses used for Chilean rodeo and comparison with the estimated back load maximum capacity, in accordance with the Beltrán formula (1954). *Arch. Med. Vet.* **2012**, *44*, 285–289. [CrossRef]

6. Arancibia, F.; Molina, E. "Cultura y Tiempo Libre" Informe Anual 2005. Instituto Nacional de Estadística. 2005. Available online: https://www.ine.cl/docs/default-source/sociales/cultura/cultura-y-tiempo-libre-2005.pdf?sfvrsn=6 (accessed on 13 April 2018).

7. Brunner, J.; Liesegang, A.; Weiss, S.; Wichert, B. Feeding practice and influence on selected blood parameters in show jumping horses competing in Switzerland. *J. Anim. Physiol. Anim. Nutr.* **2015**, *99*, 684–691. [CrossRef] [PubMed]

8. Frape, D. *Equine Nutrition and Feeding*, 2nd ed.; Blackwell Science: Oxford, UK, 1998; pp. 300–365.

9. Brunner, J.; Wichert, B.; Burger, D.; von Peinen, K.; Liesegang, A. A survey on the feeding of eventing horses during competition. *J. Anim. Physiol. Anim. Nutr.* **2012**, *96*, 878–884. [CrossRef] [PubMed]

10. Leahy, E.B.; Burk, A.O.; Green, E.A.; Williams, C.A. Nutrition-associated problems facing elite level three-day eventing horses. *Equine Vet. J.* **2010**, *42*, 370–374. [CrossRef] [PubMed]

11. Carroll, C.L.; Huntington, P.J. Body condition scoring and weight estimation of horses. *Equine Vet. J.* **1988**, *20*, 41–45. [CrossRef] [PubMed]

12. McKiernan, B. Estimating a Horse's Weight. Primefact 494. New South Wales Department of Primary Industries, NSW, Australia. 2007. Available online: http://www.dpi.nsw.gov.au/__data/assets/pdf_file/0008/109988/estimating-a-horses-weight.pdf (accessed on 1 October 2017).

13. Denhardt, R.M. The Chilean Horse. *Agric. Hist.* **2010**, *24*, 161–165.

14. García, M.; Cabezas, I.; Guzmán, R.; Valenzuela, S.; Merino, V.; Pérez, R. Características hipométricas, peso corporal y capacidad de carga del caballo fina sangre chileno en rodeo. *Avan. Cs. Vet.* **1997**, *12*, 45–51. [CrossRef]

15. Porte, E. El Nuevo estándar del caballo chileno. *Rev. Fed. Rodeo Chil. Asoc. Criads. Cab. Chil.* **1978**, *30*, 16–20.

16. Porte, E. Crecimiento y desarrollo del caballo chileno. *Av. Prod. Anim.* **2000**, *25*, 167–174.

17. Harris, P.A. Review of equine feeding and stable management practices in the UK concentrating on the last decade of the 20th century. *Equine Vet. J.* **1999**, *28*, 46–54. [CrossRef]

18. Werhahn, H.; Hessel, E.F.; Van den Weghe, H.F.A. Competition Horses Housed in Single Stalls (II): Effects of Free Exercise on the Behavior in the Stable, the Behavior during Training, and the Degree of Stress. *J. Equine Vet. Sci.* **2012**, *32*, 22–31. [CrossRef]

19. McGreevy, P.D.; Cripps, P.J.; French, N.P.; Green, L.E.; Nicol, C.J. Management factors associated with stereotypic and redirected behaviour in the thoroughbred horse. *Equine Vet. J.* **1995**, *27*, 82–83. [CrossRef]

20. Muñoz, L.; Torres, J.; Sepúlveda, O.; Rehhof, C.; Ortiz, R. Frequency of stereotyped abnormal behaviour in stabled Chilean horses. *Arch. Med. Vet.* **2009**, *41*, 73–76.

21. Muñoz, L.; Vidal, F.; Sepúlveda, O.; Ortiz, O.; Rehhof, C. Dental pathologies in incisors, canines and first premolar in adult Chilean horses. *Arch. Med. Vet.* **2010**, *42*, 85–90.

22. NRC. *Nutrient Requirements of Horses*, 6th Revised Edition; National Research Council of the National Academies: Washington, DC, USA, 2007; pp. 22–27.

23. Hintz, H.F.; Cymbaluk, N.F. Nutrition of the horse. *Annu. Rev. Nutr.* **1994**, *14*, 243–267. [CrossRef] [PubMed]

24. Keiper, R.R. Behavior: Social structure. *Vet. Clin. N. Am. Equine Pract.* **1986**, *2*, 465–483. [CrossRef]

25. Harris, P.; Dunnett, C. Nutritional tips for veterinarians. *Equine Vet. Educ.* **2018**, *30*, 486–496. [CrossRef]

26. Reeves, M.J.; Salman, N.D.; Smith, G. Risk factors for equine acute abdominal disease (colic): Results from a multi-center case-control study. *Prev. Vet. Med.* **1996**, *26*, 285–301. [CrossRef]

27. Williamson, A.; Rogers, C.W.; Firth, E.C. A survey of feeding, management and faecal pH of Thoroughbred racehorses in the North Island of New Zealand. *N. Z. Vet. J.* **2007**, *55*, 337–341. [CrossRef] [PubMed]

28. Owens, E. Sport horse nutrition—An Australian perspective. In *Advances in Equine Nutrition III.*; Pagan, J., Geor, R.J., Eds.; Nottingham University Press: Nottingham, UK, 2005; pp. 185–192.

29. Rosselot, P.; Mendonça, T.; González, I.; Tadich, T. Behavioral and Physiological Differences between Working Horses and Chilean Rodeo Horses in a Handling Test. *Animals* **2019**, *9*, 397. [CrossRef] [PubMed]

30. Dunnet, C.E. Dietary lipid form and function. In *Advances in Equine Nutrition III*; Pagan, J.D., Ed.; Nottingham University Press: Thrumpton, UK, 2005; pp. 37–54.

31. White, N.A. Colic prevalence, risk factors and prevention. In *Advances in Equine Nutrition IV*; Pagan, J.D., Ed.; Nottingham University Press: Thrumpton, UK, 2009; pp. 313–326.

32. Durham, A.E. The role of nutrition in colic. *Vet. Clin. N. Am. Equine Pract.* **2009**, *25*, 67–78. [CrossRef] [PubMed]

10

Differential Defecation of Solid and Liquid Phases in Horses

Katrin M. Lindroth [1,*]**, Astrid Johansen** [2]**, Viveca Båverud** [3]**, Johan Dicksved** [1]**,**
Jan Erik Lindberg [1] **and Cecilia E. Müller** [1]

[1] Department of Animal Nutrition and Management, Swedish University of Agricultural Sciences,
P.O. Box 7024, 750 07 Uppsala, Sweden; Johan.Dicksved@slu.se (J.D.); jan.erik.lindberg@slu.se (J.E.L.);
cecilia.muller@slu.se (C.E.M.)
[2] NIBIO, Norwegian Institute of Bioeconomy Research, P.O. Box 115, 1431 Ås, Norway; astrid.johansen@nlr.no
[3] National Veterinary Institute, 751 89 Uppsala, Sweden; viveca.baverud@sva.se
* Correspondence: katrin.lindroth@slu.se.

Simple Summary: Free faecal liquid is a condition in horses where faeces are voided in one solid and one liquid phase. The presence of free faecal liquid may cause management problems in equine husbandry and is potentially contributing to impaired equine welfare. Causes of free faecal liquid are not known, but nutritional factors such as the feeding of specific forages have been suggested to be of importance. Characterization of horses showing free faecal liquid and their feeding and management was, therefore, performed via an internet-based survey in order to map the condition further. Results showed that horses with free faecal liquid included a large variety of different breeds, ages, disciplines, coat colours, housing systems and feeding strategies, meaning that almost any type of horse could be affected. Horses that were reported to show free faecal liquid did so with all types of feeding strategies, but changes from wrapped forage to hay, to pasture, or to another batch of wrapped forage often resulted in diminished signs of free faecal liquid. Horses were also reported to have a comparably high incidence of colic in relation to published data for other horse populations. The results indicated that more detailed studies are required for a further understanding of the underlying cause of free faecal liquid.

Abstract: Free faecal liquid (FFL) is a condition in horses where faeces are voided in one solid and one liquid phase. The liquid phase contaminates the tail, hindlegs and area around the anus of the horse, resulting in management problems and potentially contributing to impaired equine welfare. The underlying causes are not known, but anecdotal suggestions include feeding wrapped forages or other feed- or management-related factors. Individual horse factors may also be associated with the presence of FFL. This study, therefore, aimed to characterize horses showing FFL particularly when fed wrapped forages, and to map the management and feeding strategies of these horses. Data were retrieved by a web-based survey, including 339 horses with FFL. A large variety of different breeds, ages, disciplines, coat colours, housing systems and feeding strategies were represented among the horses in the study, meaning that any type of horse could be affected. Respondents were asked to indicate if their horse had diminished signs of FFL with different changes in forage feeding. Fifty-eight percent ($n = 197$) of the horse owners reported diminished signs of FFL in their horses when changing from wrapped forages to hay; 46 ($n = 156$) of the horse owners reported diminished signs of FFL in their horses when changing from wrapped forages to pasture; 17% ($n = 58$) reported diminished signs of FFL when changing from any type of forage batch to any other forage. This indicated that feeding strategy may be of importance, but cannot solely explain the presence of FFL. The results also showed that the horses in this study had a comparably high incidence of previous colic (23%, $n = 78$) compared to published data from other horse populations. In conclusion, the results showed that FFL may affect a large variety of horse types and that further studies should include detailed data on individual horse factors including gastrointestinal diseases as well as feeding strategies, in order to increase the chance of finding causes of FFL.

Keywords: colic; equine; free faecal liquid; faecal water syndrome; feed changes; nutrition

---

1. Introduction

Free faecal liquid (FFL) is a condition in horses where faeces are voided in two physical phases; one solid and one liquid phase. The solid phase can be typical equine faecal balls, or more watery and similar to cowpat faeces. The liquid phase is a brown-coloured liquid that can be voided separately or together with the solid phase. The condition has previously been referred to as free faecal water and/or free faecal water syndrome (FWS), and cases have been described in Germany [1–3], Denmark [4] and Italy [5], but the overall incidence of FFL is not known. Horse owners in Sweden and Norway have anecdotally reported cases of FFL in horses, and have referred to the condition as "haylage intolerance" due to an assumed association with feeding wrapped forages (including grass conserved as silage and/or haylage with dry matter concentrations from 300–840 g per kg [6–8]). During the latest 25 years, wrapped forages such as grass silage and haylage have partially or totally replaced hay in equine feed rations in Nordic countries [9–11].

Horses affected with FFL may show discomfort when voiding faeces and/or faecal liquid, such as nervous trampling with hindlegs and extensive tail swishing, but no symptoms of disease have been described [1,5]. The faecal liquid may, however, cause lesions in the skin around the anus and on the inside of the hindlegs, as well as dirty tail and hindlegs of affected horses. The causes of FFL are unknown, but feeding wrapped forages instead of hay, feeding high amounts of alfalfa, being over 20 years of age, having poor dentition and endoparasitic infections have been suggested [1]. In a German study [1], associations between the presence of FFL and intrinsic horse factors such as being a gelding, paint-coloured and low in the social hierarchy in a group of horses were found. Improvement of the condition have been reported in association to changes in the diet in a one-horse case study [5] and after faecal transplantation performed in a study including 10 horses with FFL and twelve horses assessed to be clinically healthy [4]. However, no clear associations to feed- or management-related factors [1,3,12] has been reported. Systematic collection of data on horse characteristics as well as on feeding and management of horses affected with FFL is scarce in the scientific literature. The aim of the present study was, therefore, to characterize horses showing FFL (when fed wrapped forages) and to map the feeding and management strategies of these horses. Such information is required for further studies of causes for the condition.

2. Materials and Methods

An online survey directed to owners and/or caretakers of horses showing FFL when fed wrapped forages was performed. The inclusion criteria were that horses should be >2 years old, showing FFL when fed wrapped forages and be located in Sweden or Norway (two countries in close proximity and with similar conditions for horse feeding and management). The inclusion criteria were given on the start page of the survey, and 3 questions in the survey were control questions ensuring that the inclusion criteria were met. The survey was created using the tool Netigate (Netigate, Stockholm) and was advertised through the website of the Department of Animal Nutrition and Management, SLU (http://www.slu.se/sv/institutioner/husdjurens-utfodring-vard/), Norsk hestesenter (http://www.nhest.no), Norwegian institute of Bioeconomy Research (http://www.nibio.no) and the website Hästsverige (http://www.hastsverige.se), a Swedish platform communicating equine research to the public. The full survey is available in Table S1 (Supplementary Materials).

2.1. Data Collection

The survey was open from March 2016 to March 2017 and was available in both Swedish and Norwegian language. Respondents were instructed to give information about one horse per entry, even if they owned more than one horse showing FFL. If so, it was possible for the same respondent to answer the survey several times. The respondents were asked to answer questions about the horse, how it was kept and managed, current and previous feeding and history of gastrointestinal disturbances including FFL. The survey contained 50 questions in total, divided into horse characteristics (e.g., age, sex, breed, colour, body condition score [13] and temper as judged by the respondent); training (e.g., discipline, intensity); management (e.g., type of housing system, paddock use); current feeding (type and amount of feeds), feeding and watering strategies (e.g., number of feedings, time between feedings, how feed was offered in stable and paddock, type of water source and access to salt); presence of FFL (e.g., presence and symptoms of FFL, changes in faecal appearance due to feed changes, number of affected horses kept in the same housing system); and previous history of gastrointestinal tract (GIT) diseases.

2.2. Data Treatment

The data on horse breeds contained over 30 different breeds but with very few individuals in several breeds. Therefore, this variable was transformed to breed type. The reported breeds were divided into 4 breed groups; warmblood type horses (Appaloosa ($n = 3$), crossbred horses of warmblood type ($n = 64$), European warmblood riding horses ($n = 82$), Lusitano ($n = 1$), Standardbreds ($n = 15$), Paint horse ($n = 2$), Pura Raza Espaniola ($n = 11$) and Quarter horse ($n = 2$)); cold-blood type horses (Ardennais ($n = 4$), Clydesdale ($n = 1$), Cold-blooded trotter ($n = 13$), crossbred horses of cold-blood type ($n = 19$), Dølehorse ($n = 2$), Friesian horse ($n = 3$), Haflinger ($n = 3$), North-Swedish draught horse ($n = 8$), Norwegian Fjord Horse ($n = 12$), Shire ($n = 1$) and Tinker ($n = 1$)); Hot-blooded horses (Angloarabian ($n = 1$), Arabian ($n = 8$), and Thoroughbreds ($n = 2$)); and native pony breeds (Connemara ($n = 10$), crossbred ponies ($n = 25$), Gotland pony ($n = 6$), Icelandic horse ($n = 13$), New Forest ($n = 10$), Shetland pony ($n = 6$), Welsh cob ($n = 3$) and Welsh pony ($n = 8$)).

2.3. Calculations and Statistical Analysis

Statistical analysis was performed using SAS (Statistical Analysis System Institute Inc., Cary, NC, USA) version 9.4 for Windows. For continuous variables, minimum, maximum, quartiles (Q1, Q2 (median) and Q3), mean and standard deviation was calculated. The reported bodyweight (BW) of the horse and reported feeding levels were used to calculate the daily intake of feed (g or kg) per 100 kg BW and day. Descriptive analysis was performed using PROC FREQ. During data treatment, it was found that 23% ($n = 78$) of the horses had a history of colic. The horses were, therefore, further divided into one colic and one non-colic group for comparisons of type of clinical signs during FFL episodes. Each clinical sign was compared separately between the groups using a Chi^2-test (with expected model). Level of significance was set at $p < 0.05$.

3. Results

In total, 780 responses to the survey were obtained. Out of these, 12 responses represented horses younger than 2 years of age, 234 responses were for horses that did not show FFL but other types of problems when fed wrapped forages, and 195 responses were incomplete. These 441 responses were excluded from the dataset, leaving 339 full responses for further evaluation.

3.1. Horses and Signs of Free Faecal Liquid

The age of the horses in the study ranged from 2.5 to 28 years (average 11 ± 5.9 years). The majority of the horses were geldings (57%, n = 193) (Table 1) and of warmblood breed type (53%, n = 180) (Figure 1). Thirty-seven percent (n = 123) of the horses had bay coat colour followed by chestnut (19%, n = 64), grey (14%, n = 47) and black (8%, n = 27). Body condition scores (BCS) ranged from 1 to 4 (on a scale of 0 to 5 [13]), with a normal distribution around BCS 3 as median (55%, n = 186) (Table 1). The most frequently reported disciplines horses were used for were leisure riding (82%, n = 278), dressage (37%, n = 125) and show jumping (34%, n = 115) (Figure 2). A majority of the horses (63%, n = 215) were reported to perform a low-intensity exercise (Table 1). Extended information on horse characteristics is reported in Appendix A (Table A1). Twenty-nine percent (n = 98) of the horses were reported to show distinct irritation manifested by extensive tail swishing and nervous trampling of hindlegs while voiding faecal liquid and/or faeces, whereas 35% (n = 118) did not show any signs other than FFL (Figure 3). A bloated abdomen was reported in 29% (n = 98) of the horses during episodes of FFL (Figure 3). Fifty-two percent (n = 170) of the respondents reported that only their horse in the stable showed FFL, while 48% (n = 159) of the respondents stated that there were more horses in the stable that showed FFL.

Table 1. Characteristics of horses showing free faecal liquid (n = 339).

Item	No. of Horses	% of Horses
Country (stabled in)		
Sweden	191	56
Norway	148	44
Gender		
Mare	134	40
Gelding	194	57
Stallion	11	3
Coat colour		
Bay	123	37
Chestnut	64	19
Grey	47	14
Black	27	8
Paint	24	7
Palomino/Isabelline	21	6
Cremello	19	6
Other (Leopard pattern/buckskin)	14	4
Body condition score [1]		
<3	75	22
3	188	55
>3	76	22
Training intensity		
Low	215	63
Medium	63	19
High	23	7
Breaking in	23	7
No training [2]	15	4

[1] According to the scale of Carroll and Huntington, 1988. [2] No training includes horses kept as pets or for company.

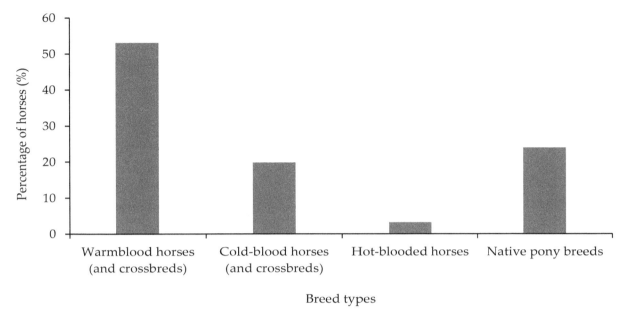

Figure 1. Distribution of breed types for horses showing free faecal liquid ($n = 339$).

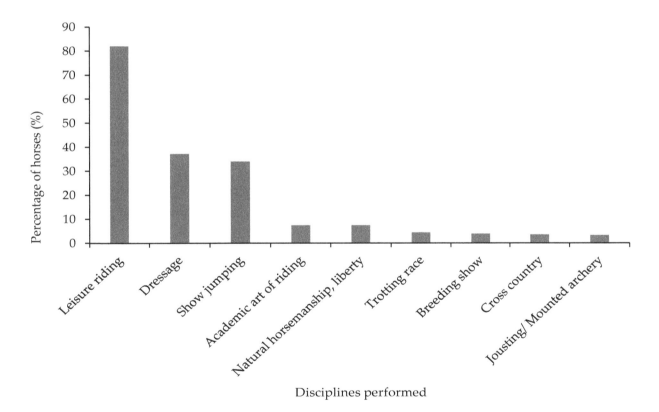

Figure 2. Distribution of disciplines performed by horses with free faecal liquid ($n = 339$), as reported by respondents. Multiple-choice question resulting in that the sum could exceed 100 percent.

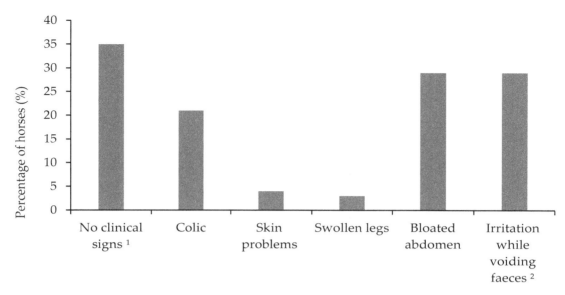

Figure 3. Percentage of horses showing different clinical signs associated with episodes of free faecal liquid, as reported by respondents ($n = 339$). Multiple-choice question resulting in a sum that could exceed 100%. [1] No clinical signs mean no signs other than free faecal liquid. [2] Including extensive tail swishing and/or trampling with hindlegs while voiding faeces and/or faecal liquid.

3.2. Management

The majority of the horses (79%, $n = 271$) were kept in individual boxes at night and outside in paddocks during the daytime, while 19% ($n = 61$) were kept in loose-housing systems (Table 2). The bedding materials used in stables and loose housing systems were a combination of straw and shavings (37%, $n = 125$), straw only (20%, $n = 68$) and shavings only (17%, $n = 58$) (Table 2). Horses that spent their daytime in paddocks were generally kept outside for 8–12 h per day (48%, $n = 163$), while 29% ($n = 98$) were kept outside for more than 12 h per day and 21% ($n = 71$) were kept outside for less than 4 h per day (Table 2). The type of paddocks were soil paddocks (39%, $n = 132$), grass paddocks (old grass during winter) (28%, $n = 94$) or sand/gravel paddocks (23%, $n = 78$) (Table 2). Forty-eight percent ($n = 163$) of the horses were kept on pasture for 9–12 weeks, while 21% ($n = 71$) were on pasture for ≤8 weeks and 8% ($n = 27$) were kept on pasture for >12 weeks (Table 2). The majority (55%, $n = 186$) of the horses were dewormed if faecal egg counts showed sufficiently high numbers to indicate deworming according to national guidelines (www.sva.se) (Table 2). Other deworming procedures included regular deworming more than one time per year (36%, $n = 122$), dewormed if considered necessary (7%, $n = 25$) or not dewormed (1%, $n = 4$) (Table 2). Horses were reported to have access to water by tubs, buckets or automatic waterers when kept in the stable, paddock and at pasture. At pasture, horses were also reported to have access to water by natural water sources (Table A1). Extended information on management factors is presented in Table A1 (Appendix A).

3.3. Feeding

The majority (74%, $n = 250$) of horses were fed forage in meals, while 26% ($n = 89$) were fed forages ad libitum. Grass haylage (defined as in Table S1, Supplementary Materials) was offered to 95% ($n = 322$) of the horses, whereas 5% ($n = 17$) of the horses were fed grass silage (Figure 4). Hay was fed to 50% ($n = 170$) of the horses (Figure 4). In general, horses were fed roughage-dominated feed rations with on average 90% roughage, and 7% concentrates in the daily feed ration (Table 3). Daily amounts of different feedstuffs are reported in Table 3. Most of the horses (67%, $n = 227$) fed forage in meals were fed forage 3 to 4 times daily, and the time between two forage feedings seldom exceeded 8 h (Table A1). Horses that were fed in their paddocks were served forage in tubs or haynets (60%, $n = 204$), or on the ground (45%, $n = 153$) (Table A1). Eight percent ($n = 27$) were not fed forage in their paddocks (Table A1). The majority (66%, $n = 224$) purchased their forage from a producer outside the farm, while the remaining proportion used forage produced on the farm (Table A1). About half (48%, $n = 163$) of the respondents stated that they did not know the forage nutritive contents (Table A1).

Table 2. Description of the management of horses showing free faecal liquid ($n = 339$, if not otherwise mentioned. Deviances in N were due to missing responses for that particular question).

Item	No. of Horses	% of Horses
Housing system ($n = 337$)		
Individual box	271	79
Loose housing system	64	19
Group housing	2	1
Bedding ($n = 336$)		
Straw	67	20
Shavings	57	17
Combination of straw and shavings	125	37
Sawdust	40	12
Wood pellets	26	8
Straw pellets	11	3
Other (paper, mix of sawdust and peat, rubber mat, raw sawdust)	10	3
Time spent per day in paddock during winter ($n = 245$)		
<4 h	5	2
4–7 h	67	20
8–12 h	163	48
>12 h	10	30
Paddock ground ($n = 332$)		
Grass (old grass during winter)	94	28
Sand/Gravel	79	24
Soil	133	40
Other	26	8
Annual time spent on pasture		
<4 weeks	5	2
4–8 weeks	67	20
9–12 weeks	163	48
>12 weeks	100	30
No pasture	4	1
Anthelmintic routines		
Regularly dewormed ≥ 1 times per year	122	36
Dewormed due to high [1] egg counts ≥ 1 times per year	154	45
Dewormed due to high [1] egg counts < 1 times per year	34	10
Dewormed if considered necessary	25	7
Not dewormed	4	1

[1] According to national guidelines (www.sva.se).

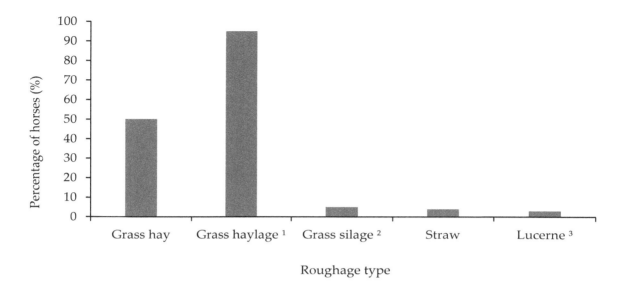

Figure 4. Different types of roughage fed to horses showing free faecal liquid ($n = 339$). Multiple roughages could be assigned in the survey, resulting in a sum of percentages exceeding 100. [1] Wrapped forage with ≥50% DM. [2] Wrapped forage with <50% DM. [3] Includes both pelleted lucerne and lucerne chaff in a dried format.

Table 3. Daily amounts of different feedstuffs (kg per 100 kg bodyweight (BW) per day) and proportion (%) of roughage and concentrate in the diet offered to horses showing free faecal liquid ($n = 339$).

Item	No. of Horses	Min	Q1	Q2	Q3	Max	Mean	SD
Roughage and concentrate feeding, Kg/100 kg BW/d [1]								
Grass hay	165	0.1	0.2	0.2	0.3	0.5	0.2	0.06
Grass haylage	251	0.2	1.7	2.0	2.3	6.0	2.0	0.67
Grass silage	4	1.2	1.2	2.9	3.8	4.7	2.9	1.73
Straw	14	0.1	0.2	0.3	0.6	2.3	0.4	0.45
Lucerne [2]	10	0.01	0.1	0.1	0.2	0.7	0.1	0.13
Total amount of roughage	217	0.3	1.5	2.0	3.1	4.8	1.8	2.17
Total amount of concentrate	190	0.01	0.1	0.2	0.3	1.0	0.2	0.18
Roughage proportion of total feed ration (%) [3]	249	20	90	100	100	100	90	0.14
Concentrate proportion of total feed ration (%) [3,4]	107	0	1	5	10	80	7	0.14
Mineral supplementation, g/100 kg BW	218	0.1	6.0	10.8	17.8	83.3	13.5	11.43

[1] Horses reported to have ad libitum access to roughage, forage or having straw as bedding material were not included. [2] Horses reported to have access to roughage ad libitum without concentrates in the diet were included. [3] Horses reported to have access to roughage ad libitum without concentrates in the diet were included. [4] Horses reported not to be fed concentrate were excluded. Min = Minimum value. Q1–Q3: First-, second- (=median) and third quartile. Max = Maximum value. SD = Standard deviation.

More than half of the horses (56%, $n = 190$) were fed concentrates, and the most common type was commercial concentrates ($n = 118$) followed by vegetable oil ($n = 104$) and molassed sugar beet pulp ($n = 22$) and (Table A1). Supplemental feeds were used for 84% ($n = 285$) of the horses in the study and mostly comprised mineral and vitamin feed (Table A1). For horses reported to be fed concentrates, 217 horses were fed concentrates 1–2 times per day and the remaining proportion was fed concentrates more often (Table A1). The presence of FFL was reported to diminish when changing from wrapped forages to hay (58%, $n = 197$), from wrapped forage to pasture (46%, $n = 156$) and from one batch of wrapped forage to another batch (17%, $n = 58$) (Table 4). However, not all horses showed any change in the presence of FFL with feed changes (7%, $n = 24$) and not all horses had been subjected to all feed changes (2%, $n = 7$) (Table 4).

Table 4. Changes in the presence of free faecal liquid in the horses in the study ($n = 339$) with diet changes as reported by respondents. "Less loose" refer to the absence and/or reduced amount of liquid phase in faeces compared to before the feed change, as reported by respondents. Not all respondents had tried all response alternatives.

Item	No. of Horses	% of Horses
Faecal appearance less loose when changing from wrapped forage to hay	198	58
Faecal appearance less loose when changing from wrapped forage to pasture	157	46
Faecal appearance less loose when changing to another batch of wrapped forage	56	17
No change in faecal appearance with any change in feeding	24	7
Faecal appearance more loose in association to changing feeds	20	6
Faecal appearance less loose when changing from primary to regrowth harvest [1]	16	5
Faecal appearance less loose when using feed additives [2]	8	4
Have not tried any change in feeding	5	2

[1] Wrapped forages. [2] Feed additives reported included yeast, linseed, psyllium seed, thiamine and various types of commercial pro- and prebiotics.

3.4. Gastrointestinal Health

Eighty-seven percent ($n = 295$) of the horses in the study had not been treated for any gastrointestinal disease within the 3 previous months prior to responding to the survey. Twenty-two percent ($n = 76$) of the horses in the study were reported to have been examined for stomach ulcers, however, only 14 were reported to have been examined with gastroscopy. Of these, 9 were diagnosed with gastric ulcers. Nearly one-quarter of the horses in the study (23%, $n = 78$) were reported to have had a previous history of colic. Therefore, horses were divided into two groups, one with a previous history of colic and one with no previous history of colic, and compared within each type of symptom they were showing during episodes of FFL (Figure 5). For horses reported to show no clinical signs during episodes of FFL ($n = 119$), there was a higher proportion of horses with no previous history of colic (74%, $n = 88$) compared to horses with a previous history of colic (26%, $n = 31$) ($p < 0.001$). For horses reported to show signs of colic during episodes of FFL ($n = 71$) there was a higher proportion of horses with a previous history of colic (87%, $n = 62$) compared to horses with no previous history of colic (13%, $n = 9$) ($p < 0.001$).

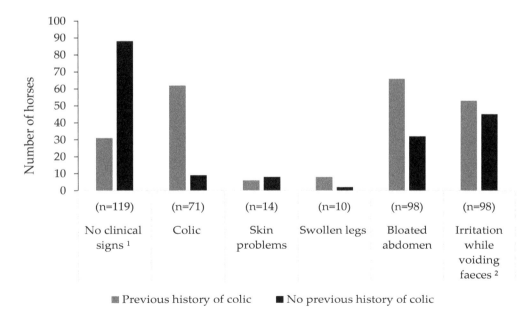

Figure 5. The horse owner reported clinical signs during episodes of free faecal liquid in horses with ($n = 77$) and without ($n = 256$) a previous history of colic. Multiple signs could be selected in the survey, resulting in the numbers of horses for all symptoms exceeding the total number of horses in the study. [1] No clinical signs mean no signs other than FFL. [2] Including extensive tail swishing and/or trampling with hindlegs.

For horses reported to show a bloated abdomen ($n = 98$), there was a tendency ($p = 0.08$) for a higher proportion of horses with a previous history of colic (67%, $n = 66$) compared to horses with no previous history of colic (33%, $n = 32$).

3.5. Stereotypic Behaviour

Nineteen percent ($n = 64$) of the horses in the study were reported to show stereotypic behaviour. The reported stereotypic behaviours included crib biting (15%, $n = 8$), wind sucking (4%, $n = 2$), weaving (7%, $n = 4$), box walking (11%, $n = 6$), wood chewing (60%, $n = 33$) and tongue rolling (7%, $n = 4$).

4. Discussion

4.1. Horses

In the present study, the typical horse showing FFL was reported to be of warmblood type, have bay coat colour, be a gelding, on average 12 years old and be used for leisure riding, dressage or show jumping. A similar distribution of horse characteristics has been described within other horse populations in Sweden and Norway [14,15], which indicates that the population in the present study is a reflection of the normal horse population. As a large variety of breeds, ages and disciplines were represented in the population in the present study, it is evident that FFL could appear in almost any type of horse. The proportion of geldings was larger compared to the proportions of mares and stallions in the current study, which is in agreement with the findings in a previous study on FFL [1] where a larger proportion of geldings was found in the case group compared to controls. However, it may be more common to keep geldings as leisure horses [16,17], compared to mares and stallions, and this could explain the higher proportion of geldings in both studies. The reported BCS was in relation to what was expected. The majority of horses were reported to have a BCS of 3 (normal BCS), which is in agreement with previous descriptions of FFL [1,5] where horses were not reported to show weight loss or loss of BCS. It has also been reported that horse owners are commonly underestimating the body condition score of their horses compared to a trained professional scorer [18]. This may indicate that the horses in the present study were not underweight or in lower than normal BCS. In a previous study, horses with the coat colour paint were over-represented (29%, $n = 12$) in a group of horses showing FFL compared to horses in two control groups, which comprised 10% ($n = 4$) and 8% ($n = 3$) paint coloured horses, respectively) [1]. In the current study, the proportion of paint horses was lower than the proportions of the bay, chestnut, grey and black horses. Therefore, an association between the coat colour paint and the presence of FFL cannot be confirmed from the results in the current study. It has been suggested that grey horses may be more prone to show FFL due to the higher risk of melanoma [19], which may cause defecation difficulties, but no such association has been identified in the literature or in this study. Social stress has been suggested to play a role for the presence of FFL, as the majority of FFL-affected horses did not defend their feed against other horses and were judged as low in the hierarchy in a previous study [1]. However, in the same study [1], high ranked horses also showed FFL, indicating that stress from being low in the hierarchy is not a sole explanation for the presence of FFL.

4.2. Management

The majority of the horses in this study were kept outside in paddocks for 4–8 h in soil or sand/gravel paddocks and almost half the horses were fed forage on the ground. Feeding forage on the ground could lead to increased ingestion of soil and sand particles causing irritation of the intestinal mucosa, which has been associated with gastrointestinal conditions such as diarrhoea, including voiding of loose and watery faeces [20], and colic [21]. Whether or not this management factor plays a role in the presence of FFL remains to be further elucidated.

4.3. Feeding

Despite the fact that about half of the horses were reported to have less loose faeces when feeding was changed from wrapped forages to hay (58%) or to pasture (46%), this was not the case for all horses in the study. In addition, about one-quarter of the horses improved by changing from one batch to another batch (including from primary to regrowth harvest) of wrapped forage. This indicates that the occurrence of FFL cannot be generally attributed to feeding wrapped forages. This finding is supported by results from a previous study in which horses displaying FFL predominantly were fed hay [1]. In addition, FFL has not been reported in controlled feeding studies with healthy horses fed silage, haylage and hay from the same grass sward and harvest [22,23]. As there may be individual variation in the gut microbiota of horses [24–26], it is possible that individuals respond differently to the same feed. Further study within this area is highly interesting and may provide more insight into factors contributing to the presence of both FFL and colic.

Although the forage conservation method may influence both chemical and microbial composition in forage [22], forages differ in a number of other factors as well. One important factor is plant maturity at harvest, which greatly influences overall digestibility in the equine GIT and the nutritive value of the forage [27]. Valle et al. [5] reported a horse with FFL to maintain a reduced or absent production of FFL with gradual changes in the nutritional plan to meet the theoretical nutritional requirements of the horse. Changes included reducing energy content in the feed ration by excluding concentrate feeds and decreasing the amount of forage and changing batch of forage in combination with increased exercise. However, simultaneously with the nutritional changes, the horse was treated with sulfasalazine, making it difficult to evaluate the effect of the changes in the feeding. Only half of the horse owners in the current study reported that they knew the nutritive content in their forage (through forage analytical reports), indicating that half of the horses may have been under- or overfed in relation to their nutritional requirements. Whether this is a factor of importance for the presence of FFL is currently not known.

The composition of the total feed ration and the ratio between forage and concentrate may be of importance for the physical appearance of faeces, as two-phase separation (liquid and solid phase) of both digesta and faeces has been reported in horses fed hay ad libitum with inclusion of grains (4.55 kg every 12 h) but not when fed only the same hay ad libitum [28]. In horses fed hay only, a clear separation between solid and liquid phases was present in the contents of the right dorsal colon (RDC), but faecal balls were well-formed and with no separation in liquid and solid phases. No or minimal gas bubbles were present in RDC content when horses were fed hay [28]. When horses were fed hay with the inclusion of grains, RDC contents were more homogenous and foamy, with less separation of phases, and the liquid phase was more viscous than in horses fed hay only. Faeces of horses fed hay and grains were, however, less formed and had a clear separation, where the liquid phase had noticeable gas bubbles and was more viscous compared to faeces of horses fed hay only [28]. One explanation for this result could be differences in the hydrophilic properties of ingesta components, which may differ when horses were fed hay only or hay with grain inclusion. The hydrophilic properties of the ingesta have been suggested as a cause of osmotic diarrhoea [29,30]. In the current study, horses were fed much smaller proportions of concentrates compared to what was described by Lopes et al. [28]. However, smaller amounts of concentrates fed daily (2.5–5 kg) have been reported to increase the risk of colic [31–33] and may affect the ingesta and its transit as well. As horses displaying FFL seldom show clinical signs of disease, FFL may be a type of osmotic diarrhoea. Further insights in causes of FFL may be provided by investigations of forage and concentrate proportions in the total feed ration of horses with FFL.

4.4. Gastrointestinal Health

In previous studies, no symptoms of disease, such as fever or weight loss or loss of body condition has been described in horses with FFL [1,5]. Nine horses were reported to have been diagnosed with gastric ulcers, but overall very few horses ($n = 14$) had been examined for gastric ulcers with gastroscopy. From the results of the current study, it cannot be ruled out that gastric ulcers may be associated with the presence of FFL, even though the incidence of gastric ulcers was lower than what has been described in other studies of gastric ulcers in leisure horses (17% to 58%, [34–36]).

In the present study, almost one-quarter of the horses were reported to have had a previous history of colic. A colic incidence between 3.5% and 10.6% has been reported for general horse populations [10,32,37,38] and of 4.8% within a German population of horses showing FFL [3]. This indicates that the incidence of colic was higher for Swedish and Norwegian horses showing FFL. In addition, in the present study, 55% of the horses did not show any clinical signs other than faecal liquid during FFL episodes, whereas the remaining proportion of horses were reported to have one or a combination of several clinical signs including e.g., colic symptoms. The latter proportion also had a higher number of horses with a previous history of colic. This indicates that causes of FFL could differ among different horses, or that FFL is a generic symptom from several different conditions of a similar nature. The number of clinical signs during an FFL episode could also depend on the severity of the condition, which could not be assessed in this study. Further studies of FFL should preferably include detailed descriptions of duration, intensity and severity of FFL episodes as well as previous disturbances in the GIT of FFL-affected horses. It is possible that the hindgut microbiota of FFL-horses is responsible for the clinical signs. Transplantation of faecal microflora in affected horses has been reported to decrease the severity of FFL in a controlled study [4]. However, in the same study [4], horses were also treated with omeprazole and psyllium seeds in addition to faecal transplant, making it difficult to evaluate the effects.

4.5. Stereotypic Behaviour

The reported incidence of stereotypic behaviour among FFL horses in the study was 19% when wood chewing was included. Tree-wood chewing may not always be a stereotypy but could be related to low-fibre diets [39], and when excluded, the incidence of stereotypies was approximately 8%, which is comparable to the previously reported incidence of 7.1–12.3% [40–44] in other horse populations. Factors reported to be associated with an increased risk of stereotypic behaviour in previous studies include low levels of social interactions with other horses [44–46], low forage availability [45,46] and low number of horses kept in the same and/or adjacent paddocks [45]. The majority of horses in the current study were kept in individual boxes when stabled and fed forage in predetermined portions during the day, which may have contributed to the incidence of stereotypic behaviour among horses in this study.

4.6. Survey Response and Limitations of the Study

The advantages of performing online surveys are the ease and low cost of data collection, the automation in data input and handling, which reduce errors, and the flexibility in survey design to make it easier for participants to respond to the questionnaire. The disadvantage includes the absence of an interviewer, and that data are reported by the horse owner, which could result in misinterpretations among respondents for some of the questions. Some of the variables have previously been shown to be difficult to estimate correctly for horse-owners, such as BCS [18]. It should be recognized that all conclusions drawn from this study were based on the perceptions of the respondents, which may vary in their knowledge of equine feeding and management.

In order to control that the respondents of the survey were within the intended group, control questions based on the inclusion criteria, such as the age of the horse, and if the horse had problems with FFL when fed wrapped forages, were asked. This resulted in the elimination of 441 responses to the present study, indicating that such controls could be of high importance to enhance the quality of data input in online surveys.

5. Conclusions

There was a large variety of horse characteristics, including breed type, age and coat colour, in horses with FFL. Many, but not all horses in this study were reported to show less separation of solid and liquid phases in their faeces when changing from wrapped forages to hay, pasture or another batch of wrapped forage. Horses with FFL were also reported to have a higher incidence of a previous history of colic compared to reports from other horse populations. Further research on FFL in horses is of interest and should include details on feeding (such as forage nutritive values, feed ration composition), and gastrointestinal tract health and function (such as the presence of stomach ulcers, colic, gastrointestinal tract response to different feedstuffs), as well as detailed descriptions of severity and duration of FFL episodes.

Author Contributions: The authors have contributed to the following sections of this work: Conceptualization, K.M.L. and C.E.M.; data curation, K.M.L.; formal analysis, K.M.L., A.J., V.B., J.D., J.E.L., and C.E.M.; funding acquisition, A.J. and C.E.M.; investigation, K.M.L. and C.E.M.; methodology, K.M.L. and C.E.M.; project administration, K.M.L., A.J., and C.E.M.; writing—original draft preparation, K.M.L.; writing—review and editing, K.M.L., A.J., V.B., J.D., J.E.L., and C.E.M. All authors have read and agreed to the published version of the manuscript.

Acknowledgments: The authors would like to thank all horse-owners who contributed to the data. In addition, thanks to Professor Ulf Olsson, Biostokastikum, Department of Energy and Technology, Swedish University of Agricultural Sciences, for very valuable and helpful statistical support and advice.

Abbreviations

The following abbreviations are used in this manuscript:

BCS	Body Condition Score
DM	Dry matter
FFL	Free faecal liquid
FMT	Faecal microbiota transplantation
GIT	Gastrointestinal tract
RDC	Right dorsal colon

Appendix A

Horse owner responses to the survey for variables not presented in tables or figures in the main text. The number (*n*) and proportion (%) of horses presented are presented in Table A1.

Table A1. Extended information on management factors and feeding strategies for horses with free faecal liquid ($n = 339$), as reported by respondents.

Variables	Total Number of Horses	% of All Horses
Horses showing FFL when fed wrapped forages	339	100
Region of stable		
Southern	126	37
Central	81	24
North	39	12
Western [1]	23	7
Eastern [1]	70	21
Horse imported		
No	253	75
Do not know	6	2
Yes	80	24
Ability of horse to keep desired BCS		
Easy keeper	97	29
Normal	187	55
Hard keeper	55	16
Type of water source in stable/loose housing system		
Frostless waterer	49	14
Frostless tub	43	13
Waterer	31	9
Tub	31	9
Bucket	78	23
Natural water source	4	1
Combination of bucket and waterer	103	30
Type of water source in paddock during winter		
Frostless waterer	34	10
Frostless tub	80	24
Waterer	4	1
Tub	124	37
Bucket	37	11
Natural water source	5	1
Other (Combination of tub and bucket, bucket and waterer)	55	16
Type of pasture		
Pasture on arable land	106	31
Natural pasture	135	40
Forest	6	2
No pasture	30	9
Other (Combination of different pasture types)	61	18
Type of water source on pasture		
Frostless waterer	45	13
Frostless tub	12	4
Waterer	10	3
Tub	165	49
Bucket	8	2
Natural water source	33	10
Other (Combination of bucket and tub)	66	19
Access to saltlick while on pasture		
Yes	220	65
No	119	35
Saltlick in stable/loose housing system		
Yes	296	87
No	43	13

Table A1. Cont.

Variables	Total Number of Horses	% of All Horses
Time from last deworming		
Not dewormed	11	3
0–3 months ago	75	22
4–6 months ago	126	37
7–12 months ago	68	20
>1 year ago	59	17
Origin of the forage		
Bought	226	67
Produced on farm, but not by the owner	42	12
Produced on farm by the owner	69	20
Other	1	0
Forages analysis		
Yes	144	42
No	163	48
Do not know	37	9
Number of feedings of forage per day		
1 time	1	0
2 times	14	4
3 times	122	36
4 times	108	32
>4 times	33	10
Free access	89	26
Storage of forage		
Indoors	223	66
Outdoors	114	34
Outdoors, covered	65	19
Outdoors, uncovered	49	14
Hay indoors, wrapped forages outdoors uncovered	1	0
Hay indoors, wrapped forages outdoors covered	1	0
Maximum time between two feedings of roughage		
0–2 h	2	1
2–4 h	35	10
4–8 h	115	34
8–12 h	101	30
>12 h	8	2
Don't know	78	24
Feeding strategy for roughage in paddock [2]		
Forage not fed in the paddock	27	8
On the ground	113	33
In the feeding rack	79	23
In a haynet	19	6
In a tub or similar	62	18
Other (combination of ground and feeding rack/haynet)	39	12
Type of concentrate fed [2]		
Grains	16	5
Molassed sugar beet pulp	22	6
Linseed/Linseed cake	4	1
Soybean meal	14	4
Potato protein	33	10
Wheat bran	3	1
Vegetable oil	104	31
No concentrate fed	149	44
Commercial concentrate	118	63

Table A1. *Cont.*

Variables	Total Number of Horses	% of All Horses
Number of concentrate feedings per day		
0 times	149	44
1 time	116	34
2 times	107	32
3 times	49	14
4 times	7	2
>4 times	1	0
Type of supplemental feeds fed [2]		
Mineral feeds	237	70
Multivitamins	58	17
B-vitamins (and Biotin)	10	3
Selenium and Vitamin E	9	3
Garlic	13	4
Herbs	8	2
Other (Yeast, magnesium)	93	27
Not fed supplemental feeds	51	16
Storage of concentrates		
In covered/closed containers indoors	255	66
In uncovered/open containers indoors	11	3
In paper bags/original package indoors	18	5
No concentrate	51	15
Other	4	1
Previous treatment of other gastro-intestinal diseases		
No	296	87
Don't know	11	3
Yes	32	9

[1] Only for Norwegian respondents. [2] Multiple choice question.

References

1. Kienzle, E.; Zehnder, C.; Pfister, K.; Gerhards, H.; Sauter-Louis, C.; Harris, P. Field study on risk factors for free fecal water in pleasure horses. *J. Equine Vet. Sci.* **2016**, *44*, 32–36. [CrossRef]

2. Ertelt, A.; Gehlen, H. Free fecal water in the horse-an unsolved problem. *Pferdeheilkunde* **2015**, *31*, 261–268. (In German) [CrossRef]

3. Zehnder, C. Field Study on Risk Factors for Free Faecal Water in Leisure Horses. Ph.D. Thesis, Ludwig-Maximilians-Universität München, Munich, Germany, 6 February 2009. (In German).

4. Laustsen, L.; Edwards, J.; Smidt, H.; van Doorn, D.; Luthersson, N. Assessment of faecal microbiota transplantation on horses suffering from free faecal water. In Proceedings of the 9th European Workshop on Equine Nutrition, Swedish University of Agricultural Science, Uppsala, Sweden, 16–18 August 2018.

5. Valle, E.; Gandini, M.; Bergero, D. Management of chronic diarrhea in an adult horse. *J. Equine Vet. Sci.* **2013**, *33*, 130–135. [CrossRef]

6. Gordon, C.H.; Derbyshire, J.C.; Wiseman, H.G.; Kane, E.A.; Melin, C.G. Preservation and feeding value of alfalfa stored as hay, haylage and direct-cut silage. *J. Dairy Sci.* **1961**, *44*, 129–131. [CrossRef]

7. Finner, M.F. Harvesting and handling low-moisture silage. *Trans. ASAE* **1966**, *9*, 377–381. [CrossRef]

8. Jackson, N.; Forbes, T.J. The voluntary intake by cattle of four silages differing in dry matter content. *Anim. Sci.* **1970**, *12*, 591–599. [CrossRef]

9. Enhäll, J.; Nordgren, M.; Kättström, H. *Horses in Sweden*; Report 2012:1; The Swedish Board of Agriculture: Jönköping, Sweden, 2012. (In Swedish)

10. Larsson, A.; Müller, C.E. Owner reported management, feeding and nutrition-related health problems in Arabian horses in Sweden. *Livest. Sci.* **2018**, *215*, 30–40. [CrossRef]

11. Vik, J.; Farstad, M. *Horse, Horse Management and Feeding: The Status of Horse Husbandry in Norway*; Norsk senter for Bygdeforskning: Trondheim, Norway, 2012. (In Norwegian)

12. Gerstner, K.; Liesegang, A. Effect of a montmorillonite-bentonite-based product on faecal parameters of horses. *J. Anim. Physiol. Anim. Nutr. (Berl.)* **2018**, *102*, 43–46. [CrossRef]
13. Carroll, C.L.; Huntington, P.J. Body condition scoring and weight estimation of horses. *Equine Vet. J.* **1988**, *20*, 41–45. [CrossRef]
14. Heldt, T.; Macuchova, Z.; Alnyme, O.; Andersson, H. Socio-Economic Effects of the Equine Industry. Falun: Dalarna University. (Report 2018:04). Available online: https://hastnaringen-i-siffror.se/files/Hästnäringens_samhällsekonomi_rapport2018_4_2.pdf (accessed on 5 November 2018). (In Swedish).
15. Hartmann, E.; Bøe, K.E.; Jørgensen, G.H.M.; Mejdell, C.M.; Dahlborn, K. Management of horses with focus on blanketing and clipping practices reported by members of the Swedish and Norwegian equestrian community. *Anim. Sci. J.* **2017**, *95*, 1104–1117. [CrossRef]
16. Ross, S.E.; Murray, J.K.; Roberts, V.L.H. Prevalence of headshaking within the equine population in the UK. *Equine Vet. J.* **2018**, *50*, 73–78. [CrossRef]
17. Wylie, C.E.; Ireland, J.L.; Collins, S.N.; Verheyen, K.L.P.; Newton, J.R. Demographics and management practices of horses and ponies in Great Britain: A cross-sectional study. *Res. Vet. Sci.* **2013**, *95*, 410–417. [CrossRef] [PubMed]
18. Wyse, C.A.; McNie, K.A.; Tannahil, V.J.; Love, S.; Murray, J.K. Prevalence of obesity in riding horses in Scotland. *Vet. Rec.* **2008**, *162*, 590–591. [CrossRef] [PubMed]
19. Hofmanová, B.; Vostrý, L.; Majzlík, I.; Vostrá-Vydrová, H. Characterization of greying, melanoma, and vitiligo quantitative inheritance in Old Kladruber horses. *Czech J. Anim. Sci.* **2015**, *60*, 443–451. [CrossRef]
20. Niinistö, K.E.; Määttä, M.A.; Ruohoniemi, M.O.; Paulaniemi, M.; Raekallio, M.R. Owner-reported clinical signs and management-related factors in horses radiographed for intestinal sand accumulation. *J. Equine Vet. Sci.* **2019**, *80*, 10–15. [CrossRef] [PubMed]
21. Kilcoyne, I.; Dechant, J.E.; Spier, S.J.; Spriet, M.; Nieto, J.E. Clinical findings and management of 153 horses with large colon sand accumulations. *Vet. Surg.* **2017**, *46*, 860–867. [CrossRef]
22. Müller, C.E. Equine digestion of diets based on haylage harvested at different plant maturities. *Anim. Feed Sci. Technol.* **2012**, *177*, 65–74. [CrossRef]
23. Müller, C.E. Silage and haylage for horses. *Grass Forage Sci.* **2018**, *73*, 815–827. [CrossRef]
24. Muhonen, S.; Julliand, V.; Lindberg, J.E.; Bertilsson, J.; Jansson, A. Effects on the equine colon ecosystem of grass silage and haylage diets after an abrupt change from hay. *J. Anim. Sci.* **2009**, *87*, 2291–2298. [CrossRef]
25. Daly, K.; Proudman, C.J.; Duncan, S.H.; Flint, H.J.; Dyer, J.; Shirazi-Beechey, S.P. Alterations in microbiota and fermentation products in equine large intestine in response to dietary variation and intestinal disease. *Br. J. Nutr.* **2012**, *107*, 989–995. [CrossRef]
26. Fernandes, K.A.; Kittelmann, S.; Rogers, C.W.; Gee, E.K.; Bolwell, C.F.; Bermingham, E.N.; Thomas, D.G. Faecal microbiota of forage-fed horses in New Zealand and the population dynamics of microbial communities following dietary change. *PLoS ONE* **2014**, *9*, e112846. [CrossRef] [PubMed]
27. Müller, C.E.; Von Rosen, D.; Udén, P. Effect of forage conservation method on microbial flora and fermentation pattern in forage and in equine colon and faeces. *Livest. Sci.* **2008**, *119*, 116–128. [CrossRef]
28. Lopes, M.A.; White, N.A., II; Crisman, M.V.; Ward, D.L. Effects of feeding large amounts of grain on colonic contents and feces in horses. *Am. J. Vet. Res.* **2004**, *65*, 687–694. [CrossRef] [PubMed]
29. Clarke, L.L.; Roberts, M.C.; Argenzio, R.A. Feeding and digestive problems in horses: Physiologic responses to a concentrated meal. *Vet. Clin. N. Am. Equine Pract.* **1990**, *6*, 433–450. [CrossRef]
30. Jones, S.L.; Spier, S.T. Pathophysiology of colonic inflammation and diarrhea. In *Equine Internal Medicine*; Reed, S.M., Bayly, W.M., Eds.; WB Saunders: Philadelphia, PA, USA, 1998; pp. 678–679.
31. Tinker, M.K.; White, N.A.; Lessard, P. Descriptive epidemiology and incidence of colic on horse farms: A prospective study. In Proceedings of the 5th Equine Colic Research Symposium, The University of Georgia, Athens, GA, USA, 26–28 September 1994.
32. Tinker, M.K.; White, N.A.; Lessard, P.; Thatcher, C.D.; Pelzer, K.D.; Davis, B.; Carmel, D.K. Prospective study of equine colic risk factors. *Equine Vet. J.* **1997**, *29*, 454–458. [CrossRef]
33. Hudson, J.M.; Cohen, N.D.; Gibbs, P.G.; Thompson, J.A. Feeding practices associated with colic in horses. *J. Am. Vet. Med. Assoc.* **2001**, *219*, 1419–1425. [CrossRef]
34. Luthersson, N.; Nielsen, K.H.; Harris, P.; Parkin, T.D. The prevalence and anatomical distribution of equine gastric ulceration syndrome (EGUS) in 201 horses in Denmark. *Equine Vet. J.* **2009**, *41*, 619–624. [CrossRef]

35. Sandin, A.; Skidell, J.; Häggström, J.; Nilsson, G. Postmortem findings of gastric ulcers in Swedish horses older than age one year: A retrospective study of 3715 horses (1924–1996). *Equine Vet. J.* **2000**, *32*, 36–42. [CrossRef]

36. Niedźwiedź, A.; Kubiak, K.; Nicpoń, J. Endoscopic findings of the stomach in pleasure horses in Poland. *Acta Vet. Scand.* **2013**, *55*, 45. [CrossRef]

37. Hillyer, M.H.; Taylor, F.G.R.; Proudman, C.J.; Edwards, G.B.; Smith, J.E.; French, N.P. Case control study to identify risk factors for simple colonic obstruction and distension colic in horses. *Equine Vet. J.* **2002**, *34*, 455–463. [CrossRef]

38. Traub-Dargatz, J.L.; Kopral, C.A.; Seitzinger, A.H.; Garber, L.P.; Forde, K.; White, N.A. Estimate of the national incidence of and operation-level risk factors for colic among horses in the United States, spring 1998 to spring 1999. *J. Am. Vet. Med. Assoc.* **2001**, *219*, 67–71. [CrossRef] [PubMed]

39. Normando, S.; Canali, E.; Ferrante, V.; Verga, M. Behavioral problems in Italian saddle horses. *J. Equine Vet. Sci.* **2002**, *22*, 117–120. [CrossRef]

40. Christie, J.L.; Hewson, C.J.; Riley, C.B.; McNiven, M.A.; Dohoo, I.R.; Bate, L.A. Management factors affecting stereotypies and body condition score in nonracing horses in Prince Edward Island. *Can. Vet. J.* **2006**, *47*, 136. [PubMed]

41. Normando, S.; Meers, L.; Samuels, W.E.; Faustini, M.; Ödberg, F.O. Variables affecting the prevalence of behavioral problems in horses. Can riding style and other management factors be significant? *Appl. Anim. Behav. Sci.* **2011**, *133*, 186–198. [CrossRef]

42. McGreevy, P.D.; French, N.P.; Nicol, C.J. The prevalence of abnormal behaviors in dressage, eventing and endurance horses in relation to stabling. *Vet. Rec.* **1995**, *137*, 36–37. [CrossRef] [PubMed]

43. Luescher, U.A.; McKeown, D.B.; Dean, H.A. Cross-sectional study on compulsive behaviour (stable vices) in horses. *Equine Vet. J.* **1998**, *27*, 14 18. [CrossRef]

44. McGreevy, P.D.; Cripps, P.J.; French, N.P.; Green, L.E.; Nicol, C.J. Management factors associated with stereotypic and redirected behaviour in the thoroughbred horse. *Equine Vet. J.* **1995**, *27*, 86–91. [CrossRef]

45. Bachmann, I.; Audige, L.; Stauffacher, M. Risk factors associated with behavioural disorders of crib-biting, weaving and box-walking in Swiss horses. *Equine Vet. J.* **2003**, *35*, 158–163. [CrossRef]

46. Redbo, I.; Redbo-Torstensson, P.; Odberg, F.O.; Henderdahl, A.; Holm, J. Factors affecting behavioural disturbances in racehorses. *Anim. Sci.* **1998**, *66*, 475–481. [CrossRef]

11

Equine Milk Production and Valorization of Marginal Areas

Nicoletta Miraglia [1], Elisabetta Salimei [1,*] and Francesco Fantuz [2]

[1] Dipartimento Agricoltura, Ambiente e Alimenti, Università degli Studi del Molise, Campobasso 86100, Italy; miraglia@unimol.it

[2] Scuola di Bioscienze e Medicina Veterinaria, Università degli Studi di Camerino, Camerino MC 62032, Italy; francesco.fantuz@unicam.it

* Correspondence: salimei@unimol.it.

Simple Summary: The revaluation of equine milk for human consumption is showing an increased interest from a scientific point of view. As practical relapse of the peculiar characteristics of horse and donkey milk, and their potentialities as food products, the dairy equine enterprise is developing worldwide. The milk production can therefore contribute to the whole equine industry, but crucial factors still need to be elucidated. Aiming to promote advances of knowledge on the dairy equine enterprise, aspects of management of the dairy horse and donkey are reviewed in the frame of marginal areas, with a special focus on dam and foal feeding, and welfare, besides milk quality.

Abstract: The equine dairy chain is renewing the interest toward horse and donkey breeding for the production of milk with potential health promoting properties. The dairy equine chain for human consumption could contribute to the rural eco-sustainable development for the micro-economies of those areas threatened by marginalization. As a part of the whole equine industry, and its possible impact in the modern and future society, the main traits of the equine dairy enterprise are reviewed with a special focus on management of animals and milk. Equine milk compositional and nutritional peculiarities are described as also related to milk hygiene and health issues. Scientific and technical aspects of the feeding management are considered in the frame of the emerging dairy equine enterprise, where pasture is an essential element that allows to match production goals for horses and donkeys, biodiversity preservation, as well as landscape safeguard.

Keywords: equine milk; dairy equine chain; dairy equine management and feeding; biodiversity; landscape; pasture

1. Introduction

Equine breeding represents one of the most promising activities in rural development, which is considered a key strategy for restructuring the agriculture sector by means of diversification and innovation [1]. The equine species are involved not only in activities concerning their use for work and tourism, but also in niche activities related to the production of food and non-food products [2,3]. The high versatility of the equine species represents a strong argument for the conservation of endangered equine breeds and populations [4,5]. Many breeds occupy special niches and contribute to the biodiversity due to their own genetic characteristics, coming from adaptive mechanisms developed in centuries of evolution in specific local environments [6,7]. Consequently, policies for the safeguard of endangered equine breeds and autochthonous populations have been developed, also considering the recovery of the relationship among humans, animals, and territory, as a 'system integrator' of the rural eco-sustainable development [8,9]. The renewed interest toward equine milk and derivatives is today sustained by the emerging dairy equine enterprise, which is developing in France, Italy, Mongolia, China, Kazakhstan, Kirgizstan, Greece, Germany, and many other countries [10–12].

As a part of the whole equine industry and its potential impact in the modern and future society, the equine dairy enterprise is described in its main traits. Based on peculiarities of equine milk for human consumption, the essential features of management of animals and milk are reviewed in the context of marginal areas. Aspects of nutrition of the dairy equids are examined in the frame of those areas where pasture and natural meadows represent the main land use, as a further contribution to landscape safeguard.

2. Equine Milk: Properties, Potentials, and Benefits

The nutritional and therapeutic peculiarities of equine milk are known since ancient times, as Hippocrates [13] and Herodotus [14] described in the 5th century BC. Moreover, the consumption of koumiss (or airag), i.e., a traditional drink made in Central Asia, according to a nomads' recipe [15], is reported in literature, not only as an ingredient of the traditional "white diet" of the Mongolian steppes population [16], but also as a popular remedy for a variety of diseases [17,18]. The traditional use of donkey milk is also reported in China and South America for the treatment of many illnesses [19].

Recent scientific findings on the equine milk compositional peculiarities and their potential health promoting properties have increased interest toward its use for human consumption, especially for sensitive consumers, such as children with allergies to cow's milk protein, as well as immunocompromised or debilitated people [10,20]. In Europe, the dairy equine enterprise started up in France as part of a project on animal diversity preservation [21], and spread out in many marginal areas of the world where these monogastric herbivores are well adapted to difficult environments, with scarce availability of forages, often of poor-quality. Today, equine milk is mainly marketed for human consumption as raw, pasteurized, or freeze-dried [22], and as fermented derivatives [15]. In Italy, the price of donkey milk ranges from 9 to 15 €/L of raw milk, 14 to 17.5 €/L of pasteurized milk, and 27.5 to 36 €/100 g of powdered milk, either spray dried or lyophilized [23,24]. Equine milk is also used in the non-food sector, as an ingredient in cosmetic products [25]. Data on the worldwide production of equine milk are not available, but equine milk has been reported to be consumed by 30 million people [26]. It should also be considered that consumer cognizance of equine milk and derivatives is so far limited, as well as common awareness of its local availability [18,27]. Besides communication gaps to be overcome, the emerging niche market of equine milk raises questions on appropriate management strategies of dam and foals, as mainly related to animal nutrition as well as environmental issues, besides food security and animal welfare.

2.1. Equine Milk Compositional and Nutritional Features

Table 1 summarizes the average horse and donkey milk gross composition and energy content from the recent literature. Values from human and cow milk are given for comparison. It should be considered that equine milk components are mainly affected by nutrition, length of lactation, and health status of the mammary gland, besides genetics. Equine milk has a high water content and shows a lower fat content than human and cow milk (Table 1). The milk fat globules diameter, likely related to lipid digestibility, is reported to be lower than in human and bovine milk [28].

Horse and donkey milk are closer to human milk in terms of both protein and lactose content than cow milk (Table 1). On this regard, it is worth noting that, although about 50% of the world population is lactose intolerant, the daily intake of 14 g of lactose is usually well-tolerated [28]. Moreover, the content of lactose is lower in fermented milk. From a nutritional point of view, it is also important to highlight the hypocaloric content (Table 1) that makes equine milk an inadequate food for infants, when not supplemented with vegetal oil (about 40g L^{-1}) [28].

Table 1. Average milk gross composition and energy content from different species [1].

Item	Horse	Donkey	Human	Cow
Total solids, g kg^{-1}	103.1	95.3	125	127
Fat, g kg^{-1}	10.3	7	35	41
Protein, g kg^{-1}	16.8	16	12	34
Lactose, g kg^{-1}	63	66	64	48
Ash, g kg^{-1}	4.2	4.1	1.9	7
Gross energy, MJ kg^{-1}	1.98	1.75	2.69	3.19

[1] Sources: [15,26,28–31].

The ash content of equine milk (Table 1), which is intermediate between human and cow milk, shows a decline throughout the lactation consistent with Ca and P concentrations in milk [15,20]. Although the absolute values of Ca and P in equine milk are reported variable, and in average higher than in human milk, the Ca:P ratio is reported to be in average 1.3 and 1.72, respectively, for donkey and horse milk, while it accounts for 1.7 and 1.23 in human and bovine milk, respectively [31–33]. As a further dietary consideration, the mineral content of milk is not reported to be influenced by the maternal diet in mammalians, except for Se and I [34,35].

Pieszka et al. [36] reviewed the level of fat-soluble vitamins (A, D, E) in mare milk and found them consistent with values reported for bovine milk. Donkey milk is reported to contain a higher level of vitamin D [37] but it displays very low contents of vitamins A and E, as probably related to the low-fat content [28]. Among the water-soluble vitamins, pyridoxine, pantothenic acid, cobalamin, and vitamin C have so far been detected at high levels only in mare milk [28,36].

After the first clinical evidences on the successful use of equine milk in children with multiple food allergies, reported in 1992 and 2000, respectively, for donkey and horse milk [38,39], donkey milk has mainly been the subject of numerous studies about its use in the diets of children affected by cow's milk protein allergy, thanks to its high palatability, due to the high lactose content, and low allergenicity, related to the nitrogenous components [28,40]. Equine milk with high hygiene characteristics, and properly supplemented from a nutritional point of view, has been confirmed as a promising alternative in the dietary treatment of children affected not only by Immunoglobulin E-mediated cow's milk protein allergy, but also by food protein-induced enterocolitis, occurring in the first six months of life [40,41]. However, findings on the efficacy of equine milk use in the fulfilment of nutrient requirements of children cannot be so far considered conclusive, as they need to be confirmed by larger studies [27]. For these reasons, the use of donkey milk is nowadays considered an ingredient in a solid-food diet, or after the first year of life for children [42].

In regards to the allergenicity of equine milk, the proteomic profile of equine milk has been extensively studied in recent years [43,44], and microheterogeneity is displayed due to genetic variants and post-translational modification [28]. A significant effect of the breed and stage of lactation on gene expression and milk composition, and the association among genetic polymorphisms, gene expression, and milk protein and fat contents have also been observed in mare milk [45–47]. This leads to the relevant role of the dairy equine enterprise in the survival of equine breeds, e.g., Lipizzan, Icelandic, German Warmblood, Akhal-Teke, Franches-Montagnes, Comtois, Italian Heavy Draught, Russian Heavy Draft, and Polish Coldblood among horses; and Poitou, Zamarano Leonés, Burro de Miranda, Ragusano, and Amiata among donkeys, and in the preservation of environment, landscape, and vegetal diversity of areas where they are adapted to live [2,24].

2.2. Functional and Bioactive Compounds

Milk, besides allergens, is a source of many bioactive and functional compounds, i.e. metabolites, enzymes, hormones, trophic, and protective factors that are involved in proper growth and nutrition in newborns [48], or in proper secretion of the mammary gland [49].

Among the bioactive and functional proteins detected in milk, there are enzymes active in protection against protozoa, bacteria, and viruses, e.g., lysozyme and lactoferrin [48]. Lysozyme accounts for 10.5% and 21% of whey proteins, respectively, in horse and donkey milk, but only 5.5% of whey protein in human milk; on the contrary, a higher level of lactoferrin (26.6% of whey protein) is detected. Lactoferrin in horse and donkey milk accounts, on average, for only 7% and 4.48% of whey protein, respectively [11,50,51]. Lysozyme activity was found unaffected by thermal treatment at 72 °C up to 3 min [52,53].

Other enzymes in milk are of technological relevance, such as alkaline phosphatase representing an index of pasteurization efficiency, with activity reported to be about 100 mU L^{-1} in thermal treated equine milk [22].

In mammalian milk, hormones and growth factors have also been detected and classified as bioactive peptides [54] derived from the maternal metabolism. Among them, there are leptin, insulin, ghrelin, Insulin-like growth factor 1 (IGF-1), and thyroid hormones that are involved in the central regulation of food intake and in the maintenance of energy balance. Their role in milk may be related to the regulation of growth, to the development and maturation of the neonatal gut, and of the immune and neuroendocrine system of the newborn [48]. Considering the species-specificity of many proteins, it is worth noting that leptin has been measured as human equivalent in both horse and donkey milk, and human-like ghrelin, IGF-1, and triiodothyronine (T3) were measured in donkey milk [15,20]. It is worth noting that the milk T3 content was affected by the diet in lactating donkeys [55]. The role of variations in the maternal hormone status of equids, as related to both physiological status, and how intensive husbandry strategies might interact with their adaptive capacities in the farming environment, deserves attention and needs to be further considered.

Bioactive peptides are also encrypted in the sequence of milk proteins and are released from them following enzymatic proteolysis, under gastrointestinal digestion or during fermentation. These dietary components exert health promoting, i.e., antimicrobial, antihypertensive, antioxidant, antithrombotic, immunomodulatory, antiproliferative, and opioid activities in the organism, beyond their nutritive value [56–59]. It should be noted, however, that technological treatments carried out to prolong milk shelf life could considerably affect structure, as well as the functional and nutritive properties of milk components, especially peptides and proteins, that might lead to a greater susceptibility to infection and/or the development of allergies [60].

In immunonutrition, the antioxidant properties of nutrients, such as alpha-tocopherol and beta-carotene are known, but increasing scientific evidence suggests the role of dietary lipids in the regulation of neonatal immune function and in the severity of symptoms of allergies [61]. Recent studies show the interesting free fatty acids profile of equine milk, with saturated fatty acids content (50%) lower than that reported for goat and sheep milk, and a higher proportion of monounsaturated fatty acids and polyunsaturated fatty acids (PUFA) than ruminant milk [28]. A balanced ratio between n3PUFA and n6PUFA, respectively considered anti-inflammatory and pro-inflammatory nutrients, is reported for horse and donkey milk [62]. Moreover, in regards to the variability of the lipid fraction, Martini et al. [63] observed an increased content of oleic, palmitoleic, and vaccenic acids considered with a positive effect on human health, and a lowered concentration of stearic acid in donkey milk samples collected in the winter.

The atherogenic and thrombogenic indices, calculated on fatty acid composition, candidate equine milk as an interesting food for people with allergic and inflammatory conditions [28,64]. Moreover, the fatty acids profile detected after in vitro digestion shows significant differences depending on the milk sources, with a prevalence of saturated fatty acids released from both human and donkey milk [65]. Heat damages have been observed in donkey's milk on functional lipid compounds, which may also directly and indirectly influence gut environment and immunoinflammatory functions [66,67].

The recent advances of knowledge on the claimed nutraceutical properties, here summarized, suggest that, when scientifically demonstrated, the added value of the equine milk should be properly exploited in the dairy equine enterprise.

3. Dairy Equine Management and Nutrition

3.1. Equine Milk Yield and Management of the Dairy Equine Enterprise

The core of the dairy equine enterprise is related to the management of dams and foals, and of the milking practice, showing important differences from the conventional dairy species. Firstly, dams and foals live together until weaning, which occurs at 7 months (for foals) or later; dams won't start to be milked before 20 d from foaling [10,68]. Secondly, since the equine mammary gland is characterized by small volume, and milk is mainly alveolar [69], milk harvesting can be carried out many times per day. In the Steppes of Central Asia, mares are milked 4–5 times per day [70], while in more intensive dairy farms located in Europe, mares and jennies are frequently milked depending on consumer demand, up to eight times a day [10,62,71]. Milking is carried out at least 2 hours after foal separation from the mother [70,72]. This distinctive trait of the dairy equine enterprise introduced the neologism "milking session", i.e., the interval from foal separation up to the end of each milking [62]. It must be noted that milk ejection is not reported to be affected by the presence of the foal during milking in the dairy donkey farm [68], while it is recommended in the dairy horse farm for a complete oxytocin release [10]. In this regard, the selection for milkability of mares would greatly improve the milking routine, reducing the labor costs [73].

Milk harvested per milking session is reported to range within 500–2000 mL and 200–900 mL for mares and jennies, respectively [62,70,74–77], regardless of the milking technique used (mechanical or manual).

The available literature data on daily equine milk yield have been obtained under different methodological approaches, which partially explains the high variability of values reported in Table 2. The daily milk production is estimated to be 15–35 g kg^{-1} bodyweight [10,29,78,79]. However, literature data are inconclusive, as the value recently estimated for the dairy donkey, i.e., 12 g milk kg^{-1} body weight, shows [80]. Todini et al. [81] reported an average milk yield per milking of 2.68 mL kg^{-1} bodyweight.

Table 2. Daily milk yield (kg/d) reported in literature for horse and donkey from d30 to d180 of lactation [1].

Item	Horse	Donkey
Mean value	11.66	2.68
s.d.[2]	5.3	1.96
Min	3.9	0.72
Max	17.2	6

[1] Sources: Horse: [15,80,82,83], Donkey: [23,25,30,31,72,76–80,84]; [2] standard deviation.

Milk yield is affected by many factors, including the farming system, nutrition and feeding, strategy and type of milking (manual or mechanical), individual milkability, stage of lactation, and size and body condition of animals, besides genetics [11]. Because of the lack of standardized methodologies in equine milking studies, the effect of the breed on dairy performances of mares and jennies is not currently defined. According to Doreau and Martin Rosset [10], any breed can be milked, provided the animals accept the milking procedure.

The farming system is a major cause of the observed variability in equine milk production, as reported for pastoralist areas of the Steppes of Central Asia [70,85], or for more intensive systems, described for both koumiss and dairy donkey farms. In the latter, shelters are available on pasture, and milking is usually carried out in dedicated areas or facilities [72,83]. Donkeys raised under temperate conditions are reported to need more protection in rainy and windy weather than horses, as the results of the adaptation of donkeys to semi-arid environments of Africa vs. continental climate, and Eurasian Steppe environments where horses evolved [86]. The grazed area must be close to the milking site [70,79] so that the proximity of pasture represents a constraint in the dairy equine enterprise and management of milking influences the feeding strategy.

In intensive farming systems, the dairy mare and jenny are milked in ad hoc facilities equipped with sheep milking machines adapted to the equine mammary characteristics [15,68]. With trained animals and skilled operators, no difference was observed in the amount of milk harvested manually or mechanically per milking session, but milk microbial contamination can be reduced by the proper use of milking machine. This introduces a crucial aspect of the equine milk production and its commercialization, related to consumer safety.

3.2. Equine Milk: Hygiene and Health Issues

In Europe, equine milk is mainly commercialized at farm or by means of vending machines (raw milk), but it is also available at shops and supermarkets (pasteurized milk) or online (pasteurized and powdered milk) [22,71,87,88].

The risk associated with equine milk consumption is considered reasonably low when compared to bovine milk. The presence of pathogens, such as *Escherichia coli* O157, *Salmonella* spp., *Campylobacter* spp., *Yersinia enterocolitica*, *Brucella* spp., *Mycobacterium* spp., *Bacillus cereus*, *Cronobacter sakazakii*, *Streptococcus equi* subsp. *zooepidemicus*, *Rhodococcus equi*, *Streptococcus dysgalactiae* subsp. *equisimilis*, *Clostridium difficile,* and *Burkholderia mallei* is reported to be low [87]. However, the variable level of microbial contamination of equine raw milk, ranging from 3.0 to 5.87 log CFU mL^{-1} milk, warns against ineffective sanitization of equipment and facilities, as well as packaging and storing conditions of milk, even after thermal treatments [22,29,87].

For these reasons, while alternative processing for equine milk sanitation and shelf life extension are studied [22], thermal treatment is always recommended before consumption [87].

In regards to the mammary gland health status, the somatic cell count is reported to be below 50,000 cells mL^{-1} milk and mastitis is rarely observed in the dairy equine farm [29,87,89]. However, injuries or improper milking procedures reported for more intensive farming systems can affect the mammary health status [31,87].

Equine milk is gaining interest as an alternative food for sensitive consumers, so that high hygiene standards represents an important issue in the dairy equine enterprise, and it affects the labor costs for cleanliness of facilities, and areas frequented by the animals.

3.3. Feeding the Dairy Equine and Pasture Management

The nutritive value and the potential health-promoting properties of equine milk are related to the horse and donkey's metabolic utilization of the diet. These monogastric species and hindgut fermenter herbivores are reported to be better utilizers of metabolizable dietary energy than ruminants, at high levels of cell wall [90]. It is also well known that the dietary influence on milk composition is more direct in the equine species than in ruminants [10,90]. Regardless of the farming system, as already mentioned, the common denominator in the diet is the presence of forages and pasture, whose management is crucial for dairy equine production and welfare [71,72,74,91,92]. Because of the evolutionary history of the two equine species, their different feeding behavior and metabolism should also be considered in relation to nutrient requirements, management of feeding, and their impact on land preservation.

3.3.1. Feeding the Dairy Horse

According to Doreau and Martin-Rosset [10], no different approaches are required in feeding the dairy or nursing horse, as far as the energy and nitrogen requirements are concerned. The nutritional requirements of the lactating mare (600 kg body weight) are summarized in Table 3 [93]. To sustain the milk production, forages account on average for 50 to 80 percent of the dry matter of the diet, and they can supply 40 to 70 percent of the mare's annual nutritional requirements [5,94]. The dry matter intake of mares (Table 3) depends on the quality of the diet at foaling [93]. At the onset of lactation, the voluntary intake of mares is reported to be high (20–30 g dry matter per kg body weight). However, the dry matter intake is scarcely a limiting factor for the mare to meet nutritional requirements [93].

Table 3. Recommended nutrient requirements and intake for lactating mares (600 kg body weight) [93].

Lactation, Month	Milk Yield, kg d^{-1}	Horse Feed Units *, n d^{-1}	Horse Digestible Crude Protein **, g d^{-1}	Dry Matter Intake, kg d^{-1}
1st	18	10.1	1131	13.5–18.0
2nd	19.8	10.3	1091	15.0–19.0
3rd	19.2	9.6	1030	15.0–19.0
4th	17.4	9.1	844	13.5–18.0
5th	13.2	7.9	629	12.5–15.0
6th	12	7.6	603	10.5–13.0

* Horse Feed Units (UFC) = 9.42 MJ Net Energy; ** Horse Digestible Crude Protein (MADC).

The diet composition varies according to quality and availability of pasture and forages. In case of good grassland conditions, dairy mare foals generally in spring, just before they turn out, and use natural or sown pasture during the grazing season [95]. They are generally dried up in autumn (early October), after 190–210 days of lactation. During winter (110–120 days), the mares are fed a limited amount of hay of medium quality (organic matter digestibility, OMD = 50–55%) [93], and cereals, or a mixed diet based on straw, ad libitum, and hay of good quality (OMD = 55–60%). In case of harsh conditions, mare foals generally in early spring, one month before turning out. They graze pastures of uplands. They are dried off in autumn (late October), and grazed resources meet the requirements of animals over 9 months of lactation. In case of low productive areas, mares graze for about 60–70% of the total grazing season. In late autumn and early winter, mares graze refusals of cattle and sheep in the lowlands [93].

In the dairy horse enterprise, the strategy of the feeding system is based on pasture availability throughout the year and consists, generally, in matching the highest requirements of the animals with the maximum biomass production [96]. It must also include provisions of preserved feedstuffs to be used in case of particularly adverse climatic conditions. The main aim of the feeding strategy, notwithstanding the horse breed, is that dairy mares gain body weight in early lactation to nurse adequately the foal and to be rebred as soon as possible, to achieve a 12-month interval between two subsequent foalings [94]. Foals live with their mothers at pasture and they are allowed to nurse when mares are not milked. Table 4 shows the nutrient requirements of foals (600 kg of adult body weight) performing an optimal or moderate growth rate.

Table 4. Recommended nutrient requirements and intake for foals (600 kg adult body weight) at 3–6 months of age with a growth rate optimal or moderate [93].

Body Weight, kg	Gain, g d^{-1}	Horse Feed Units *, n d^{-1}	Horse Digestible Crude Protein **, g d^{-1}	Dry Matter Intake, kg d^{-1}
249	1000–1200	6	647	6.0–8.0
207	800–900	4.8	497	5.5–7.5

* Horse Feed Units (UFC) = 9.42 MJ Net Energy; ** Horse Digestible Crude Protein (MADC).

Mares should be managed at pasture with the aim to recover body weight and a proper body condition at drying off in late summer or fall, to ensure good nutritional conditions in pregnancy during winter [93]. Mares increase their body weight (+6–8%) during the last three months of pregnancy, as they are usually fed from 100% to 120% of their energy requirements, and during the first month after foaling (+3%) when they turn out in spring [93].

In extensive farming systems, grazing dairy mares should meet 80% of the total requirements over the 7-month lactation period. The animals use the vegetation regrowth from September until December. As already mentioned, in more intensive farming systems, hays or maize silage (30–35% dry matter content, 0.80–0.84 Horse Feed Unit per kg dry matter) and low concentrate supplementation are offered during winter.

3.3.2. Feeding the Dairy Donkey

Nutrient requirements and suggested allowances, nowadays available specifically for donkeys, are mainly devoted to working animals, i.e., used for transportation, small agricultural works, and equine therapy, and to animals at maintenance, i.e., companion animals, often castrated, hosted in international animal rescue charities, e.g., Donkey Sanctuary [72,79,97,98]. The available nutrient requirements of the dairy donkey are either based on results from one study on foal growth, or they are derived from domestic horse data, whose behavior and physiology are known to differ from those of the donkey. Consequently, they cannot be considered conclusive and need further investigation [20,72,79,98,99]. As reported by the US Research Council on equids, donkeys maximize their dry matter intake when good quality hay is offered [100]. For donkeys at maintenance, Raspa et al. [79] report a maximum dry matter intake per kg of body weight, declining from 32 to 12 g, with ad libitum diets based on either legume forages or barley straw. In order to prevent nutritional diseases, e.g., hyperlipemia and obesity, a diet high in fiber is suggested for companion donkeys at maintenance [99].

Because the mentioned lack of information in the specific literature on protein and energy requirements of lactating donkeys [20,72,79,98], the common strategy is represented by ad libitum administration of forage-based diets associated to a monthly evaluation of the body condition score [31,79].

In the dairy equine farming system, pasture should be always available for its positive effect on animal welfare and milk quality [79]; however, when jennies are milked, grazing time and quanti-qualitative availability of grazed areas are limited, as also observed for dairy mares [70]. According to preliminary results on the grazing behavior of Miranda breed jennies in mountain pastures, Couto et al. [101] observed that the activities in searching and prehension lasted, on average, 16 hours per day with a preferential intake of herbaceous species. However, up to 30% of the intake was represented by shrubs, probably because of a low grass availability. This suggests the interesting role of these autochthonous donkeys in preserving the pasture areas from degradation and fire risks [101]. It is also worth considering that grazing time does not significantly affect the daily dry matter intake of donkeys at maintenance, when they also have free access to preserved forages [102].

Results of a survey carried out on 12 dairy donkey farms in Italy confirm the inclusion of pasture in the lactating donkey diets always associated to hay administration [72], likely due to the limited availability of pasture. Cereals and/or mixed feeds, commercial or not, are also administered to lactating donkeys, and diets are frequently salt supplemented [31,72].

Other data from on field studies about milk production report a high feed intake (30–32 g dry matter per kg body weight) of dairy jennies at the first 3–4 months of lactation. Moreover, diets are characterized on average (on a dry matter basis) by a 70:30 forage-to-concentrate ratio, a protein content of 10–13 g per 100 g, and a digestible energy value of 8.5–10.0 MJ per kg [20].

After digestion, dietary fats, soluble carbohydrates, and proteins are mainly absorbed by the small intestine of equids. Due to the negligible biohydrogenation before absorption, the direct influence of the diet on the fatty acid composition of milk is expected. In this regard, the supplementation of the mares' diet in late pregnancy, and early lactation with eicosapentaenoic acid (EPA) and/or docosahexaenoic acid (DHA) did not affect the linoleic and linolenic milk content, but it increased the arachidonic acid, EPA, and DHA milk concentrations [103]. However, in jennies, the transfer of n3 polyunsaturated fatty acids (PUFAs) from blood to milk is reported to be more efficient than that of n6 PUFAs [20]. For a nutritionally correct ratio of n3:n6 PUFA in equine milk, dietary lipid sources should be evaluated with regard to the fatty acid profile. Dietary factors can also influence the palatability of donkey milk. 'Green' aromatic notes and related compounds have been identified in milk when jennies were fed fresh forage [20].

The survey by Dai et al. [72] reported that the diet always includes hay (100% of farms) and pasture (about 92% farms) in non-lactating jennies. Concentrates and salt supplements are also administered, but in a lower percentage than during lactation [72]. Stallions, which are either grouped with females or housed individually, are mainly fed hay supplemented with mixed feeds, cereals, and salt or additives. Pasture availability is reported for about 70% of farms [72].

Nutrient requirements for foals are not defined and only rare data are available on growing rates in donkeys [104]. The administration of milk replacement formulas to foals is not common in the dairy donkey farm [72]. However, a highly digestible creep feed is usually distributed to nursing foals until one month of age, when dams are not milked [20]. Later, complementary feeds are administered to foals until weaning (7–12 months of age) [72]. Constant access to clean water and salt blocks is highly recommended for both foals and dams [20].

The welfare status of the animals needs to be constantly monitored by recommended indicators, such as body condition score, and hydration score [105]. Vaccinations and deworming are also recommended in all animals, along with regular hoof, dental, and health care treatments [105], even if they are not reported to be common practices in dairy donkey farms [72].

3.4. Pasture in the Dairy Equine Enterprise

Different systems of grazing management are possible: extensive vs. semi-extensive, associated or not with ruminants [106]. Depending on the grazing species and their nutrient requirements, the correspondence between animals and characteristics of the forage availability (in quality and quantity) is crucial for a sustainable use of the landscape [107]. As herbivore species, horses and donkeys have the ability to exploit large amounts of fibrous forages, often of low nutritive value in less favored areas, available for grazing and/or foraging. The ability of the equids to produce in high forage feeding systems is mainly explained by their distinctive features in selecting, consuming, and digesting forages and grazed resources [102], so that seasonal variations in grazing behavior and diet selection have been observed in mares [108]. In particular, equids show several adaptive abilities in harsh conditions when the total nutrient requirements can be achieved on a long-term period [102].

In free-ranging conditions, horses spend up to 70% of their time to consume available food resources and only 30% for other activities. This ingestive activity is usually distributed over several meals during the day and grazing also occurs during the night [70,109]. Moreover, grazing time can increase in autumn and in winter, and the length of grazing is in relation to the cell wall content of the sward [110].

In high forage systems, pasture is the major source of nutrients for dams and foals along the breeding cycle. Based on the type and composition of the grazing species, as well as on the carrying capacity of the pasture, the sustainable grazing period ranges from 100 to 130 days (Northern Europe), to 230 days (Central Europe) [5]. Especially in marginal areas, the most relevant management of horses and donkeys at pasture implies the evaluation of the nutritive value of forages available for the optimal animal response to match economic profits, technical feasibility, and ecological sustainability [107].

Pasture productivity varies according to the geographical zone and the climatic conditions. In Europe, the grazing period is usually limited in Northern countries by short summers. In countries of Central Europe, generally characterized by extensive grazing lands and high quality forages, long grazing periods are observed, while the grass growth is usually depressed by summer dryness in Southern Europe [5]. Table 5 shows the average chemical composition of pasture in selected areas of Europe.

Table 5. Chemical components and estimated energy content of pasture in European areas. Values expressed on a dry matter basis [5,111].

Country	Crude Protein, g kg^{-1}	Crude Fiber, g kg^{-1}	Horse Feed Units *, n kg^{-1}
Finland	200–230	180–200	0.69–0.73
France, lowlands	131–168	244–276	0.76–0.82
France, uplands	111–166	223–304	0.66–0.92
Italy, lowlands	85–159	242–325	0.67–0.90
Italy, uplands	117–155	285–345	0.63–0.85

* Horse Feed Units (UFC) = 9.42 MJ Net Energy.

In addition, it should be noted that climate changes may affect the forage population dynamics, its nutritive value, as well as the growing and grazing seasons, so that different approaches in the management of land, animals, and forage resources may be required [112]. The forage intake depends on the quality of plant resources and their ingestibility, the time of grazing, the grazing activity, and the stocking rate, especially in multispecies herding situations. On this purpose, practical and flexible models have been studied for the assessment of a grazing pressure compatible with the conservation of pasture in less favored areas [113]. Moreover, the adoption of appropriate strategies is also recommended, such as rotational grazing, control of infesting species, safeguard from parasites diffusion in the sward, and fertilization (180 kg N ha^{-1}) [107,108,114].

In extensive systems, characterized by low quality and poor productivity of natural pastures, the total requirements can be met using low grazing intensity, with a stocking rate of 0.3–0.7 animal ha^{-1}, depending on the grass availability [106,114]. In more intensive systems of Central Europe, a concentrate supplementation is offered to horses, depending on their activity. Grass is plentiful until the beginning of July (beginning of the third vegetation cycle), then the production declines from mid-July to the end of August [94]. In Mediterranean regions, depending on the geographical area, the grazing season starts between April and May. A considerable reduction of the grass production is observed, depending on the variable rainfall in July, August, and early September. Then, up to the end of October, a regrowth of the grass can occur, offering the availability of fresh forages to foals in the weaning period [5].

The average growth of pasture grass during the grazing period in Italy is shown in Figures 1 and 2, respectively, for Central and Southern Italy pastures. It is interesting to note that in the pasture area located in Southern Italy, and in a mixed grazing system, including cattle, sheep goat, and horse, the sustainable stocking rate simulated in two subsequent years varied from 1.14 to 1.35 Adult Bovine Unit ha^{-1}, due to different climatic conditions and carrying capacities of the pasture [107].

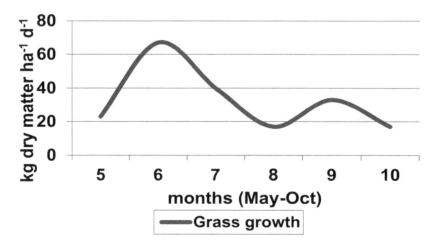

Figure 1. The average growth of pasture grass in Central Italy during the grazing period (modified from [94]).

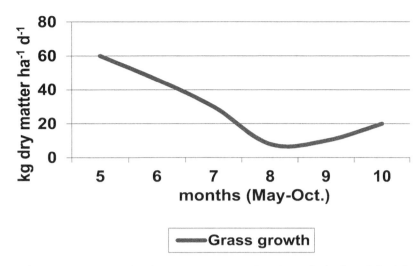

Figure 2. The average growth of pasture grass in Southern Italy (modified from [94]).

In order to achieve biodiversity and production goals in sustainable grazing systems of less favored areas, further management strategies include the reduction of the stocking rate, the periodic exclusion of the more degraded areas from grazing, the administration of complementary hay and concentrates, as well as the use of autochthonous breeds [115,116].

As a final consideration, the dairy equine enterprise, here described in its essential and promising traits, is the result of different environmental conditions, management strategies, and socio-economical aspects. Furthermore, no data on the evaluation of the economic impact of the dairy equine milk production are available in literature. However, besides labor, feeding, housing, and milking facilities, the evaluation of costs should also include those related to availability of infrastructures on pastures and marginal areas, and social costs of labor and bureaucracy, whose incidence can be relevant and different among countries [92]. Moreover, among the immaterial benefits, the impact of the dairy equine enterprise to environmental issues, such as landscape safeguard and biodiversity preservation, should also be included in a costs-to-benefits ratio evaluation [117], as also reported for horses used for tourism and work [3]. Alternatively, a price premium, based on environmental standards and labels, should be recognized to the products of the dairy equine enterprise.

4. Conclusions

The dairy enterprise involving equids, here discussed in its essential traits, represents a promising activity for the micro-economies of marginal areas around the world, because of its potentialities in human nutrition, biodiversity, and landscape preservation. Notwithstanding the advances of knowledge on milk nutritional and safety characteristics, as well as the improvement of technical skills in milk management, in depth studies are still required, especially in terms of animal nutrition and feeding. A better understanding on nutrient requirements of the dairy equid at pasture in heterogeneous and marginal areas will boost the interest toward endangered equine breeds, their milk, and their habitat. Positive relapses would in fact include the protection of plant diversity in the achievement of a productive and sustainable use of the landscape. Among the innovations for sustainable agriculture, the production of equine milk and derivatives with high nutritional value and health promoting properties should be therefore considered a promising extension of the equine industry for the modern and future society.

Author Contributions: Conceptualization, N.M., E.S. and F.F.; methodology, E.S., N.M. and F.F.; writing—original draft preparation, E.S., F.F. and N.M.; writing—review and editing, E.S. and F.F.; supervision, E.S., F.F. and N.M. All authors have read and agreed to the published version of the manuscript.

References

1.   Martin-Rosset, W. Research, development and transfer in Equine Science. In *EAAP Leroy Fellowship Award, Barcelona, Spain, 24–27 August 2009*; Wageningen Academic Publishers: Wageningen, The Netherlands, 2009.

2.   Miraglia, N. Sustainable development and equids in rural areas: An open challenge for the territory cohesion. In *EAAP Scientific Series*; Vial, C., Evans, R., Eds.; Wageningen Academic Publishers: Wageningen, The Netherlands, 2015; Volume 136, pp. 167–176. ISBN 978-90-8686-279-5.

3.   Rzekec, A.; Vial, C.; Bigot, G. Green assets of equines in the european context of the ecological transition of agriculture. *Animals* **2020**, *10*, 106. [CrossRef] [PubMed]

4.   Salimei, E.; Fantuz, F.; Coppola, R.; Chiofalo, B.; Polidori, P.; Varisco, G. Composition and characteristics of ass's milk. *Anim. Res.* **2004**, *53*, 67–78. [CrossRef]

5.   Miraglia, N.; Saastamoinen, M.; Martin-Rosset, W. Role of pastures in mares and foals management in Europe. In *Nutrition and Feeding of the Broodmare*; Miraglia, N., Martin-Rosset, W., Eds.; Academic Publishers: Wageningen, The Netherlands, 2006; Volume 120, pp. 279–297.

6.   Fleurance, G.; Duncan, P.; Mallevaud, B. Daily intake and the selection of feeding sites by horses in heterogeneous wet grasslands. *Anim. Res.* **2001**, *50*, 149–156. [CrossRef]

7.   Edouard, N.; Fleurance, G.; Martin-Rosset, W.; Duncan, P.; Dulphy, J.P.; Dubroeucq, H.; Grange, E.; Baumont, R.; Perez-Barberia, F.J.; Gordon, I.J. Voluntary intake and digestibility in horses: effect of forage quality with emphasis for individual variability. *Animal* **2008**, *2*, 1526–1533. [CrossRef]

8.   Van der Zijpp, A.; Boyazoglu, J.; Renaud, J.; Hoste, C. *Research Strategy for Animal Production in Europe in the 21st Century*; Wageningen Press: Wageningen, The Netherlands, 1993; Volume 64, p. 163.

9.   Yachi, S.; Loreau, M. Biodiversity and ecosystem productivity in a fluctuating environment: The insurance hypothesis. *Proc. Natl. Acad. Sci. USA* **1999**, *96*, 1463–1468. [CrossRef]

10. Doreau, M.; Martin-Rosset, W. Animals that produce dairy foods – horse. In *Encyclopedia of Dairy Sciences*, 2nd ed.; Fuquay, J.W., Fox, P.F., McSweeney, P.L.H., Eds.; Elsevier Academy Press: London, UK, 2011; Volume 1, pp. 358–364.

11. Salimei, E.; Fantuz, F. Horse and donkey milk. In *Milk and Dairy Products in Human Nutrition: Production, Composition and Health*; Park, Y.W., Haenlein, G.F.W., Eds.; John Wiley & Sons Ltd: Oxford, UK, 2013; pp. 594–613.

12. McLean, A.K.; Navas Gonzalez, F.J. Can scientists influence donkey welfare? Historical perspective and a contemporary view. *J. Equine Vet. Sci.* **2018**, *65*, 25–32. [CrossRef]

13. Adams, F. *The Genuine Works of Hippocrates*; The Sydenham Society: London, UK, 1849; Volume 1.

14. Mustoxidi, A. *Le Nove Muse di Erodoto Alicarnasseo*, 2nd ed.; Sonzogno Publisher: Milan, Italy, 1822.

15. Salimei, E.; Park, Y.W. Mare milk. In *Handbook of Milk of Non-Bovine Mammals*, 2nd ed.; Park, Y.W., Haenlein, G.F.W., Wendorff, W.L., Eds.; John Wiley & Sons Ltd.: Oxford, UK, 2017; pp. 369–408.

16. Langlois, B. The history, ethnology and social importance of mare's milk consumption in Central Asia. *J. Life Sci.* **2011**, *5*, 863–872.

17. Bimbetov, B.; Zhangabylov, A.; Aitbaeva, S.; Benberin, V.; Zollmann, H.; Musaev, A.; Rakhimzhanova, M.; Esnazarova, G.; Bakytzhanuly, A.; Malaeva, N. Mare's milk: Therapeutic and dietary properties. *Bull. Natl. Acad. Sci. Rep. Kazakhstan* **2019**, *3*, 52–58. [CrossRef]

18. Romaniuk, K.; Majszyk-Świątek, M.; Kryszak, K.; Danielewicz, A.; Andraszek, K. Alternative use of mare milk. *Folia Pomer. Univ. Technol. Stetin.* **2019**, *348*, 121–130. [CrossRef]

19. Li, L.; Liu, X.; Guo, H. The nutritional ingredients and antioxidant activity of donkey milk and donkey milk powder. *Food Sci. Biotechnol.* **2018**, *27*, 393–400. [CrossRef]

20. Salimei, E. Animals that Produce Dairy Foods: Donkey. In *Reference Module in Food Sciences*, 1st ed.; Elsevier Academic Press: London, UK, 2016; pp. 1–10.

21. Drogoul, C.; Prevost, H.; Maubois, J.L. Le lait de juments un produit. Une filiere a developer? In *Quoi de Neuf en Matiere d'Etudes de Recherches sur le Cheval, 18eme Journee d'Etude, 4 Mars*; CEREOPA: Paris, France, 1992; pp. 37–51.

22. Giacometti, F.; Bardasi, L.; Merialdi, G.; Morbarigazzi, M.; Federici, S.; Piva, S.; Serraino, A. Shelf life of donkey milk subjected to different treatment and storage conditions. *J. Dairy Sci.* **2016**, *99*, 4291–4299. [CrossRef] [PubMed]

23. Bordonaro, S.; Dimauro, C.; Criscione, A.; Marletta, D.; Macciotta, N.P.P. The mathematical modeling of the lactation curve for dairy traits of the donkey (*Equus asinus*). *J. Dairy Sci.* **2013**, *96*, 4005–4014. [CrossRef] [PubMed]

24. Camillo, F.; Rota, A.; Biagini, L.; Tesi, M.; Fanelli, D.; Panzani, D. The current situation and trend of donkey industry in Europe. *J. Equine Vet. Sci.* **2018**, *65*, 44–49. [CrossRef]

25. Cosentino, C.; Paolino, R.; Musto, M.; Freschi, P. Innovative use of jenny milk from sustainable rearing. In *The Sustainability of Agro-Food and Natural Resource Systems in the Mediterranean Basin*; Vastola, A., Ed.; Springer International Publishing: Cham, Switzerland, 2015; pp. 113–132.

26. Uniacke-Lowe, T.; Fox, P.F. Milk, Equid milk. In *Encyclopedia of Dairy Sciences*, 2nd ed.; Fuquay, J.W., Fox, P.F., McSweeney, P.L.H., Eds.; Elsevier Academy Press: London, UK, 2011; Volume 3, pp. 518–529.

27. Mansueto, P.; Iacono, G.; Seidita, A.; D'Alcamo, A.; Iacono, S.; Carroccio, A. Ass's milk in allergy to cow's milk protein: A review. *J. Food Allergy* **2012**, *1*, 181–191.

28. Fantuz, F.; Salimei, E.; Papademas, P. Macro- and micronutrients in non-cow milk and products and their impact on human health. In *Non-Bovine Milk and Milk Products*, 1st ed.; Tsakalidou, E., Papadimitriou, K., Eds.; Elsevier Academic Press: London, UK, 2016; pp. 209–261.

29. Kaić, A.; Luštrek, B.; Simčič, M.; Potočnik, K. Milk quantity, composition and hygiene traits of routinely machine milked Lipizzan mares. *Slov. Vet. Res.* **2019**, *56*, 115–123.

30. Aroua, M.; Jemmali, B.; Said, S.B.; Kbaier, H.B.H.; Mahouachi, M. Physicochemical properties of north African donkey milk. *Agric. Res. Tech. Open Access J.* **2019**, *21*, 1–3.

31. Malacarne, M.; Criscione, A.; Franceschi, P.; Bordonaro, S.; Formaggioni, P.; Marletta, D.; Summer, A. New insights into chemical and mineral composition of donkey milk throughout nine months of lactation. *Animals* **2019**, *9*, 1161. [CrossRef]

32. Fantuz, F.; Maglieri, C.; Lebboroni, G.; Salimei, E. Ca, Mg, Zn, Fe, Cu and Mn content of ass's milk. *It. J. Anim. Sci.* **2009**, *8*, 703–705. [CrossRef]

33. Fantuz, F.; Ferraro, S.; Todini, L.; Piloni, R.; Mariani, P.; Salimei, E. Donkey milk concentration of Calcium, Phosphorus, Potassium, Sodium and Magnesium. *Int. Dairy J.* **2012**, *24*, 143–145. [CrossRef]

34. Fantuz, F.; Ferraro, S.; Todini, L.; Mariani, P.; Piloni, R.; Salimei, E. Essential trace elements in milk and blood serum of lactating donkeys as affected by lactation stage and dietary supplementation with trace elements. *Animal* **2013**, *7*, 1893–1899. [CrossRef]

35. Fantuz, F.; Ferraro, S.; Todini, L.; Piloni, R.; Mariani, P.; Malissiova, E.; Salimei, E. Minor and potentially toxic trace elements in milk and blood serum of dairy donkeys. *J. Dairy Sci.* **2015**, *98*, 5125–5132. [CrossRef] [PubMed]

36. Pieszka, M.; Luszczyński, J.; Zamachowska, M.; Augustyn, R.; Dlugosz, B.; Hędrzak, M. Is mare milk an appropriate food for people?—A review. *Ann. Anim. Sci.* **2016**, *16*, 33–51. [CrossRef]

37. Martini, M.; Altomonte, I.; Licitra, R.; Salari, F. Short communication: Technological and seasonal variations of vitamin D and other nutritional components in donkey milk. *J. Dairy Sci.* **2018**, *101*, 1–5. [CrossRef] [PubMed]

38. Iacono, G.; Carroccio, A.; Cavataio, F.; Montaldo, G.; Soresi, M.; Balsamo, V. Use of ass's milk in multiple food allergy. *J. Ped. Gastroent. Nutr.* **1992**, *14*, 177–181. [CrossRef]

39. Businco, L.; Giampietro, P.G.; Lucenti, P.; Lucaroni, F.; Pini, C.; Di Felice, G.; Iacovacci, P.; Curadi, C.; Orlandi, M. Allergenicity of mare's milk in children with cow's milk allergy. *J. Allergy Clin. Immunol.* **2000**, *105*, 1031–1034. [CrossRef]

40. Sarti, L.; Martini, M.; Brajon, G.; Barni, S.; Salari, F.; Altomonte, I.; Ragona, G.; Mori, F.; Pucci, N.; Muscas, G.; et al. Donkey's milk in the management of children with cow's milk protein allergy: nutritional and hygienic aspects. *It. J. Ped.* **2019**, *45*, 102–110. [CrossRef]

41. Restani, P.; Ballabio, C.; Di Lorenzo, C.; Tripodi, S.; Fiocchi, A. Molecular aspects of milk allergens and their role in clinical events. *Anal. Bioanal. Chem.* **2009**, *395*, 47–56. [CrossRef]

42. Giovannini, M.; D'Auria, E.; Caffarelli, C.; Verduci, E.; Barberi, S.; Indinnimeo, L.; Dello Iacono, I.; Martelli, A.; Riva, E.; Bernardini, R. Nutritional management and follow up of infants and children with food allergy: Italian Society of Pediatric Nutrition/Italian Society of Pediatric Allergy and Immunology Task Force Position Statement. *It. J. Ped.* **2014**, *40*, 1–9. [CrossRef]

43. Uniacke-Lowe, T.; Huppertz, T.; Fox, P.F. Equine milk proteins: chemistry, structure and nutritional significance. *Int. Dairy J.* **2010**, *20*, 609–629. [CrossRef]

44. Cunsolo, V.; Saletti, R.; Muccilli, V.; Gallina, S.; Di Francesco, A.; Foti, S. Proteins and bioactive peptides from donkey milk: the molecular basis for its reduced allergenic properties. *Food Res. Int.* **2017**, *99*, 41–57. [CrossRef]

45. Cieslak, J.; Wodas, L.; Borowska, A.; Sadoch, J.; Pawlak, P.; Puppel, K.; Kuczynska, B.; Mackowski, M. Variability of lysozyme and lactoferrin bioactive protein concentrations in equine milk in relation to LYZ and LTF gene polymorphisms and expression. *J. Sci. Food Agric.* **2017**, *97*, 2174–2181. [CrossRef] [PubMed]

46. Wodas, Ł.; Maćkowski, M.; Borowska, A.; Pawlak, P.; Puppel, K.; Kuczyńska, B.; Czyżak-Runowska, G.; Wójtowski, J.; Cieślak, J. 5′-flanking variants of the equine α-lactalbumin (LALBA) gene–relationship with gene expression and mare's milk composition. *J. Anim. Feed Sci.* **2018**, *27*, 317–326. [CrossRef]

47. Cieslak, J.; Wodas, L.; Borowska, A.; Pawlak, P.; Czyzak-Runowska, G.; Wojtowski, J.; Puppel, K.; Kuczynska, B.; Mackowski, M. 5′-flanking variants of equine casein genes (CSN1S1, CSN1S2, CSN2,CSN3) and their relationship with gene expression and milk composition. *J. Appl. Genet.* **2019**, *60*, 71–78. [CrossRef] [PubMed]

48. Ballard, O.; Morrow, A.L. Human milk composition: nutrients and bioactive factors. *Pediatr. Clin. N. Am.* **2013**, *60*, 49–74. [CrossRef]

49. Bonnet, M.; Delavaud, C.; Laud, K.; Gourdou, I.; Leroux, C.; Djiane, J.; Chilliard, Y. Mammary leptin synthesis, milk leptin and their putative physiological roles. *Reprod. Nutr. Dev.* **2002**, *42*, 399–413. [CrossRef]

50. Inglingstad, R.A.; Devold, T.G.; Eriksen, E.K.; Holm, H.; Jacobsen, M.; Liland, K.H.; Rukke, E.O.; Vegarud, G.E. Comparison of the digestion of caseins and whey proteins in equine, bovine, caprine and human milk by human gastrointestinal enzymes. *Dairy Sci. Technol.* **2010**, *90*, 549–563. [CrossRef]

51. Tidona, F.; Criscione, A.; Devold, T.G.; Bordonaro, S.; Marletta, D.; Vegarud, G.E. Protein composition and micelle size of donkey milk with different protein patterns: Effects on digestibility. *Int. Dairy J.* **2014**, *35*, 57–62. [CrossRef]

52. Martini, M.; Salari, F.; Licitra, R.; La Motta, C.; Altomonte, I. Lysozyme activity in donkey milk. *Int. Dairy J.* **2019**, *96*, 98–101. [CrossRef]

53. Yvon, S.; Schwebel, L.; Belahcen, L.; Tormo, H.; Peter, M.; Haimoud-Lekhal, D.A.; Eutamene, H.; Jard, G. Effects of thermized donkey milk with lysozyme activity on altered gut barrier in mice exposed to water-avoidance stress. *J. Dairy Sci.* **2019**, *102*, 7697–7706. [CrossRef]

54. Claeys, W.L.; Verraes, C.; Cardoen, S.; De Block, J.; Huyghebaert, A.; Raes, K.; Dewettinck, K.; Herman, L. Consumption of raw or heated milk from different species: an evaluation of the nutritional and potential health benefits. *Food Control* **2014**, *42*, 188–201. [CrossRef]

55. Todini, L.; Salimei, E.; Malfatti, A.; Ferraro, S.; Fantuz, F. Thyroid hormones in milk and blood of lactating donkeys as affected by stage of lactation and dietary supplementation with trace elements. *J. Dairy Res.* **2012**, *79*, 232–237. [CrossRef] [PubMed]

56. Brumini, D.; Criscione, A.; Bordonaro, S.; Vegarud, G.E.; Marletta, D. Whey proteins and their antimicrobial properties in donkey milk: A brief review. *Dairy Sci. Technol.* **2016**, *96*, 1–14. [CrossRef]

57. Fotschki, J.; Szyc, A.M.; Laparra, J.M.; Markiewicz, L.H.; Wróblewska, B. Immune-modulating properties of horse milk administered to mice sensitized to cow milk. *J. Dairy Sci.* **2016**, *99*, 9395–9404. [CrossRef] [PubMed]

58. Kushugulova, A.; Kozhakhmetov, S.; Sattybayeva, R.; Nurgozhina, A.; Ziyat, A.; Yadav, H.; Marotta, F. Mare's milk as a prospective functional product. *Funct. Food Health Dis.* **2018**, *8*, 537–543. [CrossRef]

59. Aspri, M.; Leni, G.; Galaverna, G.; Papademas, P. Bioactive properties of fermented donkey milk, before and after in vitro simulated gastrointestinal digestion. *Food Chem.* **2018**, *268*, 476–484. [CrossRef]

60. Roth-Walter, F.; Berin, M.C.; Arnaboldi, P.; Escalante, C.R.; Dahan, S.; Rauch, J.; Jensen-Jarolim, E.; Mayer, L. Pasteurization of milk proteins promotes allergic sensitization by enhancing uptake through Peyer's patches. *Allergy* **2008**, *63*, 882–890. [CrossRef]

61. Newburg, D.S.; Walker, W.A. Protection of the neonate by the innate immune system of developing gut and of human milk. *Pediatr. Res.* **2007**, *61*, 1–8. [CrossRef]

62. Salimei, E.; Fantuz, F. Equid milk for human consumption. *Int. Dairy J.* **2012**, *24*, 130–142. [CrossRef]

63. Martini, M.; Altomonte, I.; Manica, E.; Salari, F. Changes in donkey milk lipids in relation to season and lactation. *J. Food Compos. Anal.* **2015**, *41*, 30–34. [CrossRef]

64. Jirillo, F.; Magrone, T. Anti-inflammatory and anti-allergic properties of donkey's and goat's milk. *Endocr. Metab. Immune* **2014**, *14*, 27–37. [CrossRef]

65. Santillo, A.; Figliola, L.; Ciliberti, M.G.; Caroprese, M.; Marino, R.; Albenzio, M. Focusing on fatty acid profile in milk from different species after in vitro digestion. *J. Dairy Res.* **2018**, *85*, 257–262. [CrossRef] [PubMed]

66. Laiho, K.; Ouwehand, A.; Salminen, S.; Isolauri, E. Inventing probiotic functional foods for patients with allergic disease. *Ann. Allergy Asthma Immunol.* **2002**, *89*, 75–82. [CrossRef]

67. Sorrentino, E.; Salimei, E.; Succi, M.; Gammariello, D.; Di Criscio, T.; Panfili, G.; Coppola, R. Heat treatment of ass's milk, a hypoallergenic food for infancy. In *Technological Innovation and Enhancement of Marginal Products*; Severini, C., DePilli, T., Giuliani, R., Eds.; Claudio Grezi Editore: Foggia, Italy, 2005; pp. 569–574.

68. Salimei, E. Animals that produce dairy foods – donkey. In *Encyclopedia of Dairy Sciences*, 2nd ed.; Fuquay, J.W., Fox, P.F., McSweeney, P.L.H., Eds.; Elsevier Academy Press: London, UK, 2011; Volume 1, pp. 365–373.

69. Dzidic, A.; Knopf, L.; Bruckmaier, R.M. Oxytocin release and milk removal in machine-milked mares. *Milchwiss. Milk Sci. Int.* **2002**, *57*, 423–424.

70. Bat-Oyun, T.; Ito, T.Y.; Purevdorj, Y.; Shinoda, M.; Ishii, S.; Buho, H.; Morinaga, Y. Movements of dams milked for fermented horse milk production in Mongolia. *Anim. Sci. J.* **2018**, *89*, 219–226. [CrossRef] [PubMed]

71. Naert, L.; Vandevyvere, B.; Verhoeven, G.; Duchateau, L.; De Smet, S.; Coopman, F. Assessing heterogeneity of the composition of mare's milk in Flanders. *Vlaams Diergenskund. Tijds.* **2013**, *82*, 23–30.

72. Dai, F.; Segati, G.; Dalla Costa, E.; Burden, F.; Judge, A.; Minero, M. Management practices and milk production in dairy donkey farms distributed over the italian territory. *Mac. Vet. Rev.* **2017**, *40*, 131–136. [CrossRef]

73. Brinkmann, J.; Jagannathan, V.; Drogenmuller, C.; Rieder, S.; Leeb, T.; Thaller, G.; Tetens, J. Genetic variability of the equine casein genes. *J. Dairy Sci.* **2016**, *99*, 5486–5497. [CrossRef]

74. Bayle-Labouré, J. Approche Technico-économique de l'opportunité de Développement d'une Filière "Lait de Jument Comtoise". Thesis Ingénieure Spécialité Agriculture, Établissement National d'Enseignement Supérieur Agronomique de Dijon (ENESAD), Dijon, France, 2007; p. 154.

75. Mazhitova, A.T.; Kulmyrzaev, A.A.; Ozbekova, Z.E.; Bodoshev, A. Amino acid and fatty acid profile of the mare's milk produced on Suusamyr pastures of the Kyrgyz Republic during lactation period. *Procedia Soc. Behav. Sci.* **2015**, *195*, 2683–2688. [CrossRef]

76. De Palo, P.; Maggiolino, A.; Albenzio, M.; Caroprese, M.; Centoducati, P.; Tateo, A. Evaluation of different habituation protocols for training dairy jennies to the milking parlor: effect on milk yield, behavior, heart rate and salivary cortisol. *Appl. Anim. Behav. Sci.* **2018**, *204*, 72–80. [CrossRef]

77. Salari, F.; Ciampolini, R.; Mariti, C.; Millanta, F.; Altomonte, I.; Licitra, R.; Auzino, B.; D' Ascenzi, C.; Bibbiani, C.; Giuliotti, L.; et al. A multi-approach study of the performance of dairy donkey during lactation: Preliminary results. *It. J. Anim. Sci.* **2019**, *18*, 1135–1141. [CrossRef]

78. Centoducati, P.; Maggiolino, A.; De Palo, P.; Tateo, A. Application of Wood's model to lactation curve of Italian Heavy Draft horse mares. *J. Dairy Sci.* **2012**, *95*, 5770–5775. [CrossRef] [PubMed]

79. Raspa, F.; Cavallarin, L.; McLean, A.K.; Bergero, D.; Valle, E. A review of the appropriate nutrition welfare criteria of dairy donkeys: nutritional requirements, farm management requirements and animal-based indicators. *Animals* **2019**, *9*, 315. [CrossRef] [PubMed]

80. De Palo, P.; Maggiolino, A.; Centoducati, P.; Calzaretti, G.; Milella, P.; Tateo, A. Equid milk production: evaluation of Martina Franca jennies and IDH mares by Wood's model application. *Anim. Prod. Sci.* **2017**, *57*, 2110–2116. [CrossRef]

81. Todini, L.; Salimei, E.; Malfatti, A.; Brunetti, V.L.; Fantuz, F. Thyroid hormones in donkey blood and milk: correlations with milk yield and environmental temperatures. *It. J. Anim. Sci.* **2015**, *14*, 596–601. [CrossRef]

82. Akimbekov, A.R.; Baymukanov, D.A.; Yuldashbaev, Y.A.; Iskhan, K.Z. Productive qualities of the Seleti factory-type Kazakh horse of the toad. *Bull. Natl. Acad. Sci. Rep. Kazakhstan* **2017**, *3*, 100–110.

83. Iskhan, K.Z.; Akimbekov, A.R.; Baimukanov, A.D.; Aubakirov, K.A.; Karynbayev, A.K.; Rzabayev, T.S.; Mukhatai, G.; Dzhunusova, R.Z.; Apeev, K.B. Dairy productivity of the kazakh horse mares and their cross breeds with roadsters. *Bull. Natl. Acad. Sci. Rep. Kazakhstan* **2019**, *3*, 22–35. [CrossRef]

84. Muhatai, G.G.; Chen, L.; Rugoho, I.; Xiao, G.; Chen, G.; Hodge, S.; Zhou, S. Effect of parity, milking time and stage of lactation on milk yield of Jiangyue donkey (*Equus asinus*) in North West China. *J. Dairy Res.* **2017**, *84*, 23–26. [CrossRef]

85. Minjidgorj, N.; Baldorj, O.; Austbø, D. Chemical composition of Mongolian mare milk. *Acta Agric. Scand. Sect. A-Anim. Sci.* **2012**, *62*, 66–72.

86. Proops, L.; Osthaus, B.; Bell, N.; Long, S.; Hayday, K.; Burden, F. Shelter-seeking behavior of donkeys and horses in a temperate climate. *J. Vet. Behav.* **2019**, *32*, 16–23. [CrossRef]

87. Colavita, G.; Amadoro, C.; Rossi, F.; Fantuz, F.; Salimei, E. Hygienic characteristics and microbiological hazard identification in horse and donkey raw milk. *Vet. It.* **2016**, *52*, 21–29.

88. Dai, F.; Dalla Costa, E.; Burden, F.; Judge, A.; Minero, M. The development of guidelines to improve dairy donkey management and welfare. *It. J. Anim. Sci.* **2019**, *18*, 189–193. [CrossRef]

89. Cieslak, J.; Mackowski, M.; Czyzak-Runowska, G.; Wojtowski, J.; Puppel, K.; Kuczynska, B.; Pawlak, P. Screening for the most suitable reference genes for gene expression studies in equine milk somatic cells. *PLoS ONE* **2015**, *10*, e0139688.

90. Van Soest, P.J. *Nutritional Ecology of the Ruminant*, 2nd ed.; Cornell University Press: Ithaca, NY, USA; London, UK, 1994; p. 476.

91. Cavallarin, L.; Giribaldi, M.; Soto-Del Rio, M.D.; Valle, E.; Barbarino, G.; Gennero, M.S.; Civera, T. A survey on the milk chemical and microbiological quality in dairy donkey farms located in NorthWestern Italy. *Food Control* **2015**, *50*, 230–235. [CrossRef]

92. Salimei, E.; Malissiova, E.; Papademas, P.; Colavita, G.; Galaverna, G.; Fletouris, D.; Manouras, A.; Habib, I.; Šarić, L.; Budak S.O.; et al. Donkey milk and dairy donkey farming in Mediterranean Countries: current situation, challenges and prospects. In Proceedings of the 7th IDF International Symposium on Sheep, Goat and Other Non-Cow Milk, Limassol, Cyprus, 23–25 March 2015.

93. INRA. *Equine Nutrition. INRA Nutrient Requirements, Recommended Allowances and Feed Tables*; Martin-Rosset, W., Ed.; Wageningen Academic Press: Wageningen, The Netherlands, 2015; p. 691.

94. Miraglia, N.; Burger, D.; Kapron, M.; Flanagan, J.; Langlois, B.; Martin-Rosset, W. Local animal resources and products in sustainable development: Role and potential of equids. In *Product Quality Based on Local Resources Leading to Improve Sustainability*; Rubino, R., Scpc, L., Dimitriadou, A., Gibon, A., Eds.; Wageningen Academic Publishers: Wageningen, The Netherlands, 2006; pp. 217–233.

95. Martin-Rosset, W.; Trillaud-Geyl, C.; Jussiaux, M.; Agabriel, J.; Loiseau, P.; Beranger, C. Exploitation du pâturage par le cheval en croissance ou à l'engrais. In *Le Cheval. Reproduction, sélection, Alimentation, Exploitation*; Jarrige, R., Martin-Rosset, W., Eds.; INRA: Paris, France, 1984; pp. 583–599.

96. Moulin, C. Le pâturage du cheval: questions posées par les pratiques d'éleveurs. (Grazing by horses: From farm practices to technical questions). *Fourrages* **1997**, *149*, 37–54.

97. Burden, F.; Thiemann, A. Donkeys Are Different. *J. Equine Vet. Sci.* **2015**, *35*, 376–382. [CrossRef]

98. Martin-Rosset, W. Donkey nutrition and feeding: Nutrient requirements and recommended allowances—A review and prospect. *J. Equine Vet. Sci.* **2018**, *65*, 75–85. [CrossRef]

99. Burden, F.A.; Bell, N. Donkey nutrition and malnutrition. *Vet. Clin. Equine* **2019**, *35*, 469–479. [CrossRef]

100. NRC. Donkeys and other equids. In *Nutrient Requirements of Horses*, 6th ed.; National Research Council, Ed.; National Academies Press: Washington, DC, USA, 2007; pp. 268–279.

101. Couto, M.; Santos, A.S.; Laborda, J.; Nóvoa, M.; Ferreira, L.M.; Madeira de Carvalho, L.M. Grazing behavior of Miranda donkeys in a natural mountain pasture and parasitic level changes. *Livest. Sci.* **2016**, *186*, 16–21. [CrossRef]

102. Wood, S.J.; Smith, D.G.; Morriss, C.J.; Oliver, J.; Cuddeford, D. The effect of pasture restriction on dry matter intake of foraging donkeys in the United Kingdom. In *Forages and Grazing in Horse Nutrition*; Saastamoinen, M., Fradinho, M.J., Santos, A.S., Miraglia, N., Eds.; Wageningen Academic Publishers: Wageningen, The Netherlands, 2012; pp. 163–176.

103. Kouba, J.M.; Burns, T.A.; Webel, S.K. Effect of dietary supplementation with long-chain n-3 fatty acids during late gestation and early lactation on mare and foal plasma fatty acid composition, milk fatty acid composition, and mare reproductive variables. *Anim. Reprod. Sci.* **2019**, *203*, 33–44. [CrossRef]

104. De Palo, P.; Maggiolino, A.; Milella, P.; Centoducati, N.; Papaleo, A.; Tateo, A. Artificial suckling in Martina Franca donkey foals: effect on in vivo performances and carcass composition. *Trop. Anim. Health Prod.* **2016**, *48*, 167–173. [CrossRef]

105. Awin Welfare Assessment Protocol for Donkeys, University of Milan (Italy) and Donkey Sanctuary (UK). Available online: https://air.unimi.it/retrieve/handle/2434/269100/384805/AWINProtocolDonkeys.pdf (accessed on 15 October 2019).

106. Micol, D.; Martin-Rosset, W. Feeding systems for horses on high forage diets in the temperate zone. In *Recent*

Developments in the Nutrition of Herbivores; Journet, M., Grenet, E., Farce, M.-H., Thériez, M., Demarquilly, C., Eds.; INRA Editions: Paris, France, 1995; pp. 569–584.

107. Fatica, A.; Circelli, L.; DiIorio, E.; Colombo, C.; Crawford, T.W.; Salimei, E. Stresses in pasture areas in South-Central Apennines, Italy, and evolution at landscape level. In *Handbook of Plant & Crop Stress*, 4th ed.; Pessarakli, M., Ed.; CRC Press Taylor and Francis Group: Boca Raton, FL, USA, 2019; pp. 271–291.

108. Ferreira, L.M.M.; Celaya, R.; Benavides, R.; Jáuregui, B.M.; García, U.; Santos, A.S.; García, R.R.; Rodrigues, M.A.M.; Osoro, K. Foraging behavior of domestic herbivore species grazing on heathlands associated with improved pasture areas. *Livest. Sci.* **2013**, *155*, 373–383. [CrossRef]

109. Doreau, M.; Moretti, C.; Martin-Rosset, W. Effect of quality of hay given to mares around foaling on their voluntary intake and foal growth. *Ann. Zootech.* **1990**, *39*, 125–131. [CrossRef]

110. Duncan, P. *Horses and Grasses: The Nutritional Ecology of Equids and Their Impact on the Camargue*; Springer: New York, NY, USA, 1992; p. 287.

111. Parrini, S. Caratterizzazione nutrizionale delle erbe dei pascoli naturali e impiego di metodi di valutazione innovativi. Ph.D. Thesis (28th cycle), Università degli Studi di Firenze, Florence, Italy, December 2015.

112. Hristov, A.N.; Degaetano, A.T.; Rotz, C.A.; Hoberg, E.; Skinner, R.H.; Felix, T.; Li, H.; Patterson, P.H.; Roth, G.; Hall, M.; et al. Climate change effects on livestock in the Northeast US and strategies for adaptation. *Clim. Chang.* **2018**, *146*, 33–45. [CrossRef]

113. Pulina, G.; Salimei, E.; Masala, G.; Sikosana, J. A computerised spreadsheet model for the assessment of sustainable stocking rate in semiarid and subhumid regions. *Livest. Prod. Sci.* **1999**, *61*, 287–299. [CrossRef]

114. Miraglia, N.; Polidori, M.; Salimei, E. A review on feeding strategies, feeds and management of equines in Central-Southern Italy. In *Working Animals in Agriculture and Transport*; Pearson, R.A., Lhoste, P., Saastamoinen, M., Martin-Rosset, W., Eds.; Wageningen Academic Publishers: Wageningen, The Netherlands, 2003; pp. 103–112.

115. Dumont, B.; Rook, A.J.; Coran, C.; Rover, K.U. Effects of livestock breed and grazing intensity on biodiversity and production in grazing systems. 2. Diet selection. *Grass and Forage Sci.* **2007**, *62*, 159–171. [CrossRef]

116. Dumont, B.; Garel, J.P.; Ginane, C.; Decuq, F.; Farruggia, A.; Pradel, P.; Rigolot, C.; Petit, M. Effect of cattle grazing a species-rich mountain pasture under different stocking rates on the dynamics of diet selection and sward structure. *Animal* **2007**, *1*, 1042–1052. [CrossRef]

117. Bele, B.; Norderhaug, A.; Sickel, H. Localized agri-food systems and biodiversity. *Agriculture* **2018**, *8*, 22. [CrossRef]

12

The Effect of Diet Composition on the Digestibility and Fecal Excretion of Phosphorus in Horses: A Potential Risk of P Leaching?

Markku Saastamoinen [1,*], Susanna Särkijärvi [1] and Elisa Valtonen [2]

[1]   Production Systems, Natural Resources Institute Finland (Luke), FI-31600 Jokioinen, Finland;
       susanna.särkijärvi@luke.fi
[2]   Department of Animal Science, University of Helsinki, FI-00790 Helsinki, Finland;
       elisa.mj.valtonen@gmail.com
*   Correspondence: markku.saastamoinen@luke.fi

Simple Summary: This study aimed to examine phosphorus utilization and excretion in feces when typical feeds and forage-based diets are fed. The hypothesis was that feeding regimes might influence phosphorus digestibility and excretion in feces, and therefore the environmental impact of horse husbandry. We also studied the nutrient digestibilities of the diets, as well as the proportion of the soluble fraction of P of the total phosphorus. Horse dung may pose a potential risk of P run off into the environment if not properly managed. Supplementation with inorganic P should be controlled in the diets of mature horses in light work to decrease the excretion of P in feces.

Abstract: The main horse phosphorus excretion pathway is through the dung. Phosphorus originating from animal dung and manure has harmful environmental effects on waters. The number of horses has increased in many countries, and several studies have pointed that leaching of P from horse paddocks and pastures are hotspots for high P leaching losses. The hypothesis was that feeding regimes might influence phosphorus digestibility and excretion in feces, and therefore the environmental impact of horse husbandry. A digestibility experiment was conducted with six horses fed six forage-based diets to study phosphorus utilization and excretion in feces. The study method was a total collection of feces. The experimental design was arranged as an unbalanced 6 × 4 Latin Squares. Phosphorus intake increased with an increasing concentrate intake. All studied diets resulted in a positive P balance and, the P retention differed from zero in all except the only-hay diet, in which the intake was lower compared to the other diets. The digestibility of P varied from 2.7 to 11.1%, and supplementing forage-diets with concentrates slightly improved P digestibility ($p = 0.024$), as it also improved the digestibilities of crude protein ($p = 0.002$) and organic matter ($p = 0.077$). The horses excreted an average of 20.9 ± 1.4 g/d P in feces. Excretion was smallest (20.0 g) in horses on a hay-only diet ($p = 0.021$). The average daily phosphorus excretion resulted in 7.6 kg P per year. The soluble P part of the total P in feces accounted for about 88% of the P excreted in feces, and is vulnerable to runoff losses and may leach into waters. Thus, horse dung may pose a potential risk of P leaching into the environment if not properly managed, and is not less harmful to the environment than that from other farm animals. Supplementation with inorganic P should be controlled in the diets of mature horses in light work to decrease the excretion of P in feces.

Keywords: environment; horse nutrition; phosphorus loss; phosphorus supplementation; phosphorus retention

1. Introduction

Macro-mineral phosphorus plays an important role in bone formation as a constituent of phosphoproteins, phospholipids, and nucleic acids, and in energy and fat metabolism [1–3]. In animals, 80–85% of phosphorus is stored in the bones and teeth, and the remainder in soft tissues and body fluids [3]. In the skeleton, phosphorus forms hydroxyapatite with calcium. In growing animals, the need for phosphorus is greater than in adult animals, since developing bones require more phosphorus than already developed bones in adult animals [4].

Phosphorus deficiency is found throughout the world in areas with soil poor in phosphorus. Deficiency symptoms include, but are not limited to, rice disease, osteomalacia, nervous system symptoms, stiff joints, muscle weakness, poor fertility, impaired ovarian function and consequent irregular rotation, poor growth in juvenile animals, and impaired weight gain in adult animals [1]. To avoid these detrimental effects of phosphorus malnutrition and ensure efficient intake, phosphorus is usually routinely supplemented in horses' diets.

Phosphorus absorption is influenced by the intake, source, and composition of the feed ration [5]. In adult horses, which mainly eat roughage, absorption efficiency is 35%, and in lactating mares and growing horses it is 45% as their diets are often supplemented by larger amounts of concentrates [1]. There may be some improvement in phosphorus absorption as the need for phosphorus increases, for example, through exercise or when the phosphorus content of the diet increases [6]. Fowler et al. [7] concluded that yearlings can utilize organic P as well as mature horses.

The main site of the gastrointestinal tract of phosphorus absorption in horses is the dorsal colon, but some phosphorus is also absorbed from the small intestine [8,9]. Fowler et al. [7] suggested that degradation either occurs after the site of P absorption, or liberated P is recycled back into the gastrointestinal tract.

The main phosphorus excretion pathway is through the dung. The phosphorus content of feces is directly proportional to the phosphorus content of the diet [10]. Especially in a diet rich in forage, the horse often gets too much supplemented phosphorus to meet its needs, and excess P is excreted from the body in the feces [11]. A very low proportion (about 1%) of the phosphorus is excreted in the urine [8,10,12], and is thus usually ignored in studies dealing with P digestibility. In the gastrointestinal tract, endogenous phosphorus secretion occurs in the small intestine and in the cecum [8,13]. Endogenous phosphorus secretion is due to the presence of phosphorus compounds in gastrointestinal fluids such as saliva and gastric, pancreatic, and biliary fluids [8,14].

The digestibility of phosphorus is influenced by its form and amount in feed, and its interaction with other feed components and minerals, e.g., Ca and Ca:P ratio [1,15]. Cereals are good sources of phosphorus [3], but a significant proportion of the phosphorus in cereals is bound to phytic acid, which is poorly digestible in monogastric animals [16]. However, the horse is able to digest phytate phosphorus [7,17,18]. The content of phosphorus in grasses approximately equates to that in cereals (about 3 g/kg DM), but the phosphorus content of grasses is significantly influenced by the age of the crop at the time of harvest [19,20]. Mineral supplements usually contain inorganic forms of phosphorus like monocalcium phosphate, dicalcium phosphate, or phosphorite as phosphorus sources [4].

Phosphorus and nitrogen originating from animal manure are the main environmental and water pollutants from agriculture. There is imbalance of N and P in the manure, and P is considered more harmful, because when it is in excess of crop requirements, soil becomes saturated resulting in P runoff [21]. Because the number of horses has increased in many countries, and horses are kept in paddocks and pastures there is risk of P leaching to the environment also from horse husbandry. In addition, horse manure is widely utilized in agriculture. However, the NRC [1] considers horse feces to be less risky to the environment compared to that from other farm animals because it assumes that horse manure contains less water-soluble phosphorus, prone to runoff. However, later studies have pointed that leaching of P from horse paddocks and pastures are hotspots for high P leaching losses [22–25].

In general, dietary strategies have been developed for many animal species to effectively reduce the total P concentration in manure. As we were interested in the possible detrimental environmental impact of horses, we studied phosphorus utilization and excretion in feces applying typical feeds and forage-based diets fulfilling the current P -intake recommendations [1,19]. We also studied the nutrient digestibilities of the diets, as well as the proportion of the soluble fraction of P of the total phosphorus. The hypothesis was that feeding regimes might influence phosphorus digestibility and excretion in feces, and therefore the environmental impact of horse husbandry.

2. Materials and Methods

A digestibility experiment was conducted with six forage-based diets typically fed to horses in Finland. The study was conducted at the facilities of the Natural Resources Institute Finland (Luke) in Southwest Finland. In animal handling and sample collection, the European Union recommendation directives (2010/63/EU) and national animal welfare and ethical legislation set by the Ministry of Agriculture and Forestry of Finland were followed carefully. The experimental procedures were evaluated and approved by the national ethical committee for animal experiments (https://www.avi.fi/web/avi/elainkoelautakunta-ella) (ESAVI/8331/04.10.07/2013).

2.1. Horses and their Management

Six adult Finnhorse mares (5–13 years; initial BW 552 ± 32 kg, mean BCS 6 = moderately fleshy) owned by Luke were used in the study. All the experimental horses had the same managing and feeding history before the trial. The horses were individually housed in stalls (3 m × 3 m) with peat as bedding. The horses were de-wormed before the experiment. Dental care and vaccinations had been carried out regularly prior to the experiment. The horses were freely exercised daily in groups in outdoor paddocks (with sand grounds) for 2–4 h, except during the collection period, when they were led in walk by a rope in the stable corridors (consisting of two connected 32 m long corridors with concrete and asphalt floors) for 15 min. In the paddocks, the horses had masks to prevent sand eating.

The study method was a total collection of feces. The experimental design was arranged as an unbalanced 6 × 4 Latin Squares. The study consisted of six treatments and four 21-day periods. Each period started with a five-day feed change period followed by 12 days of adaptation to the new diet, and four-day period of collecting feces samples. During the collection period, the peat bedding was changed to rubber mats. The body weight (BW) (electronic animal scale Lahden Vaaka/Lahti Precision Ltd., Lahti, Finland) and body condition score (BCS) [26] of the horses was monitored after each collection period.

2.2. Experimental Feeds and Feeding

The diets were formulated and adjusted to be as isocaloric as possible. The horses were individually fed at a level of 65–75 g DM kg $^{-1}$W$^{0.75}$, corresponding to the feeding and energy level recommended in light work in accordance with the Finnish Feed Tables and Feeding Recommendations [19]. The regularly obtained BCS and BW were used to control their possible changes, and the individual energy intakes were adjusted if necessary. The horses were fed three times per day (at 6:00 a.m., 12:00 noon and 6:00 p.m.), except in the mornings of blood sampling days, when the forages were fed at 7:30 and concentrates at 8:00 o'clock.

The diets were (dry matter basis) (A) hay 100%; (B) haylage 100%; (C) hay 80% + whole oats 20%; (D) hay 65% + whole oats 35%; (E) hay 80% + commercial pelleted complete feed 20% (Lantmännen, Malmö, Sweden) and (F) hay 65% + commercial pelleted complete feed 35%. All diets except those including the complete feed were balanced with a mineral mixture (Ca:P = 3.57) in which P was in the form of monocalcium phosphate (Vilomix Ltd., Paimio, Finland) according to the P and Ca needs of the horses. The complete feed (the added P was in the form of monocalcium phosphate, Ca:P 1.43) covered the mineral requirements of the horses. Forages were fed from special hay-boxes to avoid dropping, and the concentrates were fed from feed mangers. Free water was available from float valve drinkers.

The dried hay was produced by a local farmer in Ypäjä (60°48′34″ N, 23°16′35″ E). The haylage (Prohay Ltd., Punkalaidun, Finland, 61°06′40″ N, 23°06′20″ E) was packed in 20 kg air tight plastic packages and purchased from the producer. The oats were produced by Luke.

Feed samples were collected daily and stored at –20 °C until analysis. The samples were analyzed at the Luke Laboratories (Jokioinen, Finland) for dry matter (DM), crude protein (CP), NDF, ADF, crude fiber (CF), and ash with standard wet chemical methods e.g., [27], as well as for P and Ca content using the method by Huang and Schulte [28]. The chemical composition of the feeds is presented in Table 1.

Table 1. Average chemical composition (g/kg DM) and energy value (ME MJ/kg DM) of the experimental feeds.

Composition	Hay	Haylage	Oats	Complete Feed	Mineral Mixture [1]
Dry matter	83.6	59.4	86.1	86.9	96.0
Crude protein	82.1	122.5	107.0	124.4	
Crude fiber	318.7	326.7	103.9	113.5	
NDF	596.0	615.0	260.5	305.4	
ADF	317.5	326.6	104.3	123.3	
Ash	62.3	67.6	32.2	74.0	737.8
Calcium	2.4	2.9	0.5	7.0	156.4
Magnesium	1.2	1.3	1.3	4.2	45.0
Phosphorus	2.2	2.7	3.6	4.9	43.6
Energy ME	9.7	9.6	12.0	11.0	

NDF = neutral detergent fiber; ADF = acid detergent fiber; [1] Inorganic mineral sources: monocalcium phosphate, limestone meal, magnesium oxide; ME = metabolisable energy.

2.3. Feces Sampling and Analysis

Feces were collected for four days (Monday–Friday). Before the start of the collection period, the peat bedding was removed from the boxes and replaced with rubber mats. The feces collected overnight (every day at 9:00 a.m.) was thoroughly mixed and weighed, and a representative sample was taken. Partial samples were pooled throughout the collection period and stored at −20 °C. The daily amount of sampled feces was 12% of the total daily amount collected. Feces that were entangled with foreign substances such as urine were defined as waste. The amount of waste feces was weighed but not utilized in the analysis.

Fecal samples were analyzed at the Luke Laboratories for dry matter (DM), nitrogen, NDF, ADF, CF ash as well as total and (water) soluble P content. Nitrogen was analyzed with the Khjeldal method (AOAC-984.13) using Foss Khjeltec 2400 analyzer (Foss Analytical AB, Höganäs, Sweden), and the CP content was calculated as $6.25 \times N$. NDF and ADF were analyzed with ANKOM 220 fiber analyzer (ANKOM Technology, Macedon, NY, USA) using 25 μm nylon bags [29,30]. Total P was analyzed spectrometrically (ICP-OES, Thermo Jarrel Ash Iris advantage, Franklin, MA, USA). The proportion of soluble P (phospahte-P, PO_4-P) was analyzed from 1:60 water extracts using a continuous photometric flow analyzer (Aquakem 250, Thermo Fisher Scientific Inc., Vantaa, Finland) as described by Keskinen et al. [31]. Because only a very low proportion (about 1%) of the phosphorus is excreted in the urine [8,10,12], urine was not collected in this experiment.

2.4. Blood Samples and Analysis

Blood samples were collected 90 min after the morning meal every Wednesday during the collection period. A blood sample was drawn from the jugular vein into two sample tubes (2 × 10 mL). The samples were analyzed for inorganic P and Ca photometrically using wavelengths 660 nm and 340 nm, respectively. The analyses were performed in a clinical authorized laboratory (Ellab Ltd., Ypäjä, Finland).

2.5. Statistical Analysis

Differences in digestibility, as well as intake, excretion and retention parameters between the diets were statistically analyzed using the SAS (SAS 9.3, 2008) GLM procedure (SAS Institue, Cary, NC, USA) applying the following statistical model:

$$Y_{ijk} = \mu_{ijk} + a_i + p_j + d_k + e_{ijk} \tag{1}$$

where μ_{ijk} is the overall mean, a_i is the random effect of the animal (i = 1 ... 6), p_j is the fixed effect of the period (j = 1 ... 4), d_k is the fixed effect of the diet (k = 1 ... 6), and e_{ijk} is the normally distributed error with a mean of 0 and variance δ^2. The differences between the diets were tested with orthogonal contrasts: (1) B vs. A and C-F; (2) A vs. C-F; (3) C and D vs. E and F; (4) C vs. E and D vs. F; and (5) the interactions between the type of concentrate and concentrate level C and D vs. E and F, C vs. E, and D vs. F. Concerning the retention values, it was also tested if they differ from zero.

Differences in the proportion of soluble P of the excreted P were not studied because the diets were composed of different ingredients containing various sources of P (inorganic and organic sources, phytate P) that were not analyzed.

3. Results

3.1. Feed and Nutrient Intakes

The feed, energy and nutrient intakes for each diet are presented in Table 2. The DM intake was smallest in horses eating only haylage ($p < 0.001$), but same time they had the largest CP intake ($p < 0.001$). Concerning the concentrate supplemented diets (hay + oats or complete feed), the concentrate level did not affect the DM intake. The horses maintained their BW and BCS during the experiment (mean initial BW = 552 ± 32 kg, mean final BW = 558 ± 32 kg).

Table 2. Mean daily energy (MJ ME), dry matter (g) and nutrient intakes (g) for the experimental diets.

Diet	A	B	C	D	E	F	Pooled SEM	Haylage vs. Others	Hay vs. ConS	Oats vs. Comp	ConL	ConT × ConL
Forage	Hay	Haylage	Hay	Hay	Hay	Hay						
ConL	0	0	O20	035	C20	C35						
ME	74.3	73.3	79.1	82.6	77.6	78.7	0.4	<0.001	<0.001	<0.001	<0.001	0.009
DM	7536	6702	7568	7734	7601	7595	17.0	<0.001	0.038	0.037	0.439	0.338
OM	6981	6186	7148	7250	7081	7075	47.3	<0.001	0.033	0.027	0.368	0.272
CP	607	812	647	705	685	744	17.3	<0.001	0.014	0.047	0.009	0.979
CF	2366	2155	2079	1852	2102	1873	16.4	<0.001	0.001	0.196	<0.001	0.946
NDF	4427	4060	3992	3647	4073	3753	33.9	0.055	<0.001	0.018	<0.001	0.715
ADF	2361	2155	2076	1845	2109	1891	18.8	<0.001	<0.001	0.058	<0.001	0.746
Ash	556	516	510	485	519	521	17.0	0.901	0.033	0.202	0.519	0.446

ME = metabolisable energy MJ/day; DM = dry matter; OM = organic matter; CP = crude protein; CF = crude fiber; NDF = neutral detergent fiber; ADF = acid detergent fiber; O = Oats; C = Complete feed; ConL = Concentrate level (20 or 35% of oats O or complete feed C); ConS = Concentrate supplementation; ConT = concentrate type (oats/complete feed); Comp = complete feed.

3.2. Intake, Fecal Excretion and Digestibility of Phosphorus

The average P intake was 22.0 ± 2.0 g/d. The P intake increased with an increasing concentrate intake (Table 3). Horses ingesting oats had larger daily intake of P (22.8–24.8 g) compared to those fed with the complete feed (20.9–24.1 g) ($p < 0.001$) and horses fed with hay only had smaller intakes than horses supplemented with concentrates (20.6 vs. 20.9–24.8 g) ($p < 0.001$). Horses fed with haylage had somewhat smaller P intake than those who ate hay ($p = 0.036$).

Table 3. Daily intake (g), fecal excretion (g), digestibility (%) and retention (g) of phosphorus (P).

Diet/Forage	A	B	C	D	E	F	Pooled SEM	Haylage vs. Others	Hay vs.ConS	Oats vs.Comp	ConL	ConT × ConL
	Hay	Haylage	Hay	Hay	Hay	Hay						
ConL	0	0	O20	O35	C20	C35						
Intake P	20.6	22.1	22.8	24.8	20.9	24.1	0.21	0.036	<0.001	<0.001	<0.001	0.027
Excretion P	20.0	20.2	21.5	22.1	19.9	21.5	0.42	0.125	0.021	0.025	0.033	0.251
Digestibility P	2.7	8.0	5.6	11.1	4.9	10.6	1.98	0.652	0.037	0.761	0.024	0.974
Retention	0.6	1.9	1.0	2.8	0.9	1.9	0.45	0.354	0.075	0.25	0.014	0.379

O = oats; C = complete feed; ConL = concentrate level (20 or 35% of oats O or complete feed C); ConS = concentrate supplementation; ConT = concentrate type (oats/complete feed); Comp = complete feed.

The average daily quantity of dung was 15.6 ± 2.5 kg/horse. The horses excreted an average of 20.9 ± 1.4 g/d P in feces. Excretion was smallest (20.0 g) in horses on a hay-only diet ($p = 0.021$) (Table 3). Horses supplemented with oats excreted somewhat more P (21.5–22.1 vs. 19.9–21.5 g) than those supplemented with the complete feed ($p < 0.025$), and the excretion increased with increasing concentrate intake ($p = 0.033$).

The horses were on a positive P balance in all diets (Table 3). The retention of P was largest in the diet D (with the highest complete feed level) being 2.8 g/d. The retention values were different from zero for the diets B ($p = 0.002$), C ($p = 0.05$), D ($p < 0.001$), E ($p = 0.08$), and F ($p = 0.002$). The P retention increased ($p = 0.0145$) with the increasing concentrate level. Feeding concentrates slightly improved P digestibility ($p = 0.024$). The amount of water-soluble phosphorus of the P excreted in feces was 18.3 ± 2.5 g/d, on average. This corresponds to 87.6% of the P in feces.

3.3. Intake, Fecal Excretion and Digestibility of Calcium and Magnesium

The intake of calcium was largest in the horses fed with haylage only ($p > 0.001$) (Table 4). Ingestion of concentrates decreased Ca intake, depending on type of concentrate fed. Mg intake was smallest in the horses on a haylage-only diet. The intake increased when concentrates were fed. The increase was largest with complete feed (interaction $p < 0.001$).

Table 4. Daily intake (g), excretion (g) and digestibility (%) of calcium (Ca) and magnesium (Mg).

Diet/Forage	A	B	C	D	E	F	Pooled SEM	Haylage vs. Others	Hay vs. ConS	Oats vs. Comp	ConL	ConT × ConL
	Hay	Haylage	Hay	Hay	Hay	Hay						
ConL	0	0	O20	O35	C20	C35						
Intake												
Ca	32.6	34.2	29.6	28.3	24.7	30.4	0.45	<0.001	<0.001	0.010	0.001	<0.001
Mg	9.31	8.71	9.25	9.74	13.9	17.3	0.26	<0.001	<0.001	<0.001	<0.001	<0.001
Excretion												
Ca	19.1	17.7	17.5	17.8	19.4	20.2	0.98	0.330	0.757	0.048	0.610	0.797
Mg	10.4	10.1	10.1	10.4	10.6	12.9	0.28	0.030	0.107	<0.001	0.002	0.006
Digestibility												
Ca	41.6	47.1	29.3	38.1	22.6	31.7	7.10	0.096	0.187	0.371	0.265	0.987
Mg	−13.1	−19.0	-9.0	-7.0	23.4	25.3	4.73	0.001	0.002	<0.001	0.702	0.993

O = Oats; C = Complete feed; ConL = Concentrate level (20 or 35% of oats O or complete feed C); ConS = Concentrate supplementation; ConT = concentrate type (oats/complete feed); Comp = complete feed.

Concerning the excretion of minerals, horses on the haylage-only diet excreted somewhat less Mg than fed with hay only ($p = 0.03$). Comparing the concentrates, horses supplemented with complete feed excreted more both Ca and Mg than those supplemented with oats ($p = 0.048$ and $p < 0.001$, respectively). The digestibility of Ca did not differ between the diets. Concerning Mg, digestibility was lowest in the haylage-only diet, and lower when oats was fed compared with feeding the complete feed. The variation in Mg digestibility values was large (Table 4).

3.4. Digestibility of the Diet Nutrients

The dry matter digestibility of the haylage-only diet was lower compared with the other diets ($p < 0.001$) (Table 5). Supplementing the forage diets with concentrates improved the digestibilities of crude protein ($p = 0.002$) and organic matter ($p = 0.077$) but the concentrate level fed did not affect the digestibility of the fiber fractions. The CP digestibility of the haylage-only diet was better ($p = 0.009$) compared to the other diets. Correspondingly, the CP digestibility of the hay-only diet was the lowest ($p = 0.004$).

Table 5. Apparent digestibility coefficients (%) of the diet nutrients.

Diet/Forage	A	B	C	D	E	F	Pooled SEM	Statistical Significance (*p*-Values)				
	Hay	Haylage	Hay	Hay	Hay	Hay		Haylage vs. Others	Hay vs. ConS	Oats vs. Comp	ConL	ConT × ConL
ConL	0	0	O20	O35	C20	C35						
DM	55.0	49.9	56.6	59.3	57.5	60.1	1.33	<0.001	0.046	0.543	0.090	0.998
Ash	38.4	31.6	29.8	29.6	34.9	37.7	2.73	0.437	0.105	0.033	0.666	0.591
OM	56.3	51.4	58.5	61.3	59.1	61.8	1.29	<0.001	0.023	0.699	0.077	0.972
CP	50.9	63.6	51.7	61.8	57.3	63.6	1.83	0.009	0.004	0.066	0.002	0.310
CF	49.5	45.7	46.1	43.0	49.6	56.5	2.61	0.685	0.291	0.206	0.294	0.997
NDF	49.5	46.3	46.3	42.5	49.0	47.5	2.44	0.811	0.278	0.144	0.334	0.635
ADF	47.2	41.6	41.7	41.4	45.5	40.5	3.40	0.668	0.224	0.669	0.488	0.497

DM = dry matter; OM = organic matter; CP = crude protein; CF = crude fiber; NDF = neutral detergent fiber; ADF = acid detergent fiber. O = Oats; C = Complete feed; ConL = Concentrate level (%); ConS = Concentrate supplementation; ConT = concentrate type (oats/complete feed); Comp = complete feed.

3.5. Blood Concentrates of P and Ca

The between diet variation of the blood serum P and Ca concentrations was small. The average blood serum P concentration was 1.16 ± 0.04 mmol/L. Comparing concentrate types, the concentration was larger when oats was fed than when the complete feed was fed (1.21 vs. 1.13 mmol/L) ($p < 0.031$). The mean blood Ca concentration of the horses was 3.14 ± 0.04 mmol/L. It was larger when the horses were on a hay-only diet compared with the diets containing concentrates (3.22 vs. 3.12 mmol/L) ($p < 0.001$).

4. Discussion

The main goal of this experiment was to find the differences between the diets for P utilization, rather than the actual values for specific diets. Because studying digestibilities of other nutrients was only a secondary aim of this study, the results are discussed only briefly.

4.1. Feed Values and Nutrient Intakes

The feed values corresponded to the analyzed values presented for Finnish forages produced for horses [32], the hay being of "medium nutritional quality" and the haylage of "high nutritional quality". Concerning oats, the CP content was lower than that presented in the Finnish Feed Tables and Feeding Recommendations (10.4. vs. 12–13%) [19]. The NDF, ADF, and CF values were also lower than the values presented for average Finnish oats [19]. The P, Ca, and Mg content of the forages and oats was lower than the values presented for hays and haylages [19]. The mineral content was also clearly lower than reported for Norwegian and Swedish haylage samples collected from horse farms [20].

The nutrient intakes naturally varied because of the differences in the composition of the feeds and actual intakes, although the individual diets were initially formulated and balanced to correspond the needs of each horse [19] and be as isocaloric as possible. The smaller DM intake of the horses in the haylage-only diet resulted from restriction of haylage intake because of its high CP content. This led naturally also to smaller energy (ME) intake. In addition, the size of the horse affected the individual daily portion such that larger horses had larger portions. The horses maintained their BW and BCS during the experiment, indicating that the feeds, feeding regime, and intakes applied covered the nutritional requirements of the horses (energy and protein) in the course of the experimental period

The mean daily energy intake of 77.5 MJ ME/d during the course of the experiment agreed with the recommendation for horses in light work [19].

4.2. Intake, Fecal Excretion and Digestibility of Phosphorus

The differences in the P intake between the diets were due to the differences in the feeds and diet compositions, but the forms of P were not analyzed. P concentrations in oats and complete feed were higher than in the forages. In addition, the oat supplemented diets were balanced for minerals (P and Ca) with a mineral supplement mixture which also increased the P intake. The intakes of P were in accordance with the current recommendations [1,19], except on the highest concentrate levels where they exceeded the recommendations.

The excretion of phosphorus observed here was within the ranges presented in the literature [7,12,18,33,34]. P excretion is linearly related to its intake, and the intake increases with the increasing concentrate ingestion [10,12,33]. Van Doorn et al. [35] concluded that horses can regulate P digesting and thus P balance. The extra P can be excreted in the feces.

In the present study, the digestibility of P in adult horses varied from 2.7 to 11.1%, improving with an increased concentrate intake. This is well in line with the results of van Doorn et al. [18] for adult horses (2.4–15.4%). However, higher values (4.2–28.7%) have also been reported [15,34,36] for adult horses (with a range between 4 and 25 years). In many studies the digestibilities show impaired values with increasing age. The largest digestibilities (37–42%) have been reported for young horses (8-months-olds), but they decline quickly (to 2.0–7.7%) when the horses are between one and two years old [7,12]. Elzinga et al. [36] reported digestibility of 4.2% for aged (19–28 years) horses. P digestibility is therefore influenced by the age of the horse. In this study, the horses were between 5 and 13 years old, and P digestibility seems to accord well with previous studies when the age of the horses is considered. In previous studies, higher values have also been reported when hay + concentrates were fed e.g., [6,7,15,18,34,35] compared with forage-only diets [12,28]. The digestibility of P may also improve somewhat with increasing P intake [6]. Furthermore, digestibility is affected by the components (feeds) of the diet [1].

Diets with the highest concentrate levels (35% oats or complete feed) had better digestibility compared to the other diets. All diets resulted in a positive P balance and the P retention differed from zero in all except the only-hay diet, in which the intake was lower compared to the other diets. According to previous studies [10,18], the P retention increased with P intake in adult animals. The P retention values observed in our study were smaller (less than half) than reported for adult (appr. 6-year-olds) Standardbred horses [18] fed mixed hay + concentrate diets, but in the same time, the intakes were also correspondingly smaller. According to that study, the reasons for the P retention in adult horses are not known. However, Buchholz-Bryant [37] reported a higher P retention in mature horses (7 to 11 year old) at rest compared to exercised horses. In the present study the horses were only freely exercised daily in outdoor paddocks for 2–4 h, and during the collection period, they were walked manually in the stable corridors for 15 min In the study of Van Doorn et al. [18], the horses were given 1-h walk on a treadmill, and during the collection period they walked manually for 10 min indicating rather light work level. Thus, the light work load of the horses may be one reason for the P retention in our study. In addition, because phosphorus is stored in many body tissues and fluids [1–3], some P can be accumulated to these, too. The retention values presented here includes also the urine P, because urine was not collected and analyzed for P. This was done, because according to studies [8,10,12], only a very low proportion (about 1%) of the phosphorus is excreted in the urine.

The Ca:P ratio of the feed or diet affects not only the digestibility of phosphorus but also the digestibility of calcium [15]. A high calcium intake can impair phosphorus digestibility at a Ca:P ratio of 2.58 or more [1,18]. However, the dietary Ca:P ratios in the current study were much lower, 1.14–1.58. Concerning fecal Ca:P ratios, Böswald et al. [38] found that in horses (and other large hindgut fermenters), the fecal Ca:P ratio is lower than the dietary Ca:P ratio. The present study was in accordance with this, the mean fecal ratios ranging from 0.80 to 0.95.

4.3. Intake, Fecal Excretion and Digestibility of Calcium and Magnesium

Calcium and magnesium intakes also differed between the diets because of the differences in the mineral content of the feeds. The intakes of Ca and Mg agreed with the recommendations [1,19], with the exception that Mg intake was above the recommended values when complete feed was fed. Intakes of Ca from the concentrates were small. Oats was low in Ca, and the increased proportion of oats and decreased proportion of forage in the diet resulted in a decline in Ca intake. Concerning the intakes of Mg, the larger Mg content of the complete feed compared to that of oats explains the differences in Mg intake.

The excretion of Ca in the present study was smaller than reported in the literature e.g., [7,18] because the intake was smaller. It is also likely that different Ca sources have an influence. The Mg excretion agreed with previous studies [7,18].

The amounts of excreted Ca and Mg in feces are related to intakes [39,40], but no effect of intake was observed by Nielsen et al. [33]. Meyer et al. [41,42] reported that diets containing more roughage resulted in a higher renal excretion of Ca and Mg.

The observed digestibilities of Ca not differing between the diets agreed with the literature values [2,7,18,35]. Van Doorn et al. [18] have reported that high amounts of P and phytate P may decrease Ca digestibility. The larger excretion of Mg compared with its intake in all diets, except those including the complete feed, explains, the poor Mg digestibility in these diets. The digestibility of Mg has been reported to be largely varying, and negative values have also been reported [2,7,18,35], as in the present study. Because Ca and Mg are excreted in large quantities in urine [9,40], the observed digestibilities here may be underestimated.

4.4. Digestibility of the Diet Nutrients

It is very likely that the lower DM digestibility of the haylage-only diet was due to its larger fiber content (NDF, ADF, CF) compared to the other diets. The better CP digestibility of the haylage-only diet was due to its larger CP content, and correspondingly, the poor CP digestibility of the hay-only diet was due to its low CP content. These results are supported for example by Särkijärvi and Saastamoinen [27] and Ragnarsson and Lindberg [43,44]. The positive effect of including concentrates in the forage diets agrees with previous studies, being e.g., due to the lower NDF content of the diet e.g., [45,46]. The NDF content also explains the digestibility value differences between the forage diets. The digestibility values observed in the present study are comparable with those reported previously for Finnhorses of the same age e.g., [27,47].

4.5. Blood Concentrates of P and Ca

The blood phosphorus levels are affected by phosphorus intake [48,49], which may explain why blood levels were highest in the horses whose diets were supplemented with oats. Greiwe-Crandell et al. [50] suggested that mares fed an all-forage diet marginal or low in phosphorus may mobilize P from bone.

The blood Ca concentration was largest in the horses on the hay-only diet. Meyer et al. [42] reported higher plasma Ca levels for forage fed horses than concentrate fed horses. Some other studies, however, pointed out that the blood Ca concentration does not depend on the Ca intake [48,49]. Regarding all diets, the average blood serum Ca and P concentrations were within the normal ranges used for Finnhorses (https://www.movet.fi/laboratoriokasikirja/). To maintain physiologic Ca and P blood levels, mammals can absorb them from the gastrointestinal tract or change their bone turnover [38].

4.6. Impact of Horse Diets on P Leaching

The daily quantity of dung produced by the horses was in line with the literature values [7,51,52], depending, however, on diet and feed intake. The average daily phosphorus excretion of about 21 g in feces in this study, when typical diets and current recommendations [1,19] were applied, resulted in

7.6 kg P per year. If the P is not properly absorbed in stable beddings, or if the dung in paddocks is altered by rain and water from melting snow, P in feces and manure may present an environmental risk when leaching into waters.

The soluble P part of the total P in feces available for the utilization of plants accounted for about 88% of the P excreted in feces in this study. Ögren et al. [12] reported a proportion of 80% of soluble P. The P that is unavailable is vulnerable to runoff losses. According to Chapuis-Lardy et al. [53], excess dietary P is excreted in feces in water-soluble forms. Dougherty et al. [54] pointed out that around 90% of P losses occurred in water-soluble form. The leaching P from the dung of horses is mainly inorganic [55]. Consequently, the argument of the NRC [1] that horse manure is less harmful to the environment compared with that from other farm animals because of its low proportion of water-soluble P, is not correct in the light of the results of this and previous studies.

It is possible that the composition of the diet affects the solubility of P, as reported for dairy cows and pigs [56,57]. However, in this study it was irrelevant to compare the diets because of their composition, i.e., the inclusion of various P sources in the same diet. Further studies can be suggested to be carried out concerning this issue also in horses. Ögren et al. [12] concluded that soluble P has a strong positive relationship to P intake in horses.

As P loss is linearly related to its intake in various animal species [12,53,58], it is impossible to conclude how polluting horse industry is compared with other forms of animal production. However, in the study of Ögren et al. [12], the high proportion of inorganic P in horse feces indicated that P overfeeding of horses might be more harmful to the environment than P overfeeding of dairy cows. Previously, several other authors have also reported that horse paddocks may pose a high risk of extensive P loss [22–25,55,59]. Regular removal of dung from paddocks is recommended to minimize this risk [25,55,60,61]. How often this should be done naturally depends on the time the horses spend in the paddocks and livestock density/ha. Phosphorus sorbing materials (e.g., Fe containing) [22,59], filtering materials (geotextile-gravel) [61], or organic (bedding) materials [62] can also be used on paddock surfaces to reduce leaching loss.

Ögren et al. [12] concluded that an increase in the P requirement for growing horses is not justified. The present study shows that it is unnecessary to supplement the diets of mature horses, especially those in light work, with inorganic phosphorus, when the diets are supplemented with concentrates. According to Fowler et al. [7], the organic P in feeds may fulfill the needs of horses in light work, and no supplementation with inorganic P is needed. Balancing the diets for P intake can be estimated to save both money and environment in dairy production [63]. There may also be economic motives to catch the P in feces and absorb it in bedding materials, because the use of horse manure may reduce fertilizing costs. When horse manure is composted, its nutrients can be recycled and utilized [31], which reduces the use of inorganic fertilizers.

In addition, optimizing the proportions of the diet components, for example by supplementing the forage diets with concentrates, may improve the digestibility of phosphorus. However, it is necessary to analyze the feeds for the mineral concentrations because of the large variation [20,32,64]. In complete feeds for horses, P (and other mineral) concentrations are usually formulated to cover the requirements of an "average horse" when "medium-quality" forages are fed. They thus do not take into account the true mineral concentrations in the other components of the diet. This may result in over- or undernutrition in practical feeding. As in this study, they also contain the added P in inorganic form. In addition, when increasing the proportion of concentrates in the daily ration, the possible detrimental effects of starch [65] have to be considered. In the present study, the concentrate levels fed were not very large, and the diets were based on forages.

5. Conclusions

Horse dung may pose a potential risk of P leaching into the environment, and is not less harmful to the environment than that from other farm animals if not properly managed, because most of the P in feces is in soluble form. Supplementation with inorganic P should be controlled in the diets of adult horses in light work to decrease the excretion of P in feces. Supplementing forage diets with concentrates may improve the digestibility of phosphorus and, thus, improves the availability of P to horses. More research especially into cost effective feeding strategies and their applications for horses is essential, e.g., concerning diet composition and ingredients, to reduce horse industry's harmful impacts on and risks to water quality.

Author Contributions: M.S. and S.S. contributed methodology and investigation; S.S. contributed formal analysis; M.S. contributed resource, writing—original draft preparation, supervision; E.V. contributed experimental part and investigation, and data analyzing. All authors have read and agreed to the published version of the manuscript.

References

1. NRC. *Nutrient Requirements of Horses*, 6th ed.; National Research Council of the National Academies: Washington, DC, USA, 2007.
2. Lavin, T.E.; Nielsen, B.D.; Zingsheim, J.N.; O'Connor-Robison, C.I.; Link, J.E.; Hill, G.M.; Shelton, J. Effects of phytase supplementation in mature horses fed alfalfa hay and pelleted concentrate diets. *J. Anim. Sci.* **2013**, *91*, 1719–1727. [CrossRef] [PubMed]
3. McDonald, P.; Edwards, R.A.; Greenhalgh, J.F.D.; Morgan, C.A.; Sincliar, L.A.; Wilkinson, R.G. *Animal Nutrition*, 7th ed.; Pearson Education Limited: Harlow, UK, 2011.
4. Ammermann, C.B.; Baker, D.P.; Lewis, A.J. *Bioavailability of Nutrients for Animals: Amino Acids, Minerals, Vitamins*; Academic Press: San Diego, CA, USA, 1995.
5. Geor, R.J.; Harris, P.A.; Coenen, M. *Equine Applied and Clinical Nutrition: Health, Welfare and Performance*, 1st ed.; Saunders/Elsevier: Edinburgh, UK, 2013.
6. van Doorn, D.A.; Everts, H.; Wouterse, H.; Homan, S.; Beynen, A.C. Influence of high phosphorus intake on salivary and plasma concentrations, and urinary phosphorus excretion in mature ponies. *J. Anim. Physiol. Anim. Nutr.* **2011**, *95*, 154–160. [CrossRef] [PubMed]
7. Fowler, A.; Hansen, T.; Strasinger, L.; Davis, B.; Harlow, B.E.; Lawrence, L. Phosphorus digestibility and phytate degradation by yearlings and mature horse. *J. Anim. Sci.* **2015**, *93*, 5735–5742. [CrossRef] [PubMed]
8. Schryver, H.F.; Hintz, H.F.; Craig, P.H.; Hogue, D.E.; Lowe, J.E. Site of phosphorus absorption from intestine of the horse. *J. Nutr.* **1972**, *102*, 143–147. [CrossRef] [PubMed]
9. Schryver, H.F.; Hintz, H.F.; Lowe, J.E. Calcium and phosphorus in the nutrition of the horse. *Cornell Vet.* **1974**, *64*, 493–514. [PubMed]
10. Schryver, H.F.; Hintz, H.F.; Craig, P.H. Phosphorus metabolism in ponies fed varying levels of phosphorus. *J. Nutr.* **1971**, *101*, 1257–1263. [CrossRef]
11. Fowler, A.; Strasinger, L.; Hansen, T.; Davis, B.; Hayes, S.; Lawrence, L. The availability of dietary phosphorus to long yearlings and mature horses. *J. Equine Vet. Sci.* **2013**, *33*, 342–343. [CrossRef]
12. Ögren, G.; Holtenius, K.; Jansson, A. Phosphorus balance and fecal losses in growing Standardbred horses in training fed forage-only diets. *J. Anim. Sci.* **2013**, *91*, 2749–2755. [CrossRef]
13. Cehak, A. In vitro studies on intestinal calcium and phosphate transport in horses. *Comp. Biochem. Physiol. Part A* **2012**, *161*, 259–264. [CrossRef]
14. Pagan, J.D. Nutrient digestibility in horses. In *Advances in Equine Nutrition*; Pagan, J., Ed.; Nottingham University Press: Nottingham, UK, 1998; pp. 77–83.
15. Wilson, J.A.; Babb, C.W.; Prince, R.H. Protein and mineral digestibility of three pelleted equine feeds and subsequent nitrogen and phosphorus waste excretion. *Prof. Anim. Sci.* **2006**, *22*, 341–345. [CrossRef]
16. Warren, L.; Weir, J.; Harris, P.; Kivipelto, J. Effect of total phosphorus and phytate –phosphorus intake on phosphorus digestibility in horses. *J. Equine Vet. Sci.* **2013**, *33*, 352. [CrossRef]
17. Hainze, M.T.M.; Muntifering, R.B.; Wood, C.W.; McCall, C.A.; Wood, B.H. Faecal phosphorus excretion from horses fed typical diets with and without added phytase. *Anim. Feed Sci. Technol.* **2004**, *117*, 265–279. [CrossRef]
18. van Doorn, D.; Everts, H.; Wouterse, H.; Beynen, A.C. The apparent digestibility of phytate phosphorus and the influence of supplemental phytase in horses. *J. Anim. Sci.* **2004**, *82*, 1756–1763. [CrossRef] [PubMed]

19. Luke. *Finnish Feed Tables and Feeding Recommendations*; Natural Resources Institute Finland; Available online: http://urn.fi/URN:ISBN:978-952-326-054-2 (accessed on 21 November 2019).

20. Zhao, X.; Müller, C.E. Macro and micromineral content of wrapped forages for horses. *Grass For. Sci.* **2015**, *71*, 195–207. [CrossRef]

21. Knowlton, K.F.; Radcliffe, J.S.; Novak, C.L.; Emmerson, D.A. Animal management to reduce phosphorus losses to the environment. *J. Anim. Sci.* **2004**, *82*, E173–E195. [CrossRef] [PubMed]

22. Uusi-Kämppä, J.; Närvänen, A.; Kaseva, J.; Jansson, H. Phosphorus and faecal bacteria in runoff from horse paddocks and their mitigation by the addition of P-sorbing materials. *Agr. Food Sci.* **2012**, *21*, 247–259. [CrossRef]

23. Parvage, M.M.; Kirchmann, H.; Kynkäänniemi, P.; Ulén, B. Impact of horse grazing and feeding on phosphorus concentrations in soil and drainage water. *Soil Use Manag.* **2011**, *27*, 367–375. [CrossRef]

24. Parvage, M.M.; Ulén, B.; Kirchmann, H. A survey of soil phosphorus (P) and nitrogen (N) in Swedish horse paddocks. *Agric. Ecos. Environ.* **2013**, *178*, 1–9. [CrossRef]

25. Parvage, M.M.; Ulén, B.; Kirchmann, H. Are horse paddocks threating water quality through exess loading of nutrients. *J. Environ. Manag.* **2015**, *147*, 306–313. [CrossRef]

26. Henneke, R.R.; Potter, G.D.; Kreider, J.L.; Yeates, B.F. Relationship between condition score, physical measurements and body fat percentage in mares. *Equine Vet. J.* **1983**, *14*, 371–372. [CrossRef]

27. Särkijärvi, S.; Saastamoinen, M. Feeding value of various processed oat grains in equine diets. *Livest. Sci.* **2006**, *100*, 3–9. [CrossRef]

28. Huang, C.Y.L.; Schulte, E.E. Digestion of plant tissue for analysis by ICP Emission Spectroscopy. *Commun. Soil Sci. Plant Anal.* **1985**, *16*, 943–958. [CrossRef]

29. Robertson, J.; Van Soest, P. The detergent system of analysis and its application to human foods. *Anal. Diet. Fiber Food.* **1981**, *3*, 123.

30. Van Soest, P.J.; Robertson, J.B.; Lewis, B.A. Methods for dietary fiber, neutral detergent fiber, and nonstarch polysaccharides in relation to animal nutrition. *J. Dairy Sci.* **1991**, *74*, 3583–3597. [CrossRef]

31. Keskinen, R.; Saastamoinen, M.; Nikama, J.; Särkijärvi, S.; Myllymäki, M.; Salo, T.; Uusi-Kämppä, J. Recycling nutrients from horse manure: effects of bedding type and its compostability. *Agric. Food Sci.* **2017**, *26*, 68–79. [CrossRef]

32. Saastamoinen, M.T.; Hellämäki, M. Forage analyses as a base of feeding of horses. In *Forages and Grazing in Horse Nutrition*; Saastamoinen, M., Fradinho, M.J., Santos, A.S., Miraglia, N., Eds.; Wageningen Academic Publishers: Cambridge, MA, USA, 2012; pp. 304–314.

33. Pösö, A.R.; Soveri, T.; Oksanen, H.E. The effect of exercise on blood parameters in Standardbred and Finnish-bred horses. *Acta Vet. Scand.* **1983**, *24*, 170–184. [PubMed]

34. Patterson, D.P.; Cooper, S.R.; Freeman, D.W.; Teeter, R.G. Effects of varying levels of phytase supplementation on dry matter and phosphorus digestibility in horses fed common textured ration. *J. Equine Vet. Sci.* **2002**, *22*, 456–459. [CrossRef]

35. van Doorn, D.; van der Spek, M.E.; Everts, H.; Wouterse, H.; Beynen, A.C. The influence of calcium intake on phosphorus digestibility in mature ponies. *J. Anim. Physiol. Anim. Nutr.* **2004**, *88*, 412. [CrossRef]

36. Elzinga, S.; Nielsen, B.D.; Schott, H.C.; Rapson, J.; Robinson, C.I.; McCutcheon, J.; Harris, P.A.; Geor, R. Comparison of nutrient digestibilitys between adult and aged horses. *J. Equine Vet. Sci.* **2014**, *34*, 1164–1169. [CrossRef]

37. Buchholz-Bryant, M.A.; Baker, L.A.; Pipkin, J.L.; Mansell, B.J.; Haliburton, J.C.; Bachman, R.C. The effect of calcium and phosphorus supplementation, inactivity, and subsequent aerobic training on mineral balance in young, mature, and aged horses. *J. Equine Vet. Sci.* **2001**, *21*, 71–77. [CrossRef]

38. Böswald, L.F.; Dobenecker, B.; Clauss, M.; Kienzle, E. A comparative meta-analysis on the relationship of faecal calcium and phosphorus excretion in mammals. *J. Anim. Physiol. Anim. Nutr.* **2018**, *102*, 370–379. [CrossRef] [PubMed]

39. Schryver, H.F.; Craig, P.H.; Hintz, H.F. Calcium metabolism in ponies fed varying levels of calcium. *J. Nutr.* **1970**, *100*, 955–964. [CrossRef]

40. Hintz, H.F.; Schryver, H.F. magnesium metabolism in the horse. *J. Anim. Sci.* **1972**, *35*, 755–759. [CrossRef] [PubMed]

41. Meyer, H.; Heilemann, M.; Perez Noriega, H.; Gomda, Y. Postprandiale renale Ausscheidung von Calcium, Magnesium und Phosphorus bei ruhenden und arbeitenden Pferden. In *Advances in Animal Physiology and*

Animal Nutrition; Contributions to water and mineral metabolism of the horse; Springer: Berlin/Heidelberg, Germany, 1990; pp. 78–85.

42. Meyer, H.; Stadermann, B.; Schnurpel, B.; Nehring, T. The influence of type of diet (roughage or concentrate) on the plasma level, renal excretion, and apparent digestibility of calcium and magnesium in resting and exercising horses. *J. Equine Vet. Sci.* **1992**, *12*, 233–239. [CrossRef]

43. Ragnarsson, S.; Lindberg, J.E. Nutritional value of timothy haylage in Icelandic horses. *Livest. Sci.* **2008**, *113*, 202–208. [CrossRef]

44. Ragnarsson, S.; Lindberg, J.E. Nutritional value of mixed grass haylage in Icelandic horses. *Livest. Sci.* **2010**, *131*, 83–87. [CrossRef]

45. Martin Rosset, W.; Vermorel, M.; Doreau, M.; Tisserand, J.L.; Andrieu, J. The French horse feed evaluation systems and recommended allowances for energy and protein. *Livest. Prod. Sci.* **1994**, *40*, 37–56. [CrossRef]

46. Palmgren Karlsson, C.; Lindberg, J.E.; Rundgren, M. Associative effects on total tract digestibility in horses fed different ratios of grass hay and whole oats. *Livest. Prod. Sci.* **2000**, *65*, 143–153. [CrossRef]

47. Särkijärvi, S.; Sormunen-Cristian, R.; Heikkilä, T.; Rinne, M.; Saastamoinen, M. Effect of grass species and cutting time on in vivo digestibility of silage by horses and sheep. *Livest. Sci.* **2012**, *144*, 230–239. [CrossRef]

48. Breidenbach, A.; Schlumbohm, C.; Harmeyr, J. Pecularities of vitamin D andof the calcium and phosphate homeostatic system in horses. *Vet. Res.* **1998**, *29*, 173–186.

49. van Doorn, D.A.; Schaafstra, F.J.; Wouterse, H.; Everts, H.; Estepa, J.C.; Aguilera-Tejero, E.; Beynen, A.C. Repeated measurements of P retention in ponies fed rations with various Ca: P ratios. *J. Anim. Sci.* **2014**, *92*, 4981–4990. [CrossRef] [PubMed]

50. Greiwe-Crandell, K.M.; Morrow, G.A.; Kronfeld, D.S. Phosphorus and selenium depletion in Thoroughbred mares and weanlings. *Pferdeheilkunde*, Sonderausgabe 1. *Europäissche Konferenz über die Ernährung des Pferdes* **1992**, 77 80.

51. Jansson, A.; Dahlborn, K. Effects of feeding frequency and voluntary salt intake on fluid and electrolyte regulation in athletic horses. *J. Appl. Physiol.* **1999**, *86*, 1610–1616. [CrossRef] [PubMed]

52. Wartell, B.A.; Krumins, J.A.; Kang, K.; Schwab, B.J.; Fennell, D. Methane production from horse manure and stall waste with softwood bedding. *Bioresour. Technol.* **2012**, *112*, 42–50. [CrossRef] [PubMed]

53. Chapuis-Lardy, L.; Fiorini, J.; Toth, J.; Dou, Z. Phosphorus concentration and solubility in dairy feces: Variability and affecting factors. *J. Dairy Sci.* **2004**, *87*, 4334–4341. [CrossRef]

54. Dougherty, W.J.; Nicholls, P.J.; Milham, P.J.; Havilah, E.J.; Lawrie, R.A. Phosphorus Fertilizer and Grazing Management Effects on Phosphorus in Runoff from Dairy Pastures. *J. Environ. Qual.* **2008**, *37*, 417–428. [CrossRef]

55. Keskinen, R.; Nikama, J.; Närvänen, A.; Uusi-Kämppä, J.; Särkijärvi, S.; Saastamoinen, M. Reducing nutrient runoff from horse paddocks by removal of dung. In Proceedings of the Equi-Meeting Infrastructures Horses and Equestrian Facilities, Le Lion d´ Angers, France, 6–7 October 2014; pp. 60–65.

56. Kebreab, E.; Shah, M.A.; Beever, D.E.; Humphries, D.J.; Sutton, J.D.; France, J.; Mueller-Harvey, I. Effects of contrasting forage diets on phosphorus utilisation in lactating dairy cows. *Livest. Prod. Sci.* **2005**, *93*, 125–135. [CrossRef]

57. Maguire, R.O.; Dou, Z.; Sims, T.; Brake, J.; Joern, B.C. Dietary strategies for reduced phosphorus excretion and improved water quality. *J. Environ. Qual.* **2005**, *34*, 2093–2103. [CrossRef]

58. Dou, Z.; Knowlton, K.F.; Kohn, R.A.; Wu, Z.; Satter, L.D.; Zhang, G.; Toth, J.D.; Ferguson, J.D. Phosphorus characteristics of dairy feces affected by diets. *J. Environ. Qual.* **2002**, *31*, 2058–2065. [CrossRef]

59. Närvänen, A.; Jansson, H.; Uusi-Kämppä, J.; Jansson, H.; Perälä, P. Phosphorus load from equine critical source areas and its reduction using ferric sulphate. *Boreal. Environ. Res.* **2008**, *13*, 266–274.

60. Airaksinen, S.; Heiskanen, M.-L.; Heinonen, H. Contamination of surface run-off water and soil in two horse paddocks. *Bioresour. Technol.* **2007**, *98*, 1762–1766. [CrossRef] [PubMed]

61. von Wachenfelt, H.E. A field test of all-weather surfaces for horse paddocks. *J. Food Sci. Eng.* **2016**, *6*, 197–211. [CrossRef]

62. Parvage, M.M.; Ulén, B.; Kirchmann, H. Can organic materials reduce nutrient leaching from manure-rich paddock soils? *J. Eneviron. Qual.* **2017**, *46*, 105–112. [CrossRef] [PubMed]

63. Kebreab, E.; Odongo, B.W.; McBride, B.W.; Hanigan, M.D.; France, J. Phosphorus Utilization and Environmental and Economic Implications of Reducing Phosphorus Pollution from Ontario Dairy Cows. *J. Dairy Sci.* **2008**, *91*, 241–246. [CrossRef]

64.   Uotila, R.; Thuneberg, T.; Saastamoinen, M. The usage of forage analyses in optimizing horse nutrition in Finland. In *Forages and Grazing in Horse Nutrition*; Saastamoinen, M., Fradinho, M.J., Santos, A.S., Miraglia, N., Eds.; Wageningen Academic Publishers: Cambridge, MA, USA, 2012; pp. 331–334.

65.   Julliand, V.; Grimm, P. The impact of diet on the hindgut microbiome. *J. Equine Vet. Sci.* **2017**, *52*, 23–28. [CrossRef]

13

Answers to the Frequently Asked Questions Regarding Horse Feeding and Management Practices to Reduce the Risk of Atypical Myopathy

Dominique-Marie Votion [1], Anne-Christine François [2,*], Caroline Kruse [3], Benoit Renaud [2], Arnaud Farinelle [4], Marie-Catherine Bouquieaux [1], Christel Marcillaud-Pitel [5] and Pascal Gustin [2]

[1] Equine Pole, Fundamental and Applied Research for Animals & Health (FARAH), Faculty of Veterinary Medicine, University of Lieège, 4000 Liège 1 (Sart Tilman), Belgium; dominique.votion@uliege.be (D.-M.V.); mcbouquieaux@uliege.be (M.-C.B.)

[2] Department of Functional Sciences, Faculty of Veterinary Medicine, Pharmacology and Toxicology, Fundamental and Applied Research for Animals & Health (FARAH), University of Liège, 4000 Liège 1 (Sart Tilman), Belgium; benoit.renaud@uliege.be (B.R.); p.gustin@uliege.be (P.G.)

[3] Department of Functional Sciences, Faculty of Veterinary Medicine, Physiology and Sport Medicine, Fundamental and Applied Research for Animals & Health (FARAH), University of Liège, 4000 Liège 1 (Sart Tilman), Belgium; caroline.kruse@uliege.be

[4] Fourrages Mieux asbl, 6900 Marloie, Belgium; farinelle@fourragesmieux.be

[5] Réseau d'Epidémio-Surveillance en Pathologie Équine (RESPE), 14280 Saint-Contest, France; c.marcillaud-pitel@respe.net

[*] Correspondence: acfrancois@uliege.be

Simple Summary: Equine atypical myopathy is a severe intoxication of grazing equids resulting from the ingestion of samaras or seedlings of trees from the *Acer* species. The sycamore maple (*Acer pseudoplatanus*) is involved in European cases whereas the box elder (*Acer negundo*) is recognized as the cause of this seasonal pasture myopathy in the Unites States of America. In Europe, young and inactive animals with a thin to normal body condition and no feed supplementation, except for hay in autumn, are at higher risk. The risk is also associated with full time pasturing in a humid environment. Indeed, dead leaves piling up in autumn as well as, the presence of trees and/or woods presumably exposes the horses to the sycamore maple. This manuscript answers the most frequently asked questions arising from the equine field about feeding and management of equines to reduce the risk of atypical myopathy. All answers are based on data collected from 2006 to 2019 by the "Atypical Myopathy Alert Group" (AMAG, Belgium) and the "Réseau d'épidémiosurveillance en Pathologie équine" (RESPE, France) as well as on a review of the most recent literature.

Abstract: In 2014, atypical myopathy (AM) was linked to *Acer pseudoplatanus* (sycamore maple) in Europe. The emergence of this seasonal intoxication caused by a native tree has raised many questions. This manuscript aims at answering the five most frequently asked questions (FAQs) regarding (1) identification of toxic trees; reduction of risk at the level of (2) pastures and (3) equids; (4) the risk associated with pastures with sycamores that have always been used without horses being poisoned and (5) the length of the risk periods. Answers were found in a literature review and data gathered by AM surveillance networks. A guide is offered to differentiate common maple trees (FAQ1). In order to reduce the risk of AM at pasture level: Avoid humid pastures; permanent pasturing; spreading of manure for pasture with sycamores in the vicinity and avoid sycamore maple trees around pasture (FAQ2). To reduce the risk of AM at horse level: Reduce pasturing time according to weather conditions and to less than six hours a day during risk periods for horses on risk pasture; provide supplementary feeds including toxin-free forage; water from the distribution network; vitamins and a salt block (FAQ3). All pastures with a sycamore tree in the vicinity are at risk (FAQ4). Ninety-four percent of cases occur over two 3-month periods, starting in October and in March, for cases resulting from seeds and seedlings ingestion, respectively (FAQ5).

Keywords: equine atypical myopathy; *Acer* spp.; risk factors; environment

1. Introduction

Equine atypical myopathy (AM) is a severe pasture-associated intoxication that may occur in autumn and spring following the ingestion of certain species of maple (*Acer*) seeds and seedlings, respectively. This environmental intoxication is linked to *Acer pseudoplatanus* (sycamore maple) in Europe and *Acer negundo* (box elder) in the US [1,2]. These trees may contain several toxins [3]. The ingestion of samaras or seedlings of the incriminated trees goes with the ingestion of two cyclopropylamino acids, hypoglycin A (HGA) and methylenecyclopropylglycine (MCPG) [4]. These toxins have been confirmed to be implicated in European AM cases. Long before the discovery of the cause of AM, Fowden and Pratt (1973) [3], reported the presence of cyclopropyl derivates in seeds of the different representatives of the *Acer*'s species. Both *Acer pseudoplatanus* and *Acer negundo* seeds have been found to contain HGA and MCPG. On the contrary, other maple trees commonly found in Europe as *Acer platanoides* (Norway maple) and *Acer campestre* (field maple) tested negative for these compounds [3].

In fact, HGA and MCPG are not toxic *per se* but need to be converted into their active metabolites, i.e., methylenecyclopropylacetyl-CoA (MCPA-CoA) and methylenecyclopropylformyl-CoA (MCPF-CoA), respectively [4,5]. Both MCPA-CoA and MCPF-CoA inhibit enzymes that participate in β-oxidation and thus energy production from lipid metabolism [5,6]. The typical sign of intoxication is an acute rhabdomyolysis syndrome unrelated to exercise. This clinical picture may be seen on several horses within a group [7–10]. In more than 50% of the cases, the following clinical signs were observed: Weakness, recumbency, myoglobinuria, full bladder, stiffness, depression, muscle tremors or fasciculation, reluctance to move, sweating, normothermia, and congested mucous membranes [10–12] Atypical myopathy has a high mortality rate (i.e., 74%) that varies between countries and years (from 43% [11] to 97% [13]). The overall mortality rate of 74% average data among countries included in the study of van Galen et al., (2012) and does not take into account the different sources of variability [11]. For example, a study reports a lower mortality rate (i.e., 44%) in hospitalized animals [14]. It is hypothesized that less critical cases are driven to a hospital where appropriate symptomatic treatment is easier to provide. These two factors may contribute to the higher survival rate in an equine hospital than in the field. In any case, only the administration of vitamins and antioxidants has proven to be beneficial for survival [12,15].

In 2004, an alert group named "Atypical Myopathy Alert Group" (AMAG) was launched to warn horse practitioners and owners of the risk peaks. The alerts are released following case declarations and the AMAG regularly updates its data with the latest number of cases. Additionally, to its disease surveillance role, AMAG collects epidemiologic data about AM that has emerged in several European countries since 2006. French cases are gathered in close collaboration with the Réseau d'Épidémio-Surveillance en Pathologie Équine (RESPE) which monitors equine diseases through a network of French sentinel practitioners. Thanks to the European surveillance network, we now know that several Père David's deer (*Elaphurus davidianus*) have succumbed from this intoxication in different zoos in Germany. This indicates that ruminants pasturing in the vicinity of sycamore trees may also be intoxicated [16].

In light of the high mortality rate and the absence of specific treatment, prevention is the key to avoid intoxication of animals. Before discovering the cause of AM, epidemiological studies revealed risk factors associated with management practices of horse and pasture [15,17] In 2014, the cause of AM was discovered and linked to *Acer pseudoplatanus* in Europe [1]. Despite the cause of the intoxication being known, outbreaks have continued to occur. The development of a condition caused by a native tree has raised many questions among all actors attached to the equine sector.

The most frequently asked question (FAQ) concerns the identification of toxic trees. Other commonly-asked questions involve the feeding and management practice of equids in order to reduce the risk of intoxication. These main FAQs can be summarized as being: (FAQ1) "Which maples are toxic? Is this tree a maple and if so, is it toxic?"; (FAQ2) "How can AM be prevented (at pasture level)?"; (FAQ3) "How can AM be prevented (at horse level)?"; (FAQ4) "Our pasture is surrounded by sycamore maple trees, but no case of AM ever occurred in our grazing horses. Does this mean the pasture is safe for our animals?" and (FAQ5), "When does the risk of AM start and stop in autumn and spring?".

This manuscript answers to these FAQs regarding horse nutrition and management practices in order to prevent AM both by reviewing the most recent literature and by analyzing epidemiological data gathered since 2006.

2. Materials and Methods

2.1. Literature Review

A systematic review was performed using on the electronic databases PubMed and Scopus with « atypical myopathy » AND « horse » as keywords. In addition, abstracts of proceedings of meetings dedicated to horses were consulted (e.g., AAEP—American Association of Equine Practitioners Annual Convention, AVEF—Association des Vétérinaires Équins Français, BEPS—Belgian Equine Practitioners Society Study Days, BEVA—British Equine Veterinary Association and, WEVA—World Equine Veterinary Association Congress).

2.2. Epidemiological Data

Information regarding AM cases in Europe over a 13-year period (2006–2019) was collected via standardized questionnaires available on the AMAG (http://www.myopathie-atypique.be) and the RESPE (https://respe.net) websites. These forms were completed via email or phone contact with the owners or veterinarian whenever possible. Additionally, cases were also gathered by direct contact via mail or phone between owner/veterinarian and the principal investigator of this study. Information about the management and environment of diseased equines was obtained from the animals' owners whereas clinical data was collected from veterinary surgeons. Cases occurring between the 1 September to the end of February were classified as "autumnal cases" and those from the 1 March up to the end of August as "spring cases".

3. Results

3.1. Literature Review

Among 68 records identified in PubMed, five were rejected as they did not concern AM. From the search in Scopus, 514 results were obtained that were refined by selecting only research articles. From the 82 remaining studies, 63 were out of the scope and were discarded. In total, more than 127 documents (research articles and abstracts) were obtained.

3.2. Epidemiological Data

Epidemiologic data from 3039 cases were recorded in 14 different countries from autumn 2006 to 30 November 2019 (Figure 1, Table 1). This data set includes all cases communicated to the surveillance networks with a tentative diagnosis of AM. During this period, 14 autumnal outbreaks were encountered with a mean (± S.D.) of 164.6 ± 153.3 (median 79.5) reported cases. For spring, 12 outbreaks were recorded with a mean (± S.D.) of 56.5 ± 77.3 (median 24.0) diseased horses per outbreak. For all outbreaks together, the mean (± S.D.) number of cases is 112.5 ± 132.4 (median

41). Parts of these data (n = 824) have already been analyzed to define risk (Table 2) and protective (Table 3) factors on cases in Belgium [17], in the UK [12] and at a European level [11,15]. Equines particularly at risk for AM were found to be young (i.e., less than 3 years of age) and inactive animals with normal body condition score and receiving hay in autumn [17]. The risk of intoxication was also associated with full time pasturing in a humid environment where dead leaves pile up in autumn, with the presence of trees and/or woods and thus presumably exposed to the above-mentioned maple trees [12,15,17]. During the ten years that have passed since the last epidemiological study performed at an European level [11,15], 2433 new cases have been reported, reaching a total number of 3039 cases available for the current study. The whole database (i.e., 2006–2019) was cleaned by removing all equids that were not at pasture at the onset of clinical signs or within the week preceding these signs, equids that were diagnosed with another disease as well as equids having a low probability of intoxication (e.g., no pigmenturia and serum creatine kinase activities <10.000 IU/L; normal values 50–200: IU/L) according to van Galen et al., 2012 [11]. As opposed to the study of van Galen et al., 2012 [11], cases too poorly documented to make a definitive diagnosis have been retained in the study group. A total of 2371 cases was included in this study. The age distribution of these cases over the years are presented in Figure 2. The weekly occurrence of AM during the spring and autumnal seasons may be found in Figure 3a,b, respectively.

Figure 1. European distribution of atypical myopathy cases notified to the disease surveillance networks from autumn 2006 to November 2019.

Table 1. Total number of atypical myopathy cases in Europe notified to the disease surveillance networks from autumn 2006 to November 2019 (n = 3039).

Year / Countries	2006	2007		2008		2009		2010		2011		2012		2013		2014*		2015		2016		2017		2018		2019		Total/Country
	Autumn	Spring	Autumn	Spring	Autumn	Spring	Autumn	Spring	Autumn	Spring	Autumn	Spring	Autumn	Spring	Autumn	Spring	Autumn	Spring	Autumn	Spring	Autumn	Spring	Autumn	Spring	Autumn	Spring	Autumn	
Austria	0	0	0	0	0	0	0	0	0	0	4	0	0	0	0	0	0	0	0	0	0	0	0	4	9	0	0	17
Belgium	46	7	18	0	5	0	69	13	14	3	18	0	1	0	141	8	51	5	6	8	52	20	5	1	191	49	1	732
Czech Republic	0	0	0	0	0	0	0	0	0	0	0	0	0	0	9	0	0	0	0	0	2	17	0	0	19	2	0	49
Denmark	2	0	0	0	0	0	2	0	0	0	0	0	0	0	0	0	0	0	2	0	0	0	0	0	2	0	0	8
France	32	1	11	18	11	0	134	106	32	10	40	16	4	1	64	13	71	9	24	8	194	181	31	18	114	47	26	1216
Germany	7	0	3	5	0	0	93	21	2	2	59	2	0	0	24	1	8	0	0	0	21	20	4	0	33	4	2	311
Ireland	0	0	0	0	0	0	2	0	0	0	0	0	0	0	2	0	38	0	0	0	1	2	0	0	1	0	0	46
Luxembourg	1	0	0	0	0	0	2	0	0	0	0	0	0	0	0	0	0	0	0	0	0	0	0	0	0	0	0	3
Portugal	0	0	0	0	0	0	0	0	0	1	0	0	0	0	0	0	0	0	0	0	1	0	0	0	0	0	0	2
Spain	0	0	0	0	0	0	0	0	0	0	52	0	0	0	0	0	0	0	0	0	0	0	0	0	0	0	0	52
Sweden	0	0	0	0	0	0	0	1	0	0	1	0	0	0	0	5	0	0	0	0	0	0	0	0	0	0	0	7
Switzerland	0	0	9	0	0	0	31	3	0	0	6	0	0	1	12	0	0	0	0	0	9	3	0	0	7	0	0	81
The Netherlands	13	0	3	0	2	0	34	7	0	0	0	0	0	0	18	1	4	0	0	0	1	2	0	2	9	10	1	107
United Kingdom	1	0	13	0	0	0	39	20	3	0	33	6	2	2	52	13	154	20	2	1	11	11	3	1	13	3	5	408
Total/season	102	8	57	23	18	0	406	171	51	16	213	24	7	4	322	41	326	34	34	17	292	256	43	26	398	115	35	
Total/year	102	65		41		406		222		229		31		326		367		68		309		299		424		150		3039

Comments: This last counting replaces all previously published data; the word "autumn" should not be taken strictly since clinical series continue into early winter; (*) in 2014, atypical myopathy was linked to *Acer pseudoplatanus* (sycamore maple) in Europe.

Table 2. Risk factors identified in from former epidemiological studies [12,15,17].

Category	Risk Factors	Odds Ratio	95% CI for Odds Ratio
Demographic data	Young horses (<3 years)		
	Thin (or normal weight *)	3.08 [17] (b) (3.85 [15]/2.20 [17] (b))	1.01–9.39 [17] (b) (1.77–8.37 [15]/1.01–4.79 [17] (b))
Horse management	At pasture 24 h a day all year round	5.42 [15] 3.07 [17] (b) in winter 3.78 [17] (b) in spring 23.2 [17] (b) in summer 10.9 [17] (b) in autumn	2.577–11.42 [15] 1.45–6.50 [17] (b) in winter 1.49–9.59 [17]] (b) in spring 1.41–382 [17] (b) in summer 3.56–33.4 [17]] (b) in autumn
	Not physically active	11.8 [17] (b)	5.02–27.8 [17] (b)
Feeding practice	Hay given in autumn	4.09 [17] (a)	1.18–14.1 [17] (a)
	History of previous death of horse(s) on the pasture	4.45 [17] (b)	1.61–12.29 [17] (b)
	Lush pasture in winter	3.95 [17] (b)	1.49–10.46 [17] (b)
	Sloping pasture/steep slope	3.43 [15] 3.70 [17] (b)	1.52–7.77 [15] 1.58–8.68 [17] (b)
	Access to dead leaves piled up in autumn	11.11 [15] 10.47 [17] (b)	4.82–25.59 [15] 2.82–40.88 [17] (b)
Pasture	Presence of trees at pasture *	7.82 [15]	1.99–30.73 [15]
	Dead wood at pasture	3.12 [15]	1.42–6.84 [15]
	Humid pasture	2.63 [17] (b)	1.29–5.36 [17] (b)
	Pasture surrounded by or containing a stream/river	2.78 [17] (b)	1.24–6.19 [17] (b)
	Spreading of manure	5.73 [17] (b)	2.40–13.69 [17] (b)

(*) Parameters that are not consistent among studies [15] cases with a high probability of or confirmed atypical myopathy vs. and the cases with a low probability of having AM or with another diagnosis [17] (a) confirmed cases vs. clinically healthy co-grazing equidae [17] (b) confirmed cases vs. control horses.

Horse Nutrition and Feeding Management

Table 3. Protective factors identified in former epidemiological studies [12,15,17].

Category	Protective Factors	Odds Ratio	95% CI for Odds Ratio
Demographic data	Overweight	0.25 [17] (b)	0.09–0.69 [17] (b)
	Frequent deworming	0.11 [17] (a) 0.05 [17] (b)	0.01–0.67 [17] (a) 0.01–0.16 [17] (b)
	Regular vaccination	0.10 [17] (b)	0.05–0.21 [17] (b)
	Regular physical activity	0.08 [17]	0.03–0.19 [17]
Horse management	Weather-dependent pasturing time in spring and in autumn	0.24 spring [17] 0.10 autumn [17]	0.06–0.89 spring [17] 0.02–0.56 autumn [17]
	<6 h at pasture per day	0.04 [15] 0.62 [17]	0.01–0.19 [15] 0.16–2.36 [17]
	No access to pasture	0.03 [15]	0.00–0.22 [15]
	Supplementary feeds all year round	0.17 [15]	0.05–0.59 [15]
Feeding practice and water supply	Silage and concentrate feed in autumn + corn in winter	0.20 [17] (a) for silage 0.19 [17] (a) for concentrate feed 0.22 [17] (a) for corn	0.04–0.94 [17] (a) for silage 0.04–0.87 [17] (a) for concentrate food 0.05–0.93 [17] (a) for corn
	Salt block (all year)	3.52 [12] 0.20 [17]	1.08–11.47 [12] 0.09–0.40 [17]
	Water provision in tank/bathtub	0.25 [15]	0.09–0.69 [15]
	Water supplied by the distribution network	0.39 [17] (b)	0.17–0.88 [17] (b)
Pasture	Gentle slope	0.34 [17] (b)	0.14–0.84 [17] (b)

[12] Survivors vs. nonsurvivors [15] cases with a high probability of or confirmed atypical myopathy vs. the cases with a low probability of having AM or with another diagnosis [17] (a), confirmed cases vs. clinically healthy co-grazing equidae [17] (b), confirmed cases vs. control horses.

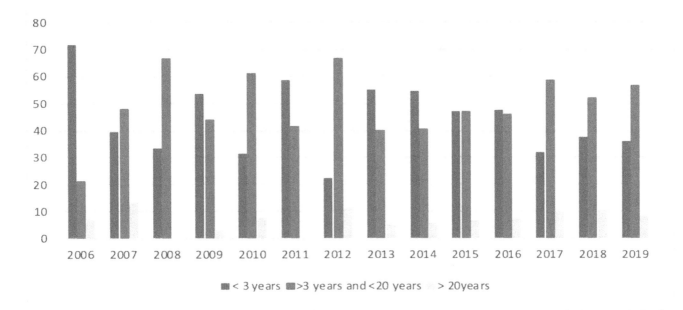

Figure 2. Frequency distribution of equids with age categories: <3 years, >3 years and <20 years and >20 years old (n = 1510) over the study period (2006–2019).

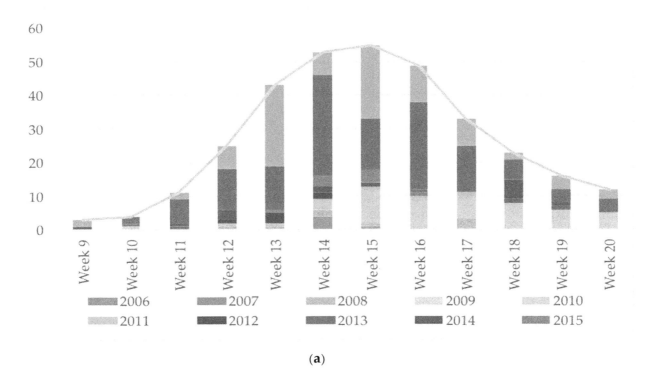

(**a**)

Figure 3. *Cont.*

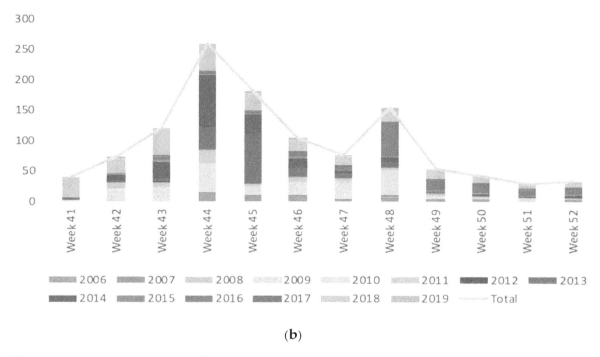

(b)

Figure 3. (a). Spring cases: weekly occurrence of atypical myopathy cases from week 9 (1 March) up to week 20 (31 May) over the study period (2006–2019); (b). Autumnal cases: weekly occurrence of atypical myopathy cases from week 41 (1 October) up to week 52 (31 December) over the study period (2006–2019).

4. Discussion

The origin of the results consists of previous epidemiological studies modified and completed by the analysis of the newest data. The results contribute to answer the FAQs regarding horse feeding and management practices to reduce the risk of AM.

4.1. FAQ1: "Which Maples Are Toxic? Is this Tree a Maple and If So, Is It Toxic?"

The question about which maple trees are toxic is often associated with a request to identify trees on the pasture. The database consultation indicates that 99% of pastures contain or are directly bordered by tree. However, looking at the data from 2014 up to now, it is observed that 20% (92/456) of AM horse owners could not answer if seeds and/or seedlings of sycamore trees were present in their meadow. Despite the educational material available on the Internet (https://en.wikipedia.org), horse owners and veterinarians are still struggling to recognize the different maple species (personal observation). This phenomenon is accentuated due to the numerous erroneous descriptions in the literature [18]. A guide from Renaud et al., (2019) is available [19], helping the different actors to differentiate the three *Acer* species commonly found in European pastures where cases have been declared [20].

The maple genus includes approximately 561 species [21]. Some of them are extensively planted as ornamental trees because of their autumnal color. As a result, there is not only a demand to distinguish non-toxic trees (*Acer platanoides* (Norway maple) and *Acer campestre* (Field maple)) from *Acer pseudoplatanus*, but also many questions regarding the potential toxicity of other maple species. Even though up to now, not all maple trees have been tested, it is one of note that almost 50 years before the discovery of the cause of AM, the incriminated toxins had already been tested in many *Acer* species [3]. Among the tested species, the following species have tested positive for HGA and/or MCPG (non-exhaustive list): *Acer palmatum*, *Acer japonicum*, *Acer macrophyllum*, *Acer spicatum*, *Acer saccharinurn*, and *Acer saccharum*. These exotics species may be found in ornamental gardens and may spread to the neighboring regions [22]. Therefore, these *Acer* species might ultimately represent a risk of intoxication for equids.

4.2. FAQ2: "How Can AM Be Prevented (at Pasture Level)?"

Atypical myopathy occurs seasonally with outbreaks starting in autumn that may continue in early winter. On the contrary, spring outbreaks usually cease before summer. At pasture level, the risk can be decreased (1) by avoiding contact with toxic plant material and (2) by favoring low-risk meadows for pasturing during autumn and spring.

4.2.1. Avoid Contact with Toxic Plant Materials

Former epidemiological studies identified access to dead leaves piled up in autumn, the presence of trees and dead wood on pastures as risk factors for AM [15,17]. This observation is presumably due to the presence of *Acer pseudoplatanus* and the ingestion of samaras and seedlings in autumn and spring, respectively [23–25]. It is therefore important to be able to recognize *Acer pseudoplatanus*, its samaras and seedlings. When in doubt, professional expertise should be sought to identify the tree (botanists and/or forestry agents might be of help). Recently, it has been suggested that flowers falling from sycamore trees after heavy rainfall and/or wind could be an additional source of intoxication [26].

Depending on weather conditions, samaras of *Acer* species may be able to spread their seeds up to several hundred meters [27]. Therefore, pasture contamination with seeds or seedlings is not necessarily linked to the presence of a tree on the pasture. In early autumn, especially after windy weather has dispersed sycamore samaras, it is recommended to equids' owners to identify contaminated areas in their pasture. The removal of seeds may help to prevent AM [28]. However, when samaras are too abundant and/or too widely dispersed within the premise, grazing in the affected area must be prohibited. Another way to limit grazing to areas free of fallen seeds and/or flowers and/or seedlings is to create parcels within the pasture [26].

Additionally, the spreading of manure and/or harrowing of pastures was found to increase the risk of AM [17]. This practice might favor the dispersal of the toxic material throughout the pasture and subsequent intoxication of horses.

Regarding prevention at the pasture level, there is growing interest in the disposal of seedlings. It is worth noting that seedlings still contain HGA after herbicidal spraying or mowing [29]. These techniques are therefore ineffective regarding the destruction of toxic material.

4.2.2. Use or Create Low-Risk Pastures

Permanent pasturing was found to be a risk factor. This is most probably due to the associated decrease in grass quantity, which leads equids to ingest the etiological agent [12,15,17]. Our database indicates that, in 64% of our cases, the pasture grass was bare or absent of grass. This observation correlates with previous epidemiological studies [11]. A good pasture management (for example pasture rotation) is advised in order to offer lush pastures. Indeed, a green meadow will limit the ingestion of seeds and seedling by horses allowing them to eat mainly grass [11,17].

Pastures particularly at risk for AM have *Acer pseudoplatanus* in their vicinity. Grazing on these meadows should be avoided during the risky seasons (see FAQ4). Furthermore, humid pastures are of particular risk for AM and grazing should therefore also be avoided on these pastures [17]. HGA is a water-soluble toxin that may pass from plants to water by direct contact [26,30]. This solubility might explain the risk associated with humidity and the protective factor linked to drinking water provided via the distribution network. For this reason, only pastures that do not contain rivers and/or freestanding water should be used during the risky seasons.

4.3. FAQ3: "How Can AM Be Prevented (at Horse Level)?"

All type of equids were affected by AM including donkeys (1.6% of the cases) and zebras (3 cases from zoological parks) with no highlighted risk factor associated with any species. The first epidemiological study highlighted that young horses, especially those <3 years of age were the primary affected group [17]. Later on, van Galen et al. (2012) [11] found that all age groups were represented. Our data *suggest* a gradual change in age distribution of cases over the years (Figure 2). In 2006, 71% of affected equids were less than 3 years old, whereas now this age group represents only 36% of individuals. This finding is unlikely to be explained by acquisition of immunity to the toxins since some survivors of AM did succumb to a second episode of the disease (unpublished data). An explanation could be that the population at risk is increasing because toxic pressure has increased over the years. The practical usefulness of this information is that all equids must be considered at risk, whatever their age. At the animal level, the risk can be decreased by (1) management and (2) feeding practices.

4.3.1. Management of Grazing Time

The intoxication is intimately linked to pasturing as van Galen et al., (2012) reported that 98% of affected horses were at pasture at onset of clinical signs [11]. Our data confirms this information with 99.8% of horses being at pasture when clinical signs declared. The few remaining cases had been stabled for less than a week which implies these animals may have been in contact with the toxin before. Up to now, not a single case of our database has been confirmed in horses that had no access to pasture and/or paddock based on HGA and MCPA-carnitine detection (unpublished data). Thus, it is advised to stable horses during autumn and spring outbreaks if seeds or seedlings are, or may be, present at pasture [26]. However, keeping a horse in the stable day and night may be difficult and is not considered as good practice in animal welfare [31,32]. Interestingly, the limitation of grazing time to less than six hours a day was found to be a protective factor [15]. Consistent with a previous study of UK cases [12], 97.5% of equids of our database had spent more than six hours per day at pasture. This observation suggests that the length of exposure to the toxins appears to be a determining factor in the risk of AM.

Specific weather conditions have been linked with AM outbreaks [9,10,33,34] whereas weather-dependent pasturing time (i.e., stabling horses when inclement weather is forecast) reduces the risk of AM [17]. Reducing pasturing time according to weather conditions was not a recommendation implemented by owners in our cases since less than 1% of them were in compliance with this preventive measure. Our data reinforces the value of this preventive measure.

4.3.2. Feed and Water Supply

Receiving supplementary feeds (hay, straw, complete mix, oats, barley, and/or corn) throughout the year decreases the risk of AM [17]. Atypical myopathy results from an energetic imbalance subsequent to HGA and MCPG poisoning. Feed provides energy substrates (especially carbohydrates) that supports the energetic metabolism and also vitamins and antioxidants known to increase the chance of survival [15]. The mitochondrial enzymes inhibited by HGA are flavin adenine dinucleotide (FAD) dependent. This cofactor originates from riboflavin (vitamin B2) suggesting that it would be useful to give this vitamin [35]. Alfalfa is an excellent natural source of riboflavin as well as, to a lesser extent, the hay from common grass [36]. In addition, well-nourished horses might be less tempted to ingest samaras and/or seedlings. Among horses receiving supplementary feeds in their daily diet (64%; n = 665), our data indicates that 61% received concentrated feed (complete mix, oats, and/or barley) and 50% had access to a salt block providing minerals. However, a salt block did not prevent these animals to be intoxicated.

Giving hay in autumn was identified as a risk factor [17] (and 40% of our cases receiving supplementary feed were fed with hay only). Indeed, hay may contain seeds and seedlings with detectable HGA concentration after several months [29] and even years of storage [30]. Gonzales et al. (2019) suggests that AM might occur in stabled horses [29] but this hypothesis is not sustained by our data since, as above-mentioned, not a single case has been confirmed by blood testing in equids with no access to pasture and/or paddock. However, we do agree that giving hay produced from contaminated pasture would increase the risk of AM in horses kept at pasture. It is probably wise not to produce hay/haylage and/or silage from pasture areas in the vicinity of sycamore trees [29]. Providing hay in autumn is controversial. Indeed, this practice was found to be a risk factor in a case-control study [17] but that appears nevertheless as a protective factor when comparing management practice in pasture of cases versus pastures of controls [15]. From this result, we can suggest that forages free from toxins should be given *at libitum*, but hay should neither be placed on the ground, nor under sycamore trees, since both practices could increase the risk of ingesting toxic material.

4.3.3. Drinking Water

Water supplied by the distribution network [17] or stored in a tank or in an bathtub [15] are protective factors. These observations suggest that water from other sources may be contaminated by the toxins. This hypothesis is reinforced by the study of Renaud et al., (2019) [30], which showed that HGA is released by stagnant flowers or seeds from *Acer pseudoplatanus* in contact with water. On the other hand, when water is dripping off flowers, no HGA is detected in collected water. However, this latest observation can be modulated by the fact that (1) the concentration of HGA can be below the limit detection threshold of the quantification method or (2) that HGA might be degraded in water. To the authors knowledge, there is no published study about the stability of HGA in water.

4.4. FAQ4: " Our Pasture Is Surrounded by Sycamore Maple Trees, but No Case of AM ever Occured in Our Grazing Horses. Does this Mean the Pasture Is Safe for Our Animals?"

Our data and former epidemiological studies indicate that unexplained sudden deaths of horse(s) had been noted on 22% of the pastures where cases were grazing [11]. In other words, 80% of cases were grazing in pastures that had no history of previous death of equid(s) (regardless of the suspected cause). Atypical myopathy is an emerging disease and a pasture surrounded by sycamore trees should not be considered as safe for pasturing horses.

4.5. FAQ5: "When Does the Risk of AM Start and Stop in Autumn and Spring?"

As previously reported, cases of AM occur more frequently in autumn (76%; n = 1801) than in spring (24%; n = 570). The expressions "autumn" and "spring" should not be taken *stricto senso*, since the autumnal clinical series are continuing in early winter and some spring cases are occurring after the 21 June. It is worth noting that 94% of "spring" cases occurred between the 1 March and the 31 May and 94% of "autumnal" cases occurred between the 1 October up to the 31 December (Figure 3a,b).

The cause of autumnal outbreaks cessation is not precisely known. Before discovering the etiology of AM, it was hypothesized that severe frost might destroy the etiological agent since outbreaks tend to cease after several days of deep freezing [10]. Now that HGA has been described as incriminated toxin, this hypothesis can be refuted since laboratory investigation showed that HGA is unaltered after several freeze–thaw cycles [37]. In our laboratory (unpublished data), we have collected samaras from the environment on a weekly basis since 2016 from now and, with very few exceptions, HGA has always been detected in seeds of sycamore tree. However, clinical series of AM usually fully stop in winter and resume with the germination of the samaras (personal observation). The analysis of HGA concentration over time in samaras fallen on the ground showed a high variability from tree to tree and from week to week thus impeding an easy interpretation of the evolution of toxicity. These field studies were complicated by the fact that the samaras collected on the ground had fallen at very different times. Therefore, the cause of the ceasing "autumnal" outbreaks in winter is not known but could result from a reduction in accessibility (e.g., adheration to the ground following rain and frost) and/or a decrease in toxins' concentration).

Regarding spring outbreaks, horse owners wonder if the case series stops because the seedlings have lost in toxicity. Actually, the end of spring outbreaks may not be explained by the disappearance of the toxicity since the seedlings remain toxic [26]. It is however hypothesized that spring outbreaks cease following a relative decrease in risk of intoxication by grazing. This reduced risk of intoxication might result from (1) a lusher meadow, (2) the observed decrease in toxicity of seedlings with their growth [26], (3) a decrease in palatability of older plants [26], (4) less frequently encountered weather conditions favoring toxicity and (5), a significant natural disappearance of seedlings. Regarding the latter, only a small percentage (<20%) of seedlings recorded in early spring on heavily contaminated pastures are still present in early summer (unpublished data). This observation added to the fact that herbicidal spraying do not reduce HGA concentration in sycamore seedlings [29] questions the benefit of herbicide treatments.

5. Conclusions

As there is no specific treatment for AM yet, prevention is the key. The risk of developing AM results from the combinations of protective and risk factors. In order to reduce the risk of AM, it is advised to avoid humid pastures, permanent pasturing, spreading of manure, and contact with sycamore plant material. During the risky periods pasturing time should be modulated according to weather conditions and limited to less than six hours a day. Grazing equids should receive supplementary feeds, with preferences for feeds containing riboflavin. When hay or silage are fed, it is necessary to exercise caution ensuring the forages are toxin-free. Also, it is advised to supply a salt block and provide drinking water from the distribution network. It is worth noting that AM is an emerging disease and equids of any age and all pastures with a sycamore tree in the vicinity must be considered at risk. These preventive measures should be implemented for a period of 3 months twice yearly, starting in March for "spring cases" then again in October to prevent "autumnal cases". As mentioned before, these are the critical seasons and samaras or seedlings are likely to be present on the pasture. A French version of this paper with additional illustrations can be found in supplementary materials (Supplementary S1).

Author Contributions: Data curation, M.-C.B.; Funding acquisition, D.-M.V.; Investigation, D.-M.V.; Resources, A.F. and C.M.-P.; Supervision, D.-M.V. and P.G.; Writing—original draft, D.-M.V. and A.-C.F.; Writing—review and editing, B.R. and C.K. All authors have read and agreed to the published version of the manuscript.

Acknowledgments: The authors thank all communicating veterinarians and owners of affected horses for their collaboration.

References

1. Votion, D.M.; Van Galen, G.; Sweetman, L.; Boemer, F.; De Tullio, P.; Dopagne, C.; Lefère, L.; Mouithys-Mickalad, A.; Patarin, F.; Rouxhet, S.; et al. Identification of methylenecyclopropyl acetic acid in serum of European horses with atypical myopathy. *Equine Vet. J.* **2014**, *46*, 146–149. [CrossRef]
2. Valberg, S.J.; Sponseller, B.T.; Hegeman, A.D.; Earing, J.; Bender, J.B.; Martinson, K.L.; Patterson, S.E.; Sweetman, L. Seasonal pasture myopathy/atypical myopathy in North America associated with ingestion of hypoglycin A within seeds of the box elder tree. *Equine Vet. J.* **2013**, *45*, 419–426. [CrossRef]
3. Fowden, L.; Pratt, H.M. Cyclopropylamino acids of the genus Acer: Distribution and biosynthesis. *Phytochemistry* **1973**, *12*, 1677–1681. [CrossRef]
4. Bochnia, M.; Sander, J.; Ziegler, J.; Terhardt, M.; Sander, S.; Janzen, N.; Cavalleri, J.M.; Zuraw, A.; Wensch-Dorendorf, M.; Zeyner, A. Detection of MCPG metabolites in horses with atypical myopathy. *PLoS ONE* **2019**, *14*, e0211698. [CrossRef]
5. Von Holt, C.; Chang, J.; von Holt, M.; Böhm, H. Metabolism and metabolic effects of hypoglycin. *Biochim. Biophys. Acta* **1964**, *0*, 611–613. [CrossRef]
6. Melde, K.; Jackson, S.; Bartlett, K.; Stanley, H.; Sherratt, H.; Ghisla, S. Metabolic consequences of methylenecyclopropylglycine poisoning in rats. *Biochem. J.* **1991**, *274*, 395–400. [CrossRef]

7. Westermann, C.M.; de Sain-van der Velden, M.G.M.; van der Kolk, J.H.; Berger, R.; Wijnberg, I.D.; Koeman, J.P.; Wanders, R.J.A.; Lenstra, J.A.; Testerink, N.; Vaandrager, A.B.; et al. Equine biochemical multiple acyl-CoA dehydrogenase deficiency (MADD) as a cause of rhabdomyolysis. *Mol. Genet. Metab.* **2007**, *91*, 362–369. [CrossRef] [PubMed]

8. Palencia, P.; Rivero, J.L.L. Short Communications Atypical myopathy in two grazing horses in northern Spain. *Vet. Rec.* **2007**, *161*, 346–348. [CrossRef] [PubMed]

9. Finno, C.J.; Valberg, S.J.; Wünschmann, A.; Murphy, M.J. Seasonal pasture myopathy in horses in the midwestern United States: 14 cases (1998–2005). *J. Am. Vet. Med. Assoc.* **2006**, *229*, 1134–1141. [CrossRef]

10. Votion, D.M.; Linden, A.; Saegerman, C.; Engels, P.; Erpicum, M.; Thiry, E.; Delguste, C.; Rouxhet, S.; Demoulin, V.; Navet, R.; et al. History and clinical features of atypical myopathy in horses in Belgium (2000–2005). *J. Vet. Intern. Med.* **2007**, *21*, 1380–1391. [PubMed]

11. Van Galen, G.; Marcillaud Pitel, C.; Saegerman, C.; Patarin, F.; Amory, H.; Baily, J.D.; Cassart, D.; Gerber, V.; Hahn, C.; Harris, P.; et al. European outbreaks of atypical myopathy in grazing equids (2006–2009): Spatiotemporal distribution, history and clinical features. *Equine Vet. J.* **2012**, *44*, 614–620. [CrossRef] [PubMed]

12. Gonzalez-Medina, S.; Ireland, J.L.; Piercy, R.J.; Newton, J.R.; Votion, D. Equine atypical myopathy in the UK: Epidemiological characteristics of cases reported from 2011 to 2015 and factors associated with survival. *Equine Vet. J.* **2017**, *49*, 746–752. [CrossRef] [PubMed]

13. Brandt, K.; Brandtt, K.; Hinrrchs, U.; Schulze, C.; Landes, E. Atypische Myoglobinurie der Weidepferde. *Pferdeheilkunde* **1997**, *13*, 27–34. [CrossRef]

14. Dunkel, B.; Ryan, A.; Haggett, E.; Knowles, E.J. Atypical myopathy in the South-East of England: Clinicopathological data and outcome in hospitalised horses. *Equine Vet. Educ.* **2018**, 1–6. [CrossRef]

15. Van Galen, G.; Saegerman, C.; Marcillaud Pitel, C.; Patarin, F.; Amory, H.; Baily, J.D.; Cassart, D.; Gerber, V.; Hahn, C.; Harris, P.; et al. European outbreaks of atypical myopathy in grazing horses (2006–2009): Determination of indicators for risk and prognostic factors. *Equine Vet. J. Vol.* **2012**, *44*, 621–625. [CrossRef]

16. Bunert, C.; Langer, S.; Votion, D.M.; Boemer, F.; Muller, A.; Ternes, K.; Liesegang, A. Atypical myopathy in Pere David's deer (Elaphurus davidianus) associated with ingestion of hypoglycin A. *J. Anim. Sci.* **2018**, *96*, 3537–3547. [CrossRef]

17. Votion, D.M.; Linden, A.; Delguste, C.; Amory, H.; Thiry, E.; Engels, P.; Van Galen, G.; Navet, R.; Sluse, F.; Serteyn, D.; et al. Atypical myopathy in grazing horses: A first exploratory data analysis. *Vet. J.* **2009**, *180*, 77–87. [CrossRef]

18. Westermann, C.M.; van Leeuwen, R.; van Raamsdonk, L.W.D.; Mol, H.G.J. "Hypoglycin A Concentrations in Maple Tree Species in the Netherlands and the Occurrence of Atypical Myopathy in Horses. *J. Vet. Intern. Med.* **2016**, *30*, 880–884. [CrossRef]

19. Renaud, B.; François, A.-C.; Dopagne, C.; Rouxhet, S.; Gustin, P.; Votion, D. Identification of the Maple Tree Responsible for Atypical Myopathy. Available online: http://hdl.handle.net/2268/242221 (accessed on 12 December 2019).

20. Van Galen, G.; Dopagne, C.; Rouxhet, S.; Pitel, C.; Votion, D. Etiologie de la myopathie atypique: Conditions de toxicité de l'agent causal—Étude préliminaire. In *40ème Journée de la Recherche Equine*; Institut français du cheval et de l'équitation (IFCE): Paris, France, 2014; pp. 101–109.

21. Search Results—The Plant List. Available online: http://www.theplantlist.org/tpl/search?q=Acer&_csv=on (accessed on 12 December 2019).

22. Hulme, P.E.; Bacher, S.; Kenis, M.; Klotz, S.; Kühn, I.; Minchin, D.; Nentwig, W.; Olenin, S.; Panov, V.; Pergl, J.; et al. Grasping at the routes of biological invasions: A framework for integrating pathways into policy. *J. Appl. Ecol.* **2008**, *45*, 403–441. [CrossRef]

23. Unger, L.; Nicholson, A.; Jewitt, E.M.; Gerber, V.; Hegeman, A.; Sweetman, L.; Valberg, S. Hypoglycin A Concentrations in Seeds of Acer Pseudoplatanus Trees Growing on Atypical Myopathy-Affected and Control Pastures. *J. Vet. Intern. Med.* **2014**, *28*, 1289–1293. [CrossRef]

24. Baise, E.; Habyarimana, J.A.; Amory, H.; Boemer, F.; Douny, C.; Gustin, P.; Marcillaud-Pitel, C.; Patarin, F.; Weber, M.; Votion, D.M. Samaras and seedlings of Acer pseudoplatanus are potential sources of hypoglycin A intoxication in atypical myopathy without necessarily inducing clinical signs. *Equine Vet. J.* **2016**, *48*, 414–417. [CrossRef] [PubMed]

25. Zuraw, A.; Dietert, K.; Kühnel, S.; Sander, J.; Klopfleisch, R. "Equine atypical myopathy caused by hypoglycin A intoxication associated with ingestion of sycamore maple tree seeds. *Equine Vet. J.* **2016**, *48*, 418–442. [CrossRef] [PubMed]

26. Votion, D.M.; Habyarimana, J.A.; Scippo, M.L.; Richard, E.A.; Marcillaud-Pitel, C.; Erpicum, M.; Gustin, P. Potential new sources of hypoglycin A poisoning for equids kept at pasture in spring: A field pilot study. *Vet. Rec.* **2019**, *184*, 740. [CrossRef] [PubMed]

27. Katul, G.G.; Porporato, A.; Nathan, R.; Siqueira, M.; Soons, M.B.; Poggi, D.; Horn, H.S.; Levin, S.A. Mechanistic analytical models for long-distance seed dispersal by wind. *Am. Nat.* **2005**, *166*, 368–381. [CrossRef] [PubMed]

28. Votion, D. Atypical myopathy: An update. *Practice* **2016**, *38*, 241–246. [CrossRef]

29. Gonzalez-Medina, S.; Montesso, F.; Chang, Y.-M.; Hyde, C.; Piercy, R.J. Atypical myopathy-associated hypoglycin A toxin remains in Sycamore seedlings despite mowing, herbicidal spraying or storage in hay and silage. *Equine Vet. J.* **2019**, *51*, 701–704. [CrossRef]

30. Renaud, B.; Francois, A.C.; Marcillaud-Pitel, C.; Gustin, P.; Votion, D. Myopathie atypique: Les différentes sources d'intoxication. Comment gérer le risque? In *Journées Sciences et Innovations Équines*; Institut français du cheval et de l'équitation (Ifce): Saumur, France, 2019; p. 9.

31. Cooper, J.J.; Mason, G.J. The identification of abnormal behaviour and behavioural problems in stabled horses and their relationship to horse welfare: A comparative review. *Equine Vet. J. Suppl.* **1998**, *30*, 5–9. [CrossRef]

32. Cooper, J.J.; Albentosa, M.J. Behavioural adaptation in the domestic horse: Potential role of apparently abnormal responses including stereotypic behavior. *Livest. Prod. Sci.* **2005**, *92*, 177–182. [CrossRef]

33. Hosie, B.D.; Gould, P.W.; Hunter, A.R.; Low, J.C.; Munro, R.; Wilson, H.C. Acute myopathy in horses at grass in east and south east Scotland. *Vet. Rec.* **1986**, *119*, 444–449. [CrossRef]

34. Harris, P.; Whitwell, K. Atypical myoglobinuria alert. *Vet. Rec.* **1990**, *15*, 603.

35. Westermann, C.M.; Dorland, L.; Votion, D.M.; De Sain-van der Velden, M.G.M.; Wijnberg, I.D.; Wanders, R.J.A.; Spliet, W.G.M.; Testerink, N.; Berger, R.; Ruiter, J.P.N.; et al. Acquired multiple Acyl-CoA dehydrogenase deficiency in 10 horses with atypical myopathy. *Neuromuscul. Disord.* **2008**, *18*, 355–364. [CrossRef] [PubMed]

36. Rooney, D.K. Applied nutrition. In *Equine Internal Medicine*; Sellon, D.C., Reed, S.M., Bayly, W.M., Eds.; W.B. Saunders: Philadelphia, PA, USA, 2004; pp. 235–272.

37. Gonzalez-Medina, S.; Hyde, C.; Lovera, I.; Piercy, R.J. Detection of equine atypical myopathy-associated hypoglycin A in plant material: Optimisation and validation of a novel LC-MS based method without derivatisation. *PLoS ONE* **2018**, *13*, 13–15. [CrossRef] [PubMed]

14

Effects of Horse Housing System on Energy Balance during Post-Exercise Recovery

Malin Connysson [1,*], Marie Rhodin [2] and Anna Jansson [2]

[1] Wången National Center for Education in Trotting, Vången 110, S-835 93 Alsen, Sweden

[2] Department of Anatomy, Physiology and Biochemistry, Swedish University of Agricultural Sciences, SE-75007 Uppsala, Sweden; marie.rhodin@slu.se (M.R.); anna.jansson@slu.se (A.R.)

[*] Correspondence: malin.connysson@wangen.se.

Simple Summary: Horse management aims to keep horses healthy and ensure good performance and animal welfare. Many horses are currently kept in individual box stalls indoors, a housing system that limits free movement, exploration, and social interaction, and may also subject horses to lower air quality. The alternative is a free-range housing system where horses are kept in groups outdoors. Anecdotal information indicates concerns among sports horse trainers that lack of rest in such systems delays recovery and impairs performance. This study examined whether recovery after competition-like exercise in Standardbred trotters was affected by housing system. The results showed that a free-range housing system did not delay recovery in Standardbred trotters, and in fact had positive effects on appetite and recovery of energy balance.

Abstract: This study examined the effects of two housing systems (free-range and box stalls) on recovery of energy balance after competition-like exercise in Standardbred horses. Eight adult geldings (mean age 11 years) were used. The study had a change-over design, with the box stall (BOX) and free-range group housing (FreeR) treatments each run for 21 days. The horses were fed forage ad libitum and performed two similar race-like exercise tests (ET), on day 7 and day 14 in each treatment. Forage intake was recorded during the last 6–7 days in each period. Blood samples were collected before, during, and until 44 h after ET. Voluntary forage intake (measured in groups with four horses in each group) was higher in FreeR horses than BOX horses (FreeR: 48, BOX: 39, standard error of the mean (SEM) 1.7 kg ($p = 0.003$)). Plasma non-esterified fatty acids (NEFA) was lower at 20–44 h of recovery than before in FreeR horses ($p = 0.022$), but not in BOX horses. Housing did not affect exercise heart rate, plasma lactate, plasma urea, or total plasma protein concentration. Thus the free-range housing system hastened recovery in Standardbred trotters, contradicting anecdotal claims that it delays recovery. The free-range housing also had positive effects on appetite and recovery of energy balance.

Keywords: NEFA; Standardbred trotters; feed intake

1. Introduction

Many horses are currently housed in individual box stalls in stables [1–4]. Box stalls facilitate supervision, individual feeding and grooming of the horses, but obviously limit their scope for free movement, exploration and social interaction. An alternative is a modern housing system where horses are kept in groups in paddocks with shelters and lying areas and with individual feeding controlled by transponders. Anecdotal information indicates concerns among sports horse trainers that lack of rest in such systems delays recovery and impairs competition performance. However, unpublished data [5] indicate that picky-eater Standardbred trotters kept in a group housing system have better body condition than when housed in box stalls, indicating better appetite and higher feed intake.

Environmental factors such as space allowance, group size and feeder characteristics have been shown to affect feed intake in pigs [6]. Little is known about how physical environment affects feed intake in horses. Keeping horses in groups may affect eating; which has been shown to be highly synchronized in group-housed horses [7,8].

In the short term, recovery involves decreasing muscle temperature, compensating for oxygen debt, and regulating acid–base balance. In the longer term, it also involves energy replenishment, fluid balance recovery, and tissue re-synthesis. Energy balance can be monitored by measuring body weight, energy expenditure (using heart rate), body condition score (BCS), and substrate usage (non-esterified fatty acids (NEFA) and urea). NEFA, originating from lipolysis of adipose tissue, have been widely shown to increase in ponies and horses during periods of insufficient energy intake [9–13]. When amino acids are used as energy substrate there is a degradation that starts with deamination, where the amino group is removed and converted into ammonia. Ammonia released by this process is removed from the body by forming urea in the liver.

This study examined whether recovery of energy balance after competition-like exercise in Standardbred horses fed ad libitum was affected by housing system. Two different systems were compared: free-range group housing (FreeR) and an individual box-stall housing system (BOX) in which activity in groups was possible for only 4–5 h/day. The hypothesis was that free-range housing hastens recovery compared with box stall housing.

2. Materials and Methods

Umeå local ethics committee approved the study (**A 54-13**) and it was performed in compliance with European Union directives on animal experiments (2010/63/EU; European Union, 2010) and with laws (Swedish Constitution, 1988:534) and regulations (Swedish Board of Agriculture Constitution, 2012:26) governing experiments on live animals in Sweden.

2.1. Horses and Management

Eight adult Standardbred geldings in training (mean age 11 years, range 9–13 years) were used. The study was performed during May–June 2015 and all horses had raced during the period 2005–2015 and had average earnings of 146,553 SEK (range 26,500–495,798 SEK). Mean bodyweight at the beginning of the trial was 509 kg (range 410–562 kg). The horses were trained under the supervision of professional trainers (licensed by the Swedish Trotting Association) according to a training program similar to that used by Swedish trotting trainers [14].

2.2. Experimental Design

The horses were randomly allocated to two groups of four and kept in the box stall (BOX) or free-range group housing system (FreeR) for 21 days, followed by a complete change-over to the other treatment. Between treatments, the horses had a three-day transition period in the new housing system. All horses performed two similar exercise tests (ET), on day 7 (ET1) and day 14 (ET2) in each treatment period. On days 3, 11, and 18 of each treatment, the horses were exercised on a track (5000 m warm-up, 3000 m at a speed of 11.1 m/s).

The horses in treatment BOX were housed individually in 3 m × 3 m boxes with wood shavings and were let out into a sand paddock (2000 m²) together with the other horses in their group for 4–5 h every day. In the boxes, the horses had ad libitum access to water in buckets. The horses in treatment FreeR were kept together in a group housing area that consisted of a paved paddock (3200 m²) with a shelter with rubber matting and automatic feeding stations. They were offered water ad libitum in water barrels.

2.3. Diet

In treatment BOX, the horses were fed forage ad libitum in their stall, but no feed was offered during the paddock stay. Feed was offered four times per day (07.00, 12.30, 17.00, and 20.00 h) and there had to be left-overs at every meal to provide ad libitum access.

In treatment FreeR, the horses were fed from automatic feeding stations that recognized the individuals by transponders. With this technique, feed allowance was regulated by time and all horses had access to the feeding station for more time than they used it (free access). All horses but one had 400 min eating time/day, while in one horse the access was 500 min/day to ensure free access.

The same forage was used in both housing systems, a haylage (dry matter (DM) 78%, 11.2 MJ metabolizable energy (ME)/kg DM, 14.3% crude protein (CP), calcium (Ca) 5.6 g/kg DM, phosphorus (P) 2.2 g/kg DM, and magnesium (Mg) 1.7 g/kg DM). The horses were also given 0.5 kg concentrate (Krafft Sport, Malmö, Sweden (12 MJ ME/kg, 11% CP)) and 80 g mineral and vitamin supplement (Krafft Miner Vit, Malmö, Sweden) (Ca 55 g/kg, P 65 g/kg, Mg 60 g/kg, salt (NaCl) 125 g/kg, copper (Cu) 900 mg/kg, selenium (Se) 15 mg/kg, vitamin A 100,000 IU/kg, vitamin D3 10,000 IU/kg, and vitamin E 5000 mg/kg) every day. In the BOX system this was fed once a day, while in FreeR the automatic feed station divided the total allowance equally between every hour of the day.

The horses were offered ad libitum access to salt blocks during the whole study and they also had extra salt mixed with beet pulp after the exercise on days 3, 11, and 18 (250 g beet pulp and 20 g NaCl).

2.4. Exercise Test

On the day of the exercise test (ET), the horses were transported 50 km to an official 1000-m oval, banked, gravel racetrack (Östersunds race track, Sweden). The horses performed the ET in groups of four, two from each treatment. The same driver drove the same horse in a harness race sulky on all test occasions and the horses raced in the same group at the same time on all occasions. The ET started with a warm-up consisting of 4000 m slow trot (6.3–6.7 m/s) and 500 m trot (10 m/s). After the warm-up, the horses walked the track for 10 min. They then trotted 2140 m in the same race field position, at 11.8–12.2 m/s for the first 1640 m and as fast as they could (free positioning) for the last 500 m. This was followed by a cool-down of 1000 m slow trot (6.3–6.7 m/s). Approximately one hour after crossing the finish line in the ET, the horses were transported back home, and three hours after the ET they were put back into their housing system.

2.5. Measurements, Sampling, and Analysis

The horses were weighed on a scale (Tru-test E2000S2, Auckland, New Zealand) every morning. Body weight measurements on the day before ET day and on the morning of ET day were used as "Rest" values for bodyweight. Rectal temperature was measured every day, at 07.00 h. Before and after each experimental period, the body condition score (BCS) of all horses was evaluated as specified by Henneke et al. [15]. During days 15–22 (in period 1) and days 14–19 (in period 2), forage intake was measured by weighing the allowance and left-overs. In treatment FreeR, the horses were fed forage from two feeding stations (four feeding places) and intake had to be measured for the whole group. Mean ambient temperature on ET days and the three following days was 8.4 °C (range 1.3–9.8 °C).

During ET days, blood samples were collected via a catheter inserted in vena jugularis after local cutaneous anesthesia (Tapin (lidocaine 25 mg/g, prilocaine 25 mg/g), Orifarm Generics, Stockholm, Sweden). The catheter was inserted approximately one hour before the first blood sampling. Blood samples were collected in 6-mL lithium-heparinized tubes (102 IU) and kept on ice until centrifuging (10 min, 920 × g, 18 °C), after which the plasma was frozen (−20 °C). Blood samples were collected at rest in the stable (Rest), at the racetrack 1 min after the finish line (FL) and after 10, 180, 240, 300, 360, and 420 min of recovery. At 20 and 44 h after the ET, blood samples were collected by venipuncture from the jugular vein.

Total plasma protein concentration (TPP) was measured in all plasma samples using a handheld refractometer (Atago, Tokyo, Japan). Plasma lactate concentration was analyzed in samples Rest, FL and after 10, 180, and 420 min of recovery, using an enzymatic (L-lactate dehydrogenase and glutamate-pyruvate transaminase) and spectrophotometric method (Boehringer Mannheim/R-Biopharm, Darmstadt, Germany), with intraassay coefficient of variation (CV) 2.2% in this study. Plasma NEFA concentration was analyzed in samples taken at Rest, FL, and after 10, 180, 240, and 420 min of recovery and also after 20 and 44 h, by quantitative determination using an enzymatic colorimetric method (Wako Chemicals GmbH, Neuss, Germany), with intraassay CV 1.8% in this study. Plasma urea concentration was analyzed with a spectrophotometric method (Urea Assay Kit, Cell Biolabs Inc., San Diego, CA, USA), with intraassay CV 1.3% in this study.

Heart rate was continuously recorded during the race and up to 420 min of recovery using a heart rate recorder (Polar CS600X Polar Electro, Kempele, Finland) and the data were analyzed using Polar ProTrainer 5 Equine Edition Software (Polar Electro, Kempele, Finland). Mean recovery heart rate was calculated using recordings from 270–410 min of recovery.

2.6. Statistical Analysis

Analysis of variance was performed with the MIXED procedure in SAS (version 9.4; SAS Institute Inc., Cary, NC, USA), using an autoregressive (AR(1)) structure. Plasma sample results were pooled into recovery 3–7 h (180–420 min), and recovery 20–44 h (22 h and 44 h). Statistical analysis was performed with a statistical model including fixed effects of housing, sample, and the interaction between them. The model for an observed variable of horse i in housing j, sample k, was:

$$Y_{ijk} = \mu + \eta_i + \pi_j + \gamma_k + (\pi\gamma)_{ik} + e_{ijk}, \tag{1}$$

where μ is the overall mean, η_i is the effect of horse, π_j is the effect of housing, γ_k is the effect of sample, $(\pi\gamma)_{ik}$ is the effect of the interaction between housing and sample, and e_{ijk} is the random error. The random part included horse, horse × housing, and period. Observations within each horse × period × housing combination were modeled as repeated measurements. For urea, race had a significant effect and was included in the model.

Forage intake, bodyweight, BCS, heart rate, and velocity data were analyzed by a statistical model including fixed effects (housing, period, and the interaction between housing and period). The model for an observed variable of horse i in period j, in housing k was:

$$Y_{ijk} = \mu + \eta_i + \pi_j + \gamma_k + (\pi\gamma)_{ik} + e_{ijk}, \tag{2}$$

where μ is the overall mean, η_i is the effect of horse, π_j is the effect of housing, γ_k is the effect of sample, $(\pi\gamma)_{ik}$ is the effect of the interaction between housing and sample, and e_{ijk} is the random error. The random part included horse and horse × housing. Observations within each horse were modeled as repeated measurements.

Post-hoc comparisons were adjusted for multiplicity using the Bonferroni method. Values are presented as least square means (LSM) with the standard error of the mean (SEM) in a parenthesis. Differences were considered statistically significant at $p < 0.05$.

3. Results

One of the horses was excluded due to a hoof crack, although it showed no clinical signs of pain. Training and ET were still performed with that horse and it was included in feed intake measurements, since these were done by group. Group forage intake was higher in treatment FreeR than treatment BOX (FreeR: 48 (1.7), BOX: 39 (1.7) kg ($p = 0.003$)) and the mean bodyweight during the whole treatment periods (21 days) tended to be higher in FreeR horses than in BOX horses (FreeR: 505 (13), BOX: 500 (13) kg ($p = 0.07$)). There was no difference in BCS between the housing systems (FreeR: 4.8 (0.4), BOX: 4.7 (0.4) ($p = 0.93$)).

3.1. Rest

Body weight was higher in FreeR horses than BOX horses at Rest (morning day before ET + morning before ET) (Table 1). Housing did not affect plasma NEFA, urea, or TPP concentration on the morning before ET (Tables 1 and 2).

Table 1. Body weight and total plasma protein in Standardbred horses kept in free-range group housing (FreeR) or box housing (BOX). SEM = standard error of the mean.

Variable	Sample	FreeR	SEM	BOX	SEM	p-Value
Body weight (kg)						
	Rest	509	13	504	13	0.038
	20–44 h recovery	504 [a]	13	498 [a]	13	0.020
Total plasma protein (g/L)						
	Rest	59.5	0.8	60.4	0.8	0.996
	Finish line	73.5 [a]	0.8	74.3 [a]	0.8	1.000
	10 min recovery	67.0 [a]	0.8	67.4 [a]	0.8	1.000
	3–7 h recovery	61.0	0.8	62.3 [a]	0.8	0.405
	20–44 h recovery	61.8 [a]	0.8	62.6 [a]	0.8	1.000

[a] Significantly different ($p < 0.05$) from Rest.

Table 2. Plasma concentrations of non-esterified fatty acids (NEFA), urea and lactate in Standardbred horses kept in free-range group housing (FreeR) or box housing (BOX). SEM = standard error of the mean.

Variable	Sample	FreeR	SEM	BOX	SEM	p-Value
Plasma NEFA (mmol/L)						
	Rest	0.26	0.04	0.21	0.04	1.000
	Finish line	0.33	0.04	0.34 [a]	0.04	1.000
	10 min recovery	0.54 [a]	0.04	0.55 [a]	0.04	1.000
	3–7 h recovery	0.24	0.04	0.22	0.04	1.000
	20–44 h recovery	0.16 [a]	0.04	0.25	0.04	0.019
Plasma urea (mmol/L)						
	Rest	5.0	0.3	4.6	0.3	0.481
	Finish line	5.1	0.3	4.9 [b]	0.3	1.000
	10 min recovery	5.2	0.3	4.9 [a]	0.3	0.818
	3–7 h recovery	5.6 [a]	0.3	5.3 [a]	0.3	0.718
	20–44 h recovery	4.8	0.3	4.7	0.3	1.000
Plasma lactate (mmol/L)						
	Rest	0.9	0.8	1.3	0.8	1.000
	Finish line	20.8 [a]	0.8	20.6 [a]	0.8	1.000
	10 min recovery	16.0 [a]	0.8	16.0 [a]	0.8	1.000
	3–7 h recovery	1.0	0.7	1.1	0.7	1.000

[a] Significantly different ($p < 0.05$) from Rest. [b] Tendency for significant difference ($p \leq 0.09$) from Rest.

3.2. Finish Line and 10 min Recovery

Housing system did not affect peak heart rate (FreeR: 224 (2), BOX: 221 (2) beats/min), plasma lactate concentration (Table 2), or velocity during the simulated race (FreeR: 12.0 (0.1), BOX: 12.0 (0.1) m/s).

Plasma lactate concentrations and TPP were higher than Rest values at the finish line and after 10 min of recovery during both FreeR and BOX (Tables 1 and 2). Plasma NEFA concentrations were higher than Rest values at 10 min of recovery during both FreeR and BOX, and for BOX treatment also at the finish line (Table 2). Plasma urea concentrations were higher than Rest values at the finish line (tendency $p = 0.071$) and after 10 min of recovery during BOX treatment (Table 2). There was no effect of FreeR compared with BOX on plasma NEFA, urea, or TPP concentration at finish line or after 10 min of recovery (Tables 1 and 2).

3.3. Short-Term Recovery (3–7 h)

As mentioned, heart rate, plasma lactate, NEFA, urea, and TPP concentrations all increased from rest to race, and therefore recovery was necessary. There was a tendency during 3–7 h recovery for higher heart rate in FreeR horses compared with BOX horses (47 (1) vs. 43 (2) beats/min; $p = 0.100$) (Figure 1). The different housing systems did not significantly affect plasma lactate, plasma NEFA, urea, or TPP concentration during 3–7 h recovery (Tables 1 and 2). During short-term recovery, plasma urea concentrations were greater than the Rest values (Table 2). During short-term recovery, TPP were higher than the Rest values during BOX housing (Table 1).

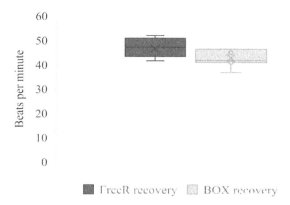

Figure 1. Box-plot of 3–7 h recovery heart rates in Standardbred horses kept in free-range group housing (FreeR) or box housing (BOX).

3.4. Long-Term Recovery (20–44 h)

Body weight did not return to the Rest values during long-term recovery (Table 1). Plasma NEFA returned to the Rest value during 20–44 h of recovery in BOX horses, while in FreeR horses it not only returned to the Rest value but fell below it (Table 2). Plasma NEFA was lower in FreeR horses than in BOX horses during 20–44 h of recovery (Table 2) and bodyweight was higher in FreeR horses than in BOX horses during this period (Table 1). Housing system had no significant effect on plasma urea or TPP concentration during 20–44 h of recovery (Tables 1 and 2). Plasma TPP concentration during 20–44 h of recovery was significantly higher than the Rest value (Table 1).

4. Discussion

In this study, there was little or no difference in short-term (3–7 h) metabolic recovery in horses kept in the free-range and box stall housing systems evaluated. In fact, the results indicated that in the long-term perspective, the free-range system may be beneficial. The lack of differences in NEFA, urea, and TPP responses during short-term recovery indicates that water and feed intake were similar in the two housing systems during the first hours after ET. During long-term recovery, the NEFA levels were very low in FreeR horses, even lower than before the ET. This indicates quick and efficient recovery of energy balance by horses in this housing system, an indication supported by the higher daily feed consumption observed in the FreeR housing system. The hormone insulin plays an important role in regulating lipolysis in adipose tissue, as increased insulin concentration results in decreased lipolysis and thereby decreased release of NEFA. An increased insulin response in FreeR (due to higher feed intake) may have lowered the NEFA response but the response may also have been influenced by different feeding intervals in the housing systems. Low post-exercise appetite has been observed in athletic horses (box-housed) and is suggested to be associated with the hormone's active ghrelin, adiponectin, and leptin, and/or gastric ulcers [16,17]. Low BCS and periods of low appetite are conditions seen in some horses during periods of intense training and racing, and our results indicate that a group housing system might counteract these problems. Our findings indicate that physical environment is important for feed intake in horses, as reported earlier for pigs [6]. The lower feed intake in box-housed horses could have been due to the daily 4–5 h without feed in the paddock, but that is close to the time span without eating observed in wild-living horses and was probably compensated for by increased feed intake. Compensatory increases in feed consumption after a period of feed restriction have been reported in ponies [18].

Although plasma NEFA concentrations indicated that energy balance was restored in the horses, bodyweight was still not back to resting values at 44 h of recovery. In French Trotters, the post racing decrease in bodyweight is reported to be on average 9.4 kg (range 0–26 kg) and bodyweight requires on average 3.3 days to recover [19]. An exercise-induced decrease in body weight also seems to be affected by whether the horses are transported [20]. It has been suggested that sweat losses are responsible for 90% of post-exercise weight losses [21], which are therefore not affected by energy balance but by water and electrolyte intake. Since total plasma protein concentrations remained elevated from resting values at the end of the study period, it is likely that the horses did not recover fluid balance within 44 h. The salt intake in the present study might have been too low to compensate for losses, since salt blocks have been shown to be an inadequate sole source of salt for athletic horses [22] and the extra salt offered (20 g) to the horses on the day after ET might not have been sufficient.

There were no differences in plasma lactate responses between the housing systems, which was probably due to events in the immediate recovery period after exercise (i.e., 1000 m slow-down trot, walk, and road transport back home) being the same in both treatments. It has been shown previously that, during the first 30 min after intense exercise, lactate removal can be increased two-fold by light exercise [23]. Interestingly, there was no significant difference in HR between the housing treatments during short-term recovery indicating that energy expenditure was similar in this period [24]. In both systems, HR seemed to be slightly higher than earlier observed [25] in box-stalled Standardbred horses at night (36–40 bpm vs. 43–47 bpm in our study). We expected HR to be higher when horses were kept in group housing compared to box housing, due to more physical activity (walking around). Earlier studies have shown that physical activity in stabled or partly stabled horses is lower than in horses housed in free-range systems [26]. In addition, keeping horses in groups in paddocks seems to increase physical activity compared with keeping them in individual paddocks [27]. In BOX horses, an elevation of HR from expected levels at rest is difficult to explain and accordingly also the lack of difference in HR between treatments. One possible reason for the lack of difference in HR is because HR was elevated in BOX horses due to horses being more alert, excited, or stressed in this system (by people and horses moving around in the stable). This assumption contradicts anecdotal claims that horses kept in a free-range group housing system are less relaxed than stabled horses.

5. Conclusions

This study found that a free-range housing system hastened recovery in Standardbred trotters, rather than delaying it as suggested by anecdotal claims. The free-range housing system also had positive effects on appetite and recovery of energy balance.

Author Contributions: Conceptualization, M.C. and A.J.; data curation, M.C., M.R., and A.J.; formal analysis, M.C. and M.R.; funding acquisition, A.J.; investigation, M.C.; methodology, M.C. and A.J.; project administration, M.C.; resources, M.C. and A.J.; supervision, M.R. and A.J.; validation, M.C., M.R., and A.J.; writing—original draft, M.C., M.R., and A.J.; writing—review and editing, M.C., M.R., and A.J.

Acknowledgments: Thanks also to Ulf Hedenström, Lars-Åke Svärdfeldt, Sara Sandqvist, Emma Folkesson, and the students at Wången for assistance during the study.

References

1. Bachmann, I.; Stauffacher, M. Haltung und nutzung von pferden in der schweiz: Eine repräsentative erfassung des status quo. *Schweizer Archiv für Tierheilkunde* **2002**, *144*, 331–347. [CrossRef] [PubMed]
2. Petersen, S.; Tolle, K.H.; Blobel, K.; Grabner, A.; Krieter, J. Evaluation of horse keeping in Schleswig-Holstein. *Züchtungskunde* **2006**, *78*, 207–217.
3. Henderson, A.J.Z. Don't fence me in: Managing psychological well being for elite performance horses. *J. Appl. Anim. Welf. Sci.* **2007**, *10*, 309–329. [CrossRef] [PubMed]
4. Hotchkiss, J.W.; Reid, S.W.J.; Christley, R.M. A survey of horse owners in Great Britain regarding horses in their care. Part 1: Horse demographic characteristics and management. *Equine Vet. J.* **2007**, *39*, 294–300. [CrossRef] [PubMed]
5. Connysson, M. (Wången National Center for Education in Trotting, Alsen, Sweden). Personal communication, 2017.
6. Li, Q.; Patience, J.F. Factors involved in the regulation of feed and energy intake of pigs. *Anim. Feed Sci. Technol.* **2017**, *233*, 22–33. [CrossRef]
7. Nicol, C.J.; Badnell-Waters, A.J.; Bice, R.; Kelland, A.; Wilson, A.D.; Harris, P.A. The effects of diet and weaning method on the behaviour of young horses. *Appl. Anim. Behav. Sci.* **2005**, *95*, 205–221. [CrossRef]
8. Sweeting, M.P.; Houpt, C.E.; Houpt, K.A. Social facilitation of feeding and time budgets in stabled ponies. *J. Anim. Sci.* **1985**, *60*, 369–374. [CrossRef] [PubMed]
9. Baetz, A.L.; Pearson, J.E. Blood constituent changes in fasted ponies. *Am. J. Vet. Res.* **1972**, *33*, 1941–1946. [PubMed]
10. Rose, R.J.; Sampson, D. Changes in certain metabolic parameters in horses associated with food deprivation and endurance exercise. *Res. Vet. Sci.* **1982**, *32*, 198–202. [CrossRef]
11. Sticker, L.S.; Thompson, D.L.; Fernandez, J.M.; Bunting, L.D.; DePew, C.L. Dietary protein and(or) energy restriction in mares: Plasma growth hormone, IGF-I, prolactin, cortisol, and thyroid hormone responses to feeding, glucose, and epinephrine. *J. Anim. Sci.* **1995**, *73*, 1424–1432. [CrossRef] [PubMed]
12. Christensen, R.A.; Malinowski, K.; Massenzio, A.M.; Hafs, H.D.; Scanes, C.G. Acute effects of short-term feed deprivation and refeeding on circulating concentrations of metabolites, insulin-like growth factor i, insulin-like growth factor binding proteins, somatotropin, and thyroid hormones in adult geldings. *J. Anim. Sci.* **1997**, *75*, 1351–1358. [CrossRef] [PubMed]
13. Connysson, M.; Essén-Gustavsson, B.; Lindberg, J.E.; Jansson, A. Effects of feed deprivation on standardbred horses fed a forage-only diet and a 50:50 forage-oats diet. *Equine Vet. J.* **2010**, *42*, 335–340. [CrossRef] [PubMed]
14. Ringmark, S. A Forage-Only Diet and Reduced High Intensity Training Distance in Standardbred Horses. Ph.D Thesis, Swedish University of Agricultural Sciences, Uppsala, Sweden, 2014.
15. Henneke, D.R.; Potter, G.D.; Kreider, J.L.; Yeates, B.F. Relationship between condition score, physical measurements and body fat percentage in mares. *Equine Vet. J.* **1983**, *15*, 371–372. [CrossRef] [PubMed]
16. Gordon, M.E.; McKeever, K.H.; Bokman, S.; Betros, C.L.; Manso-Filho, H.; Liburt, N.; Streltsova, J. Interval exercise alters feed intake as well as leptin and ghrelin concentrations in standardbred mares. *Equine Vet. J.* **2006**, *38*, 596–605. [CrossRef] [PubMed]
17. Gordon, M.E.; McKeever, K.H.; Betros, C.L.; Manso Filho, H.C. Plasma leptin, ghrelin and adiponectin concentrations in young fit racehorses versus mature unfit standardbreds. *Vet. J.* **2007**, *173*, 91–100. [CrossRef] [PubMed]
18. Glunk, E.C.; Pratt-Phillips, S.E.; Siciliano, P.D. Effect of restricted pasture access on pasture dry matter intake rate, dietary energy intake, and fecal pH in horses. *J. Equine Vet. Sci.* **2013**, *33*, 421–426. [CrossRef]
19. Leleu, C.; Miot, M.; Rallet, N. Mailliot Pivan Race-induced weight loss and its recovery time in Trotters: Factors of variation. In Proceedings of the 10th International Conference on Equine Exercise Physiology, Lorne, Australia, 12–16 November 2018.
20. Connysson, M.; Muhonen, S.; Jansson, A. Road transport and diet affect metabolic response to exercise in horses. *J. Anim. Sci.* **2017**, *95*, 4869–4879. [CrossRef] [PubMed]
21. Carlson, G.P. Haematology and body fluids in the equine athlete: A review. In *Equine Exercise Physiology*, 2 ed.; Gillespie, J.R., Robinson, N.E., Eds.; ICEEP Publications: Davis, CA, USA, 1987; pp. 393–425.
22. Jansson, A.; Dahlborn, A. Effects of feeding frequency and voluntary salt intake on fluid and electrolyte regulation in athletic horses. *J. Appl. Phys.* **1999**, *86*, 1610–1616. [CrossRef] [PubMed]

23. Marlin, D.J.; Harris, R.C.; Harman, J.C.; Snow, D.H. Influence of post-exercise activity on rates of muscle and blood lactate disappearance in the Thoroughbred horse. In *Equine Exercise Physiology*, 2 ed.; Gillespie, J.R., Robinson, N.E., Eds.; ICEEP Publications: Davis, CA, USA, 1987; pp. 686–700.

24. NRC. *Nutrient Requirements of Horses*, 6th ed.; The National Academies Press: Washington, DC, USA, 2007; p. 24.

25. Ringmark, S.; Lindholm, A.; Hedenström, U.; Lindinger, M.; Dahlborn, K.; Kvart, C.; Jansson, A. Reduced high intensity training distance had no effect on VLa4 but attenuated heart rate response in 2–3-year-old Standardbred horses. *Acta Vet. Scand.* **2015**, *57*, 17–30. [CrossRef] [PubMed]

26. Chaplin, S.J.; Gretgrix, L. Effect of housing conditions on activity and lying behaviour of horses. *Animal* **2010**, *4*, 792–795. [CrossRef] [PubMed]

27. Jørgensen, G.H.M.; Bøe, K.E. Individual paddocks versus social enclosure for horses. In *Horse Behavior and Welfare*; European Federation of Animal Science No. 122; Wageningen Academic Publishers: Wageningen, The Netherlands, 2007.

15

Horse Transport to Three South American Horse Slaughterhouses

Béke Nivelle [1,2,*], **Liesbeth Vermeulen** [3], **Sanne Van Beirendonck** [4], **Jos Van Thielen** [4,5] and **Bert Driessen** [1,2]

[1]   Laboratory of Livestock Physiology, Department of Biosystems, KU Leuven, 3001 Heverlee, Belgium; bert.driessen@dierenwelzijn.eu
[2]   Dier&Welzijn vzw, 3583 Paal, Belgium
[3]   Westvlees NV, 8840 Westrozebeke, Belgium; liesbeth_vermeulen@westvlees.com
[4]   Bioengineering Technology TC, KU Leuven, 2440 Geel, Belgium; sanne.vanbeirendonck@kuleuven.be (S.V.B.); jos.vanthielen@kuleuven.be (J.V.T.)
[5]   Thomas More, 2440 Geel, Belgium
*   Correspondence: beke.nivelle@kuleuven.be

Simple Summary: In the western world, the number of slaughtered horses is decreasing, but still about 5 million horses are slaughtered worldwide each year. The conditions in which horses are transported to the slaughterhouses are a topic of discussion. This study intended to investigate the circumstances of commercial slaughter horse transport and to detect possible risk factors for horse welfare. Therefore, 23 commercial horse transports to three South American slaughterhouses were monitored. During transport, a camera was mounted in each loading space so that horse behaviour could be analysed after transport. Fighting behaviour could not be explained by stocking density, environmental parameters, trailer characteristics, duration and distance of the journey. The temperature and relative humidity were recorded every five minutes in all loading spaces. Average temperatures exceeded the thermoneutral zone in six transports, but it is not clear if and to what extent horse welfare was impaired. Overall, loading and transporting of the horses went well, but the infrastructure of the loading area did not always promote smooth loading and can therefore be improved. At later visits, we noted that this issue was addressed.

Abstract: Between November 2016 and October 2017, 23 horse transports from 18 collection points to two slaughterhouses in Argentina and one in Uruguay were monitored. The goal of this study was to characterize the current practices in commercial horse transports and to detect potential threats to horse welfare. A total of 596 horses were transported over an average distance of 295 ± 250 km. Average transport duration was 294 ± 153 min. The infrastructure did not always promote smooth loading, but the amount of horses that refused to enter the trailers was limited. In each loading space, a camera was mounted to observe horse behaviour during the journey. Ambient temperature and relative humidity (RH) were recorded every five minutes in each loading space. In 14 of the 23 transports, the maximum temperature rose above 25 °C and the average temperature was over 25 °C during six transports. The average temperature humidity index (THI) exceeded 72 during six transports. The average stocking density was 1.40 ± 0.33 m^2 per horse, or 308 ± 53 kg/m^2. The degree of aggression differed between the front and rear loading space. Stocking density, environmental parameters, trailer characteristics, and transport duration and distance did not influence aggressiveness.

Keywords: horses; transport; slaughter

1. Introduction

In the USA, Australia and Europe, the number of slaughtered horses is decreasing. In Belgium for example, the number of slaughtered horses decreased from 21,390 slaughtered horses in 2001 to only 5895 horses in 2018 [1]. Annually, about 5 million horses are slaughtered worldwide [2]. In 2018, China was the country that slaughtered the most horses worldwide—more specifically, about 1.59 million [2]. In the same year, 389,153 horses were slaughtered in South America [2]. However, to this day, the circumstances in which horses are transported give rise to discussion [3].

Animals should be handled as carefully as possible at all times, including during the transportation process to the slaughterhouse. The animals should undergo as little stress as possible, on the one hand for welfare reasons, but also to prevent any deterioration in the quality of the horse meat. Quality loss can occur as a result of excessive stress, bruises or injuries [4–7]. Suboptimal ambient parameters such as temperature, relative humidity, ammonia and carbon dioxide concentration in the air can cause stress, but also inappropriate infrastructure and psychological stressors. Examples of psychological stressors are the determination of dominance rank, and transport activities such as loading, unloading and the transport itself. In addition, activities that are part of the management of the animals, such as the weaning of young animals, weighing of animals or changing housing, can also cause stress [8].

The conditions of (non-commercial) transports of sport and company horses and the response of these horses to transport are well studied [9,10]. On the other hand, commercial transport of horses is studied to a lesser extent. Furthermore, most studies in which horses are transported untethered in groups involve healthy horses that are used to being transported [9,11–14]. Studies investigating slaughter horse transports or horses that were sold as slaughter horses are less numerous [13,15–19]. However, from those studies it is clear that a number of transport-related factors influence horse welfare. Journey distance and time [20,21], loading density [7,15,22,23], handling [7,24,25], new environments [24], potentially re-grouping or mixing with unfamiliar animals [24], fasting and deprivation of water [24], the myriad of trailer designs [7,23], driving behaviour [7], road type and quality [7,22,24], traffic conditions [24], suspension systems and building materials of the trailers [15,23], environmental conditions in the trailer [7,22], and weather conditions [15,23,26] all affect horse welfare.

According to Morgan [27], the thermoneutral zone of a horse is on average between 5 °C and 25 °C. Another study defines the thermoneutral zone between −1 °C and 24 °C [27]. The differences in estimation of the thermoneutral zone are, among other things, probably due to acclimatization, body condition and climate [28]. For example, the upper limit of the thermoneutral zone lowers with increasing humidity. At a relative humidity (RH) of more than 50%, it gets harder for the animals to dissipate heat to the environment [29]. The temperature humidity index (THI) is a useful parameter to estimate the thermal comfort of organisms [30].

Legislation involving animal welfare differs between countries. At the same time, meat is traded from countries with less stringent welfare requirements, like Argentina [31–33] and Uruguay [34], to countries with higher welfare requirements, like members of the European Union. The European Union (EU) sets welfare requirements at the time of killing for companies willing to export animal products to the EU (Council Regulation (EC) No 1099/2009) [35], but does not impose direct requirements on the transport of those animals to the slaughterhouses in third countries. The international Horse Meat Federation [36], on the other hand, expects its members to meet the requirements set in their "Manual for the Animal Welfare of horses during transport and slaughtering", which is based on existing legislation and international guidelines [36].

The first aim of this study was to characterize the current practices of the commercial horse transport from collection points to slaughterhouses in Argentina and Uruguay. Secondly, potential risk factors for horse welfare were detected.

2. Materials and Methods

Between November 2016 and October 2017, a total of 23 horse transports in Argentina and Uruguay were monitored from loading at a collection point to one of the three selected horse slaughterhouses. A total of 596 half-bred horses with an average weight of 415 (±38) kg were picked up at 18 collection points and were transported to Lamar (Argentina), Frigorífico General Pico (Argentina) and Sarel (Uruguay). The transports were spread throughout a year, so that a number of transports were monitored in each season. A total of six transports took place in the spring (November 2016), six in the summer (March 2017), six in the fall (June 2017) and five in the winter (October 2017). Each season two transports per slaughterhouse were monitored, except for October 2017. At that time, only one transport to Frigorífico General Pico was followed up. The transports were carried out with different types of transport vehicles. In this study, three types of vehicles are distinguished, namely trucks, tractor–trailers and truck–trailers (Figure 1). A truck is a pulling vehicle with one inseparable loading space and is therefore a single unit. In the case of a tractor–trailer, the towing vehicle can be (dis)connected from the trailer via a fifth-wheel coupling. Finally, a truck–trailer is a truck, as defined above, with a trailer connected to it through a drawbar. This transport combination therefore exists of two separated loading spaces.

a   b

Figure 1. Example of different transport combinations. (**a**) A truck–trailer: a truck pulling a trailer and (**b**) A tractor–trailer: a pulling vehicle (without loading space) hauling a trailer through a fifth-wheel coupling.

The transports were monitored and supervised by the same researcher. The researcher also mounted the sensors and cameras in the trailers and recorded specific transport data. In each trailer the horses were filmed. The camera (Trophy Cam model 119437, Bushnell, China) filmed fragments of one minute at intervals of about 100 seconds. On average, 38.9% (±3.8)% of each transport was filmed. After the transports, the footage was viewed, and the behaviour and interactions of the horses were analysed. During the analysis of the videos, it was noted per trailer how many horses fell and whether the animals fought 'hardly or not', 'averagely' or 'a lot'. A fall was considered to be a loss of balance in which parts of the body other than the hooves touched the ground. If the horse could restore equilibrium without other body parts touching the ground, this was called stumbling. Furthermore, the temperature, relative humidity (RH) and dew point were automatically recorded (EL-USB-2, Lascar Electronics, Wiltshire, UK) every five minutes in each loading space. These data were automatically written to an excel file. For the analysis of the environmental parameters, the thermoneutral zone used by Morgan [27], namely between 5 °C and 25 °C, is taken as the starting point. In addition, the temperature humidity index (THI) is calculated using the following formula [37], with T, temperature in °C, RH expressed as a number between 0 and 1:

$$THI = 0.8T + RH \times (T - 14.4) + 46.4 \tag{1}$$

The timing of various operations such as loading and unloading, the duration and distance of the transport, the number of intermediate stops for (police) checks, as well as the stocking density, dimensions and characteristics of the loading spaces were recorded. The observer noted what tools the drivers, which are the persons that handle the horses, used. Furthermore, the characteristics and dimensions of loading docks were registered. The openness of the side walls was categorised as 'open' when the surroundings could be seen easily through the wall, 'half open' when the view through the side walls was limited, and 'solid' when the horses could not see anything through the side walls of the loading dock. All the parameters that were considered during this investigation are listed in the Appendix A in Tables A1 and A2.

The data were processed using SAS Enterprise Guide and SAS 9.4. Averages and standard deviations were calculated using the PROC UNIVARIATE procedure. Correlations between the environmental parameters in the front and rear trailer were calculated using the regression (PROC REG) procedure. The frequency procedure (PROC FREQ) was used for all frequency calculations and generalized linear mixed models (PROC GLIMMIX) were used to identify which parameters influenced the degree of aggression and falling of horses during transport.

3. Results

3.1. Loading Dock

In 17 of the 18 collection points, a loading dock was present to facilitate the loading of the horses. From two collection points, horses were picked up more than once. However, only one loading dock was used for two transports. In one collection point, the loading dock did not have a slope, since the transport vehicle could be parked so that surface of the loading area was at the same level as the floor of the trailer. Therefore, this loading deck was excluded from the averages (Table 1). The slope of the loading docks was on average 17.4° (±3.6)°, which equals 31.5% (±7.0)% (Table 1). Seven out of 21 loading docks had a slope steeper than 20.0° (36.4%) and the slope of all but one loading dock was steeper than 10.0° (17.6%). The length of the loading dock (measured on the surface of the loading dock) was 4.01 (±0.90) m and the height was 1.18 (±0.17) m (Table 1).

Table 1. Dimensions and slope of the loading docks in the 18 collection points in degrees and in percentages.

Parameter	Average ± Standard Deviation	N	Minimum	Maximum	Median
Length (m)	4.01 ± 0.90	21	2.60	6.50	3.89
Height (m)	1.18 ± 0.17	21	0.93	1.74	1.17
Height side walls (m)	1.59 ± 0.22	19	1.16	1.94	1.60
Slope (°)	17.4 ± 3.6	21	9.3	23.6	17.1
Slope (%)	31.5 ± 7.0	21	16.4	43.6	30.7

The surface of the loading docks consisted of only soil (26.1%), a combination of soil and wood (26.1%), only wood (13.0%), concrete (partly) covered with soil (8.7%), a combination of soil and grit (8.7%) or straw (4.4%). The side walls of the loading docks were, on average, 1.59 (±0.22)-m-high and constructed from wood—mostly planks (82.6%), but round wooden beams (4.3%) in one instance. The side walls of three (13.0%) loading docks were categorised as 'open', seven (30.4%) as 'half open' and 10 (43.5%) as 'solid'. Of three (13.0%) loading docks, the kind of side wall construction was not registered.

3.2. Loading

Loading the horses took an average of 12.2 (±7.1) minutes per transport and 0.49 (±0.27) minutes per horse (Table 2). Spread over three transports, five horses (0.84%) had to be led into the trailer with a halter: three horses in one loading did not want to enter the trailer and twice one horse refused to enter the trailer. In the end, two of these horses could not be loaded at all. The trucks left the collection points 8.3 (±4.4) minutes after the loading process was completed. In total, 73.9% of the transports departed in the morning, on average at 11:15 a.m. (±1:31; between 9:25 a.m. and 3:20 p.m.). Figure 2 shows the arrival and departure times of the transports.

Table 2. Duration of loading and standstill after loading before departure.

Parameter	Average ± Standard Deviation	N	Minimum	Maximum	Median
Loading time (minutes)	12.2 ± 7.1	23	2.0	32.0	10.0
Loading time per horse (minutes/horse)	0.49 ± 0.27	23	0.19	1.23	0.42
Duration standstill before departure (minutes)	8.26 ± 4.28	23	2.00	19.00	8.00

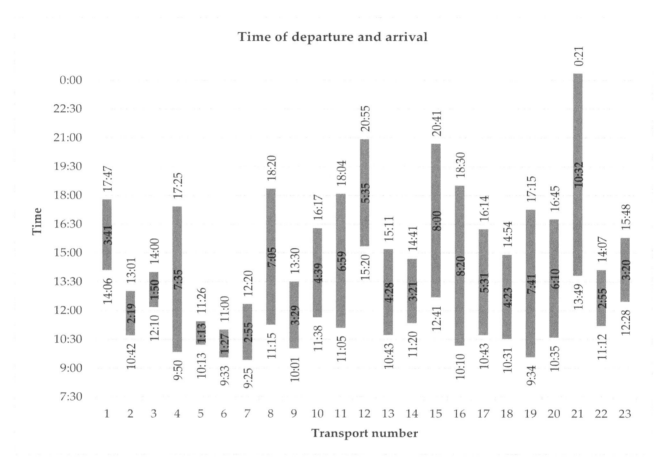

Figure 2. Arrival and departure times of the monitored transports. The duration of the transports is mentioned in the coloured bars. The transport numbers are the same as used in Tables A3 and A4 in the Appendix A.

Tools such as a flag, whip or stick were used to drive the horses on the trailers. We define a stick as a narrow, long and little or not flexible object to drive the horses with. A whip is a thin, not very flexible stick, with or without a handle on it. In this context, a flag is a stick with a piece or ribbons of textile or plastic, so that the movements of the object are more visible to the horses. Flags were used in 22 (95.7%) of the 23 transports. Four times (17.4%) a whip or a stick was used and three times (13.0%) a rider on horseback drove the slaughter horses onto the trailer. The tools were not used to hit or poke animals on sensitive body parts, but to give visual signals. Occasionally, the horses were gently touched with the stick, whip or flag, but not to the extent that the touch could cause pain or discomfort.

Divided over 13 loads, a total of 23 (3.86%) horses stumbled during loading, with a maximum of five horses during one loading. Five falling horses were noted, spread over three loadings. During a loading in which three horses stumbled, also three horses fell and during two other loads, in which respectively three and five horses stumbled, one horse fell each time. No falls were noted in other loadings. The falls were caused by pushing and/or fights among horses. One of the fallen horses was ran over by the other horses when the group abruptly turned around in the narrowing space towards the loading ramp. The abrupt turn of the group was caused by the directions of the drivers.

Horses enter the transport vehicles through trapdoors, which are guillotine-type doors. Internal doors were also trapdoors. During 13 of the 23 transports, at least one horse bumped its head against the trapdoor. In five transports, three or more horses (maximum eight) bumped their heads against the trapdoor. The height of the trapdoors where horses hit their heads varied between 1.51 and 1.81 m.

3.3. Environmental Parameters

The lowest trailer temperature observed during transport was 6.0 °C, while the maximum temperature was 35.5 °C (Table 3). In the trailers, the temperature never dropped below the lower limit of the thermoneutral zone, being 5.0 °C. The upper limit of the thermoneutral zone, 25.0 °C, was exceeded during 14 of the 23 (60.9%) transports. During six (26.1%) transports, the average temperature in the trailer was above 25.0 °C and during five (21.7%) transports, the minimum temperature in the trailer exceeded 25.0 °C. For seven (30.4%) transports, the maximum temperature was above or equal to 30.0 °C. There was a strong correlation between the front and rear load for both average, minimum and maximum temperatures (Table 4).

Table 3. The average, minimum and maximum temperature, RH and THI in the loading spaces. The average, minimum and maximum of each parameter was first calculated per transport and then per time period.

Parameter		Average ± Standard Deviation	N	Minimum	Maximum
Temperature		22.0 ± 5.0	38	6.0	35.5
	Spring	25.4 ± 3.7	11	15.0	35.5
	Summer	24.2 ± 2.2	11	17.5	32.5
	Autumn	15.3 ± 2.3	9	6.0	26.0
	Winter	21.7 ± 4.2	7	14.0	32.0
RH		57.4 ± 13.5	38	27.5	99.0
	Spring	52.6 ± 14.9	11	27.5	97.0
	Summer	59.9 ± 12.9	11	42.5	96.5
	Autumn	60.3 ± 6.2	9	43.5	80.0
	Winter	57.5 ± 16.4	7	31.5	99.0
THI		67.8 ± 6.5	37	45.5	83.0
	Spring	72.0 ± 4.6	11	59.0	83.0
	Summer	71.4 ± 2.7	11	62.6	81.5
	Autumn	59.1 ± 3.3	9	45.5	72.8
	Winter	66.5 ± 4.9	6	57.3	78.0

The RH varied between 28.0% and 99.0% (Table 3). There was a strong correlation between the front and rear loading space for both average, minimum and maximum humidity (Table 4). There was no correlation between temperature and RH within the same loading space.

Due to the lack of a reference framework with limit values for heat stress in horses [38,39], the THI is tested against the values used for dairy cattle. A THI of 72–78 is labelled as mild heat stress, while a THI between 79 and 89 stands for severe heat stress in dairy cattle [40–42] (Appendix A, Figure A1). The THI ranged from 45.5–83.0 during transports (Table 3). In 13 (57%) transports, the maximum THI exceeded 72, which is the lower limit for mild heat stress in cattle. For six (26%) transports, the average THI was above or equal to 72 and for four (17.4%) transports, the minimum THI was at least equal to 72. During six (26.1%) transports, the maximum THI value was between 78 and 89, the standard for severe heat stress in cattle. However, the average THI value always remained below 78. There was a strong correlation between the front and rear loading space for both average, minimum and maximum humidity (Table 4). The average temperature, RH, dew point temperature and THI are shown per transport in Appendix A Table A4.

Table 4. Correlations between temperature (T), relative humidity (RH) and temperature humidity (THI) in front and rear loading space. The *p*-value of "a" is the *p*-value of the correlation coefficient. T_{AF} = average temperature in the front loading space; T_{AR} = average temperature in the rear loading space; T_{MinF} = minimum temperature in the front loading space; T_{MinR} = minimum temperature rear loading space; T_{MaxF} = maximum temperature front loading space; T_{MaxR} = maximum temperature rear loading space. RH_{AF} = average RH front in front loading space; RH_{AR} = average RH in rear loading space; RH_{MinF} = minimum RH in front loading space; RH_{MinR} = minimum RH in rear loading space; RH_{MaxF} = maximum RH in front loading space; RH_{MaxR} = maximum RH in rear loading space. THI_{AF} = average THI in the front loading space; THI_{AR} = average THI in the front loading space; THI_{MinF} = minimum THI in the front loading space; THI_{MinR} = minimum THI in the rear loading space; THI_{MaxF} = maximum THI in the front loading space; THI_{MaxR} = maximum THI in the rear loading space.

Parameter	Equation (Y = aX + b)	N	r^2	*p*-Value a	*p*-Value b
T_{AF} and T_{AR}	$T_{AF} = 1.01710 \times T_{AR} - 0.26312$	14	0.9847	<0.0001	0.7465
T_{MinF} and T_{MinR}	$T_{MinF} = 1.04965 \times T_{MinR} - 0.77318$	14	0.9840	<0.0001	0.3104
T_{MaxF} and T_{MaxR}	$T_{MaxF} = 0.85989 \times T_{MiaxR} + 3.92441$	14	0.8111	<0.0001	0.2244
RH_{AF} and RH_{AR}	$RH_{AF} = 1.00838 \times RH_{AR} + 0.22294$	14	0.9861	<0.0001	0.9020
RH_{MinF} and RH_{MinR}	$RH_{MinF} = 0.86383 \times RH_{MinR} + 6.00346$	14	0.9246	<0.0001	0.0668
RH_{MaxF} and RH_{MaxR}	$RH_{MaxF} = 0.99305 \times RH_{MiaxR} + 0.78603$	14	0.9676	<0.0001	0.8120
THI_{AF} and THI_{AR}	$THI_{AF} = 1.01061 \times THI_{AR} - 0.64323$	14	0.9878	<0.0001	0.7704
THI_{MinF} and THI_{MinR}	$THI_{MinF} = 1.04821 \times THI_{MinR} - 3.03323$	14	0.9831	<0.0001	0.2441
THI_{MaxF} and THI_{MaxR}	$THI_{MaxF} = 0.92326 \times THI_{MiaxR} + 5.81791$	14	0.8453	<0.0001	0.4859

3.4. Trucks

Different types of transport vehicles were used to carry out the 23 transports (see Materials and Methods). Some vehicles were used for multiple transports. Table A3 in the Appendix A shows the type of vehicle used for each transport and the frequency of use of the vehicle during the monitoring. Only one of the trucks had a roof consisting of a black sail. The average dimensions of the loading spaces are shown in Table A3 in the Appendix A. In 10 transports (43.5%) the front loading space was divided into several compartments: eight times (34.8%) into two compartments and twice (39.1%) into three. The rear loading space was divided into two (six times; 40.0%) or three (two times; 13.3%) compartments in eight (53.3%) of the 15 transports with two loading spaces.

The floor in all loading spaces was provided with wire mesh to prevent slipping of the horses. Different types of wire mesh could be distinguished. The most common were the standard wire mesh (Figure 3a), where the rods are on top of each other and the rods do not bend between the crossings. This type of wire mesh was used in 20 (87.0%) transports. The curved wire mesh (Figure 3b), which is bent between the crossings, was used in two transports, just like the diamond-shaped wire mesh with connections in one plane (Figure 3c). In the Appendix A Table A3 shows which type of wire mesh was found per transport and per trailer. The average mesh size was 26.0 ± 4.5 cm by 23.7 ± 4.8 cm. On average, the wire mesh was 1.29 ± 0.38 cm thick and mounted at a height of 2.60 ± 0.78 cm. In our observations, both on the spot and afterwards, no shoed horses were detected in any of the transports.

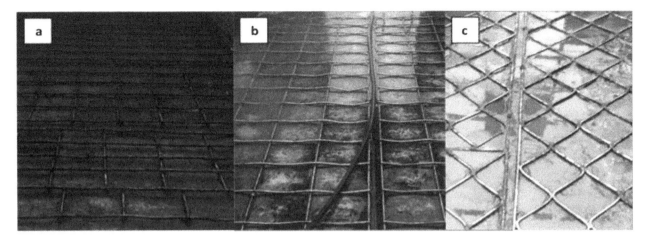

Figure 3. Different types of wire mesh that were used for the 23 observed slaughter horse transports. (**a**) Standard wire mesh—here addressed as "standard"; (**b**) standard wire mesh, which is curved between connection points—here called "curved"; (**c**) diamond shaped wire mesh, with connections in one plane—here called "diamond shaped".

3.5. Trailer Density

The average density of the trailers was 1.40 (±0.33) m^2/horse or 308 (±53) kg/m^2. Table 5 shows the average density in m^2/horse and kg/m^2 per trailer and per compartment. Average density varied between 0.94 m^2/horse and 2.45 m^2/horse. Stallions were not always separated from mares and geldings during transport. During at least five transports, one or more stallions were loaded. In at least two of these transports, the stallions were not separated from the other horses. Once, a stallion standing between mares was moved to another compartment before departure, because of his aggressive behaviour. In the other compartment however, the stallion was not separated from the other horses behind a fence or ropes either. To prevent further biting, a rope was tied tightly in the mouth. The stallion then stopped his aggressive behaviour. Several times a pony or a young horse was transported in the same compartment with significantly larger horses.

Table 5. Average available space per horse (m^2/horse) and average density (kg/m^2). T1 = front loading space; T2 = rear loading space.

Parameter		Average ± Standard Deviation	N	Minimum	Maximum	Median
Surface area of loading space (m^2)		34.70 ± 9.50	23	13.50	41.91	39.45
Average density of full loading space (m^2/horse)		1.40 ± 0.33	23	0.94	2.45	1.38
Density T1 (m^2/horse)		1.40 ± 0.36	23	0.80	2.45	1.33
	Compartment 1	1.52 ± 0.53	8	1.03	2.68	1.46
	Compartment 2	1.38 ± 0.44	7	0.97	2.25	1.33
	Compartment 3	1.25 ± 0.35	2	1.00	1.50	1.25
Density T2 (m^2/horse)		1.38 ± 0.22	15	0.99	1.88	1.38
	Compartment 1	1.36 ± 0.27	6	1.06	1.74	1.36
	Compartment 2	1.53 ± 0.49	6	1.15	2.48	1.37
	Compartment 3	1.86 ± 0.53	2	1.49	2.23	1.86
Average density (kg/m^2)		308 ± 53	21	191	402	327
Density T1 (kg/m^2)		312 ± 63	21	191	473	308
	Compartment 1	283 ± 65	6	175	352	304
	Compartment 2	322 ± 57	6	208	365	338
	Compartment 3	362	1	362	362	362
Density T2 (kg/m^2)		309 ± 42	14	227	354	326
	Compartment 1	334 ± 74	5	253	446	344
	Compartment 2	291 ± 67	5	191	362	322
	Compartment 3	212	1	212	212	212

3.6. Aggression and Falling during Transport

The degree of aggression was assessed per loading space during 22 transports, of which 15 transports with a truck–trailer, together accounting for 38 loading spaces. A loading space refers to the space in one transport component. A truck therefore has one loading space just like a tractor–trailer, while a truck–trailer has two loading spaces. In 13 (34.2%) loading spaces, hardly any or no fights were registered. In 15 (39.5%) loading spaces, an average amount of fights was recorded, and a lot of fights were recorded in 11 (28.9%) loading spaces. No relation was found between the degree of aggression and the density (front loading space: $p = 0.78$; rear loading space: $p = 0.25$), the thickness of the wire mesh on the floor (front loading space: $p = 0.23$; rear loading space: $p = 0.20$), the transport duration (front loading space: $p = 0.90$; rear loading space: $p = 0.98$) or distance (front loading space: $p = 0.93$; rear loading space: $p = 0.78$), the average temperature (front loading space: $p = 0.33$; rear loading space: $p = 0.79$), the average dew point temperature (front loading space: $p = 0.18$; rear loading space: $p = 0.99$) and the average THI (front loading space: $p = 0.28$; rear loading space: $p = 0.80$). However, the degree of aggression differed between the front and rear loading spaces ($p = 0.05$). Splitting the rear loading space into two or more compartments resulted in less fighting than when the rear loadings spaces consisted of one compartment ($p = 0.05$). In the case of the front loading space (22 transports), no relation was found between compartmentalisation and the degree of fighting ($p = 0.73$). It should be noted, however, that the fighting behaviour in the rear loading space could only be monitored for 14 transports, since only 14 out of 15 truck–trailers had the rear loading spaces successfully filmed.

In only one journey did a horse fall twice. In the 22 other journeys, no horses fell, except once before departure. That horse was then removed from the truck and not taken to the slaughterhouse. No connection could be found between the degree of fighting and the falls of the horses, but this is likely due to the limited number of horses that fell during the journey.

3.7. Transport Distance and Duration

The transport distances from collection point to slaughterhouse ranged from 37–700 km (Table 6). The transport time varied from 73–632 min (Table 6). One transport lasted 480 min or eight hours, while two (8.7%) other transports lasted longer, more specifically 500 and 632 min. The horses were not unloaded during the journey and had no ability to eat or drink. The transport duration and distance were strongly correlated ($r^2 = 0.94$; $p < 0.0001$) for the 23 observed transports: with increasing distance, the transport duration increased according to:

$$Y = 49.7 + 0.8X \tag{2}$$

(with Y = the transport duration in minutes and X = the distance in km)

Table 6. Average transport duration, distance and average number of stops and police checks per transport and per 100 km. Duration of loading, unloading and standstill after loading and before unloading.

Parameter	Average ± Standard Deviation	N	Minimum	Maximum	Median
Transport duration (minutes)	296 ± 150	23	73	632	268
Distance (km)	295 ± 177	23	37	700	250
Number of stops per transport	2.74 ± 1.33	23	0.0	5.0	3.0
Number of stops per 100 km	1.14 ± 0.90	23	0.00	4.35	1.04
Number of police checks per transport	0.87 ± 1.00	23	0.00	3.00	1.00
Number of police checks per 100 km	0.39 ± 0.58	23	0.00	1.97	0.25
Standstill between arrival and unloading (minutes)	14.8 ± 12.6	22	3.0	45.0	10.0
Duration of unloading (minutes)	12.5 ± 9.5	20	2.0	36.0	9.0
Duration of unloading per horse (minutes/horse)	0.473 ± 0.285	20	0.115	0.947	0.388

An average of 2.74 (±1.33) stops were inserted per transport, of which an average of 0.87 (±1.00) were inserted for police checks (Table 6).

3.8. Unloading

After arrival, transporters had to wait for an average of 14.8 (±12.6) minutes before unloading (Table 6). The unloading of the horses took on average 12.5 (±9.5) min per transport and 0.473 (±0.285) per horse (Table 6). There was no correlation between the duration of loading and unloading ($r^2 = 0.11$; $p = 0.08$).

4. Discussion

4.1. Loading Docks

Besides good handling, the professional federation of the international Horse Meat Sector (HoMeFe) [36] as well as the European Consortium of the Animal Transport Guides Project (CATGP) [7] stresses the importance of good loading dock design, construction and maintenance to minimize the risk of slipping, falling, injuries and stress to animals while (un)loading. Therefore, CATGP [7] sums up a number of 'good' and 'better' practices in its guidelines for transport of slaughter horses. The good practices are derived from the Council Regulation 1/2005 of 22 December 2004 on the protection of animals during transport and related operations [7,43].

The loading area should be constructed in a way that prevents distress, excitement and injury as much as possible [7,36,43]. Good practices include that the slope of the loading dock does not exceed 36.4% (or 20.0°) [7,36,43,44], better practices demand a maximum slope of no more than 10.0° (17.3%) [7]. When the slope is steeper than 10.0° or 17.6%, the loading dock must be equipped with some sort of system that improves passage of animals without the risk of slipping [36,43], such as stair steps or foot battens [7,36,45]. Argentinian legislation states that the slope of the loading dock should not exceed 30.0° (57.7%) and should be equipped with foot battens [33]. Good loading practices also include using a slip-resistant and anti-sliding surface on the loading ramp [7,36,43]. To prevent animals from falling off or escaping from the (un)loading dock, side walls should be provided [7,36,43]. Side walls which limit the view on the environment prevent animals from being distracted by what is happening around the loading dock and might thereby simplify (un)loading [7,36,44]. Grandin [46] also mentions that solid side walls are more efficient in preventing escape attempts due to the blocked vision of the animals.

4.2. Loading and Unloading

Loading duration is considered to be an indicator of the ease of loading. Since no loading times of slaughter horses are available for comparison, the observed loading durations can only be compared to these of beef cattle. María et al. [47] noted an average loading time of 1.20 (±0.86) minutes per beef bull. However, horses and cattle cannot be compared in terms of the leniency of their movements; these data support the observation that the loading of the horses in general went quite smoothly.

However, infrastructure did not always promote smooth loading. Horses regularly bumped their heads against trapdoors, clearly indicating that the trapdoors were too low. Furthermore, when stressed, for example due to rushed driving, horses carry their head higher [48] and thereby the chance of head-bumping is increased, especially when trapdoor height is rather low. Argentinian legislations states that the trapdoors should be at least 1.60 m high, which was not always the case, to prevent the horses from hurting their heads and backs [33].

In a total of seven loadings, the observing researcher noted that horses hesitated or reacted anxiously to specific elements of the infrastructure; for example, low or not fully opened trapdoors, a bar hanging too low over the passageway, a steep slope of a loading dock, or an uneven or muddy ground on or before the loading dock. Falling horses were only observed in loadings in which multiple horses stumbled. The falls were caused by pushing or fights between horses, which was clearly caused by the directions of the drivers in one instance. The directions of the drivers were not necessarily wrong in this case, but with a calmer approach the fall might not have happened. Furthermore, drivers must ensure that they do not give conflicting signals to the horses, for example by standing too close to the passageway of the horses when another driver is directing the horses to go there. These conflicting signals create confusion and thereby chaos. However, the presence of the investigators was probably—to a certain extent—an additional stress factor for the horses and loading crew.

For all transports, unloading went quite well. However, better communication between the slaughterhouse and the transporter may reduce the standstill before unloading the horses on arrival.

According to Friend [23], all horses experience stress during transport. Loading and unloading might even be more stressful to animals than the transport itself [47], but horses that were loaded before and did not have any negative experiences with loading experienced less stress than animals that were loaded for the first time [49]. It is not known whether the horses observed in this study had been loaded and transported before, and how any previous transports were perceived. It is possible that the horses that did not want to enter the trailer had had previously negative experiences during a transport, or had other bad experiences [13]. If the person who loads the horses has a good relationship with the horses, the stress level during loading decreases [49]. Since the horses do not have much contact with people during rearing, there is little evidence that there is a relationship of trust between the caretakers and the horses. Due to the lack of (positive) transport experiences of the horses and the absence of a trust relationship with the drivers, the importance of efficient and knowledgeable driving increases, in order to make loading as smooth as possible [49]. The authors of [47] state that education and training of the personnel is likely to be one of the most effective measures to improve horse welfare during loading and transport. HoMeFe demands that everyone involved in the transport of slaughter horses is trained regularly [36]. However, it is not clear whether or not the drivers and horse transporters of the observed transports had training in the handling and transporting of horses.

HoMeFe prohibits the use of electric driving aids, sticks and dogs [36]. As well as the World Organisation for Animal Health (OIE) [50], Grandin [51] mentions that the use of an electric prod should be avoided if possible, since the electric prod might cause the animals to become agitated and therefore sometimes dangerous. Furthermore, drivers should not scream, flap their arms or make sudden movements to keep the animals as calm as possible [36,50,51]. After all, besides their adverse effects on animal welfare, multiple studies demonstrate the adverse effects of the use of electric prods and incompetent handling on meat quality in pork and beef [52–54].

No electric prods were used and handlers stayed calm. Flags were used in all but one loading. Against the requirements of HoMeFe [36], a whip or stick was used in four loadings, but never to hit horses.

4.3. Environmental Parameters

About three-quarters of the transports left before noon, implying that the transports were often being carried out during the hottest moments of the day.

When the environmental temperature rises above the upper limit of the thermoneutral zone, the animal has to invest energy to keep its body temperature constant [7,26]; for example, by sweating, peripheral vasodilation, and by increasing the respiratory rate [7,26]. However, it is not clear from what point welfare is compromised. The temperature regularly rose above 25 °C, which is the upper limit of the thermoneutral zone of horses [27], and RH regularly increased above 50%. Above 50%, heat is dissipated less efficiently [29].

Based on the THI framework for producing dairy cows (Appendix A Figure A1) [42], mild and sometimes severe heat stress occurred during transports. However, THI must be interpreted carefully, since this parameter does not take solar load and wind speed into account [55]. Especially during transport in an open trailer, air displacement might enhance heat dissipation. On the other hand, the horses are standing close to each other during transport, which might limit heat dissipation. Moreover, the lack of a reference framework with limit values for heat stress in horses complicates interpretation.

Older horses are known to be more prone to heat stress than younger horses. When exposed to the same level of exercise, older horses overheat in a much shorter time than younger horses, indicating that their ability to dissipate excess heat is less compared to that of younger horses [56]. However, the age of the slaughter horses in this study was not known.

4.4. Trucks and Trailer Density

The floor of all transport vehicles was provided with wire mesh, as demanded by Argentinian legislation [33]. Different authorities and guidelines have determined the minimum and maximum density for horse transports, which explains the differences in thresholds. According to the CATGP [7], the density may vary between 1.00 m^2 per horse and 1.75 m^2 per horse, depending on the size and age of the horse (Appendix A Table A5). The available space per horse may deviate by a maximum of 10% from the directive, depending on the physical condition, the weather conditions and the probable transport time [7]. HoMeFe [36] has drafted guidelines in its specifications for the density during transport. The available space must be between 1.1 m^2 per horse and 1.4 m^2 per horse. Depending on the physical condition of the horses, the weather conditions, the travel time, the weight and the height of the horses, the actual density may deviate a maximum of 20% from the guidelines. This means that the available surface area per horse may vary between 0.88 m^2 and 1.68 m^2 [36]. In one or more compartments of six transports, the average surface area per animal deviated more than 20% from the HoMeFe guideline [36]: the surface area per animal was too large in five instances, and it was too small in one instance (Appendix A Table A6). It should be noted that the allowed stocking densities differ substantially between the CATGP [7] and HoMeFe [36] guidelines. However, comparing stocking densities expressed as surface area per horse is not evident, since adult horses can differ substantially in size. Two trailers with the same loading density expressed in m^2/horse can be, in reality, quite a different stocking density in kg/m^2 for these two loadings. On the other hand, determining the number of horses that can be loaded based on the estimated average weight of a group of horses might be prone to estimation errors.

Stull [57] compared some physiological parameters and the increase in injuries between horses transported at low (1.40–1.54 m^2/horse) and high stocking densities (1.14–1.31 m^2/horse). She concluded that it is better to provide at least 1.40 m^2/horse during transport, depending on the weight, conformation and size of the horses. Extra attention must be paid to the design of the trailer in order to prevent injuries to the horses [57].

A few times a pony or a young horse was transported in the same compartment with other, significantly larger horses. This is contrary to European Regulation EC 1/2005 [43] and the Argentinian Resolución 581/2014 [33], which states that animals of significantly different sizes or ages and sexually mature mares and stallions, must be handled and transported separately, unless the animals have been reared together, are accustomed to each other, or when the separation would cause distress. Similarly, mares accompanied by their dependent foals are not subject the above provisions. Finally, animals 'hostile to each other' and tied and untied animals should not be transported in the same compartment [43]. In at least five transports, one or more stallions were loaded. The stallions were not separated from the other horses in at least two transports, which is contrary to the abovementioned European Regulation [43], Argentinian legislation [33], and guidelines from HoMeFe [36]. One stallion that was not separated from the other horses was aggressive and was therefore placed in another compartment and had a rope tied tightly in the mouth. After putting on the rope, the stallion stopped behaving aggressively, but this was probably due to the inconvenience caused by the rope.

4.5. Aggression and Falling during Transport

In this study, no relationships could be demonstrated between environmental parameters and aggression. This suggests that the environmental parameters are not the most important factors that may or may not provoke aggression. As well as the current study (Appendix A Table A6), Iacono and colleagues [18] could not demonstrate a relationship between the degree of aggression and the density or fatigue of the horses during the transport of untied horses. According to the authors of [18], aggression is more likely to be a consequence of individual horses than stocking density. The current observations confirm this assumption (Appendix A Table A6).

4.6. Transport Distance and Duration

Argentinian legislation allows for a transport duration up to 36 hours without feeding, watering or rest [31]. In Uruguayan law, no maximum transport duration is mentioned [34]. According to the specifications of HoMeFe [36], a transport may take up to 12 h. In exceptional cases, the transport may last up to 14 h, for example if the destination can be reached by continuing the ride for a maximum of two hours [36]. This requirement differs substantially from the specifications of the European Regulation EC 1/2005 [43], that states that a transport may only take up to eight hours. In exceptional cases, the transport of trained horses may last up to 24 h if they are watered and fed every eight hours and if the transport vehicle meets some extra requirements for roof construction, presence and quality of litter, feeding and watering regime, partitions, ventilation, climate control and navigation system. However, for unbroken horses, the European Regulation does not allow any extension of the transport duration of eight hours, regardless of the transport vehicle in which the unbroken horses are transported [43].

Friend [14] claims that it is advisable to regularly provide the horses with water on the truck during long-distance transport in warm conditions, in order to reduce dehydration, stress and fatigue. CATGP [7] recommends to water horses every 4.5 h, while the European Council Regulation EC 1/2005 [43] states that Domestic Equidae have to be watered every eight hours. Notwithstanding, Friend [14] also mentions that water consumption can be highly variable among different horses in the same situation [11]. Likewise, other studies question the watering of horses during transport [17,58]. After all, it often takes some time before the horses start drinking, about 20 min to an hour after the water is offered. Some horses did not drink, possibly for fear of the new water source or because of stress associated with the transport. Furthermore, the difference in weight loss between horses that did and did not drink suggested that the horses probably did not drink a large amount of water [17,59]. Therefore, it seems especially important that the horses are sufficiently hydrated before departure and have access to sufficient fresh water immediately after arrival [14,36,60].

5. Conclusions

Our study identified the current practices of the commercial horse transport from collection points to slaughterhouses in Argentina and Uruguay. Some risk factors have been detected and could be improved. The loading and unloading of the horses generally went quite smoothly. Better training of drivers and optimized infrastructure (a level ground surface before and on the loading dock, sufficiently high trapdoors, steepness of the loading docks, provision of steps or foot battens on the loading dock, etc.) can prevent a lot of confusion and chaos for the horses, and thereby improve welfare. Driving aids were always used correctly. Most journeys started before noon, implying that the horses were often transported during the hottest moments of the day.

Still, on the one hand, interpretation of THI values is difficult because of a lack of reference framework for horses. On the other hand, not all parameters that affect thermal comfort are included in the THI. Therefore, it is not clear from what point on welfare is compromised. Stocking densities were not always according to relevant guidelines and significantly smaller horses or stallions were not always separated from the other horses. No influence of environmental parameters or transport characteristics on the degree of fighting behaviour could be demonstrated. On the contrary, the degree of aggression differed between the front and rear loading spaces of the same transport vehicle, suggesting that animal-specific factors, rather than environmental factors, determine the occurrence of aggressive behaviour. The willingness of all actors involved—slaughterhouses, transporters, loading crew, etc.—to conduct this study and to address shortcomings, underscores the growing awareness of animal welfare issues in Argentina and Uruguay.

Author Contributions: Conceptualization, B.D.; methodology, B.D. and L.V.; formal analysis, L.V. and B.N.; investigation, L.V.; data curation, L.V. and B.N.; writing—original draft preparation, B.N. and B.D.; writing—review and editing, L.V., S.V.B. and J.V.T.; supervision, B.D. and J.V.T.; project administration, L.V and B.N. All authors have read and agreed to the published version of the manuscript.

Acknowledgments: The staff of the slaughterhouses and all drivers are acknowledged for their collaboration.

Appendix A

Table A1. Observed parameters of the transport and transport vehicle.

Parameter	Unit
Transport date	
Slaughterhouse	
Date of slaughter	
Address of the collection point	
License plate of all parts of the transport combination	
Presence of a roof on the loading spaces	Yes/no
Does the roof cover the whole loading space??	Yes/no
Colour of the roof	
Internal dimensions of the loading space - length, width, height and height of the solid wall	m
Surface area of the loading spaces	m^2
Number of compartments in the front and rear loading space	
Dimensions of the compartments - length and width	m
Loading density (averagely, per loading space and per compartment)	m^2/horse kg/m^2
Presence of tread plates in the front and rear loading space	Yes/no
Number of tread plates	
Length and width of the tread plates	m
Connecting method of the trailers: connected (yes) or not connected (no). If the trailers are connected, both trailers can be loaded through the hind gate. If trailers are not connected, both trailers should be (un)loaded separately.	Yes/no
Dimensions of the gates to enter the vehicle, between loading spaces, compartments or separate trailers (width and height)	m
Presence of wire mesh on the floor of the loading spaces	Yes/no
Type of wire mesh	Standard; Standard, bent; Diamond shaped
Mesh dimensions (length and width), height of the top of the wire mesh and thickness of the wire mesh	m
Presence of fences in the loading space to create different compartments	Yes/no

Table A2. Observed parameters during loading and unloading.

Parameter	Unit
Dimensions of the loading ramp	
- length, height, width	m
- slope	degrees
Material that covers loading ramp (soil, wood, …)	
Point of time on which loading and unloading started and ended (total, per trailer and per compartment)	
Duration of loading and unloading (total, per trailer and per compartment)	Minutes
Number of horses per transport, per trailer and per compartment	
Hygiene of the loading space: is the loading space clean?	Yes/no
Number of horses that bumped their head against the upside of the gates while (un)loading	
Number of horses that stumbled during (un)loading	
Number of horses that fell during (un)loading	
Number of horses that had leg injuries at the moment of (un)loading	
Number of horses that had a belly injury at the moment of (un)loading	
Number of horses that limped while (un)loading	
Number of horses that held the mouth opened while (un)loading	
Number of horses that made noise while (un)loading (long and loud, screamy whinnying)	
Number of horses that had the nostrils widely opened while (un)loading	
Number of horses that fled back from the trailer during loading	
Number of horses that had a head injury while (un)loading	
Number of horses that heavily snort during (un)loading	
Number of horses that sweat heavily during (un)loading	
Number of horses that scrape the floor during (un)loading	
Number of horses that fight during (un)loading	
Colour of the flags (and other driving tools) at (un)loading	
Material used for (un)loading the horses	
Did the drivers leave the gate of the loading spaces opened when going back to drive the other horses on the vehicle?	
Was the gate fully opened when (un)loading the horses?	Yes/no
Were there any remarks during the (un)loading?	
Weather conditions at (un)loading	
- sunny	Yes/no
- cloudy	Yes/no
- rainy	Yes/no
- windy	Yes/no
- misty	Yes/no
- moist	Yes/no
- CO_2-concentration	ppm
- light intensity	lux
Point of time of departure, arrival, starting and finishing loading and unloading (total, per trailer and per compartment) and point of time of the end of weighing	
Duration of standstill before departure, standstill before unloading, (un)loading (total, per loading space and per compartment) and weighing	minutes
Distance of the transport	km
Average speed	km/u
Duration of transport	minutes
Number of stops during transport	
Number of police controls during transport	
Degree of aggression/fighting during transport (per loading space)	Not or barely, averagely or a lot
Number of fallen horses during transport (per trailer)	
Temperature in the loading space during transport (measurement every five minutes, calculated average, minimum and maximum) per loading space	°C
RH in the loading space during transport (measurement every five minutes, calculated average, minimum and maximum) per loading space	%
Dew point temperature in the loading space during transport (measurement every five minutes, calculated average, minimum and maximum) per loading space	°C
Dimensions of the (un)loading ramp	
- length, width, height	m
- slope	degrees
Is the gate completely opened when unloading?	Yes/no

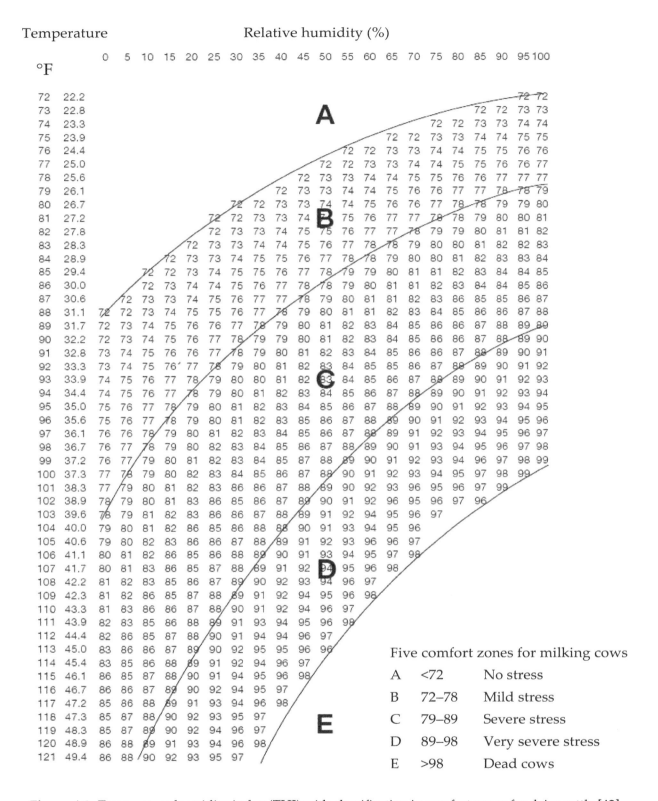

Figure A1. Temperature humidity index (THI) with classification in comfort zones for dairy cattle [42].

Table A3. Some details of the 23 observed transports. A letter in the column "Same transport vehicle" indicates that this (part of) the transport combination was also used in the other transport where the same letter is mentioned. In the column "Number of trailers" is mentioned of how many the loading spaces the transport combination consists. The internal dimensions of the loading spaces and the height of the solid wall are given in the last four columns. "Solid wall" refers to the part of the side wall of the loading space which is not made up of trellis. The function of the solid wall is to block the view of the horses and more importantly to prevent the horses from getting stuck (with their legs) in the trellis.

Number	Season	Transport Date	Slaughterhouse	Number of Trailers	Type of Vehicle	Number of Horses	Same Transport Vehicle	Duration (Minutes)	Distance (km)	Type of Wire Mesh	Length of Loading Space (m)	Width Loading Space (m)	Height Loading Space (m)	Height Solid Wall (m)
1	Spring	1/11/'16	Lamar	1	Tractor-trailer	25	a	191	206	Standard	15.76	2.48	2.05	1.30
2	Spring	3/11/'16	Lamar	2	Truck-trailer	9		135	152	Standard	5.81	2.45	2.15	1.13
						16				Standard	10.20	2.52	2.20	1.13
3	Spring	7/11/'16	General Pico	1	Truck	8		110	37	Standard	5.50	2.50	1.83	1.13
4	Spring	9/11/'16	General Pico	2	Truck-trailer	11	b	455	461	Standard	5.92	2.50	2.03	1.16
						18				Standard	10.15	2.50	2.03	1.16
5	Spring	14/11/'16	Sarel	2	Truck-trailer	21	c	73	63	Standard, bent	8.34	2.54	1.90	1.30
						21				Standard, bent	8.16	2.54	1.90	1.30
6	Spring	15/11/'16	Sarel	2	Truck-trailer	21		87	69	Standard, bent	6.70	2.50	2.05	1.40
						17				Standard, bent	7.56	2.50	2.02	1.40
7	Summer	7/03/'17	Sarel	2	Truck-trailer	16	d	175	154	Standard	7.83	2.36	1.80	1.40
						16				Standard	7.53	2.40	1.80	1.40
8	Summer	9/03/'17	Sarel	1	Tractor-trailer	38		425	407	Standard	15.00	2.67	1.83	1.35
9	Summer	14/03/'17	Lamar	2	Truck-trailer	10	e	209	192	Standard	6.15	2.50	2.20	1.30
						18				Standard	10.00	2.48	2.10	1.28
10	Summer	16/03/'17	Lamar	2	Truck-trailer	10	f	279	250	Standard	4.74	2.45	2.00	1.21
						20				Standard	10.48	2.45	2.00	1.21
11	Summer	20/03/'17	General Pico	1	Truck	11		419	470	Standard	5.40	2.50	2.00	1.30
12	Summer	20/03/'17	General Pico	2	Truck-trailer	6		335	289	Standard	5.30	2.50	1.78	1.15
						14				Standard	10.50	2.50	1.90	1.14
13	Fall	1/06/'17	Sarel	2	Truck-trailer	12		268	228	Standard	6.25	2.50	1.81	1.20
						20				Standard	10.42	2.49	1.83	1.19
14	Fall	2/06/'17	Sarel	1	Truck	12		201	185	Diamond shaped in one surface	6.20	2.30	1.89	1.30
15	Fall	6/06/'17	General Pico	1	Tractor-trailer	15	b	480	585	Standard	14.70	2.50	2.05	1.20
16	Fall	7/06/'17	General Pico	2	Truck-trailer	9	c	500	569	Standard	5.92	2.50	2.03	1.16
						17				Standard	10.15	2.50	2.03	1.16
17	Fall	12/06/'17	Lamar	2	Truck-trailer	11		331	338	Standard	6.65	2.40	1.95	1.24
						16				Standard	10.00	2.37	1.95	1.24
18	Fall	13/06/'17	Lamar	2	Truck-trailer	10		263	323	Standard	6.00	2.50	1.77	1.18
						19				Standard	10.00	2.50	2.10	1.19
19	Winter	13/10/'17	Sarel	2	Truck-trailer	10	d	461	380	Standard	6.00	2.34	1.80	1.17
						16				Diamond shaped in one surface	10.73	2.34	1.81	1.23
20	Winter	13/10/'17	Sarel	1	Truck	16	f	370	380	Standard	7.83	2.36	1.80	1.40
21	Winter	17/10/'17	General Pico	2	Truck-trailer	11		632	700	Standard	4.74	2.45	2.00	1.21
						23				Standard	10.48	2.45	2.00	1.21
22	Winter	23/10/'17	Lamar	2	Truck-trailer	11	e	175	160	Standard	6.15	2.50	2.20	1.30
						16				Standard	10.00	2.48	2.10	1.28
23	Winter	24/10/'17	Lamar	1	Tractor-trailer	26	a	200	180	Standard	15.76	2.48	2.05	1.30

Table A4. Some details of the transport conditions. If in the column "weather conditions" "rain" is mentioned, this does not necessarily mean that it rained continuously during loading, transport and/or unloading.

Number	Season	Transport Date	Slaughterhouse	Number of Trailers	Type of Vehicle	Number of Horses	Same Transport Vehicle	Duration (Minutes)	Distance (km)	Density m²/horse	Density kg/m²	Weather Conditions	Average T in Loading Space (°C)	Average RH (%)	Averagera Dew Point Temperature (°C)	Average THI
1	Spring	1/11/'16	Lamar	1	Tractor–trailer	25	a	191	206	1.58	307.8	Cloudy, rainy, misty	23.9	60.1	15.5	71.0
2	Spring	3/11/'16	Lamar	2	Truck–trailer	9		139	152	1.58	247.3	Sunny, windy	26.8	35.0	10.1	72.2
						16				1.61	243.5	Standard	26.3	35.2	9.7	71.6
3	Spring	7/11/'16	General Pico	1	Truck	8		110	37	1.72	231.6	Cloudy, rainy	15.9	92.2	14.6	60.5
4	Spring	9/11/'16	General Pico	2	Truck–trailer	11	b	455	461	1.33	346.1	Sunny, windy	23.8	46.8	11.0	69.4
						18				1.43	320.6	Standard	24.2	45.9	11.2	69.9
5	Spring	14/11/'16	Sarel	2	Truck–trailer	21	c	73	63	1.01	333.8	Sunny	28.5	47.4	16.1	75.8
						21				0.99	341.2	Standard, bent	29.1	46.4	16.3	76.5
6	Spring	15/11/'16	Sarel	2	Truck–trailer	21		87	69	0.80	473.1	Cloudy, sunny at arrival	28.8	54.7	18.8	77.3
						17				1.11	339.4	Standard, bent	28.8	54.8	18.8	77.3
7	Summer	7/03/'17	Sarel	2	Truck–trailer	16	d	175	154	1.15	336.1	Sunny	26.9	56.2	17.4	75.0
						16				1.13	341.8	Standard	26.4	51.1	17.2	74.4
8	Summer	9/03/'17	Sarel	1	Tractor–trailer	38		425	407	1.05	343.3	Cloudy, rainy, moist	22.7	86.3	19.9	71.3
9	Summer	14/03/'17	Lamar	2	Truck–trailer	10	e	209	192	1.54	308.5	Sunny, moist	27.2	49.6	15.7	74.5
						18				1.38	344.2	Standard	27.0	49.6	15.5	74.2
10	Summer	16/03/'17	Lamar	2	Truck–trailer	10	f	279	250	1.16	391.5	Cloudy, sunny at arrival	21.5	51.6	11.1	67.3
						20				1.28	354.2	Standard	21.4	52.0	11.1	67.2
11	Summer	20/03/'17	General Pico	1	Truck	11		419	470	1.23	357.0	Alternately sunny and cloudy, moist	24.6	56.1	15.2	71.8
12	Summer	20/03/'17	General Pico	2	Truck–trailer	6		335	289	2.21	192.9	Alternately sunny and cloudy	22.9	59.8	14.5	69.5
						14				1.88	227.2	Standard	22.8	59.7	14.4	69.4
13	Fall	1/06/'17	Sarel	2	Truck–trailer	12		268	228	1.30	306.0	Sunny	18.2	66.0	11.7	63.5
						20				1.30	307.2	Standard	18.1	66.1	11.6	63.2
14	Fall	2/06/'17	Sarel	1	Truck	12		201	185	1.19	335.9	Sunny	18.3	56.8	9.6	63.3
15	Fall	6/06/'17	General Pico	1	Tractor–trailer	15		480	585	2.45	191.0	Sunny, windy	14.5	56.8	5.8	57.8
16	Fall	7/06/'17	General Pico	2	Truck–trailer	9	b	500	569	1.67	267.7	Alternately sunny and cloudy, windy	15.8	61.5	8.3	59.8

Table A4. *Cont.*

Number	Season	Transport Date	Slaughterhouse	Number of Trailers	Type of Vehicle	Number of Horses	Same Transport Vehicle	Duration (Minutes)	Distance (km)	Density		Weather Conditions	Average T in Loading Space (°C)	Average RH (%)	Averagera Dew Point Temperature (°C)	Average THI
										m²/horse	kg/m²					
17	Fall	12/06/'17	Lamar	2	Truck–trailer	17	c	331	338	1.47	303.4	Standard	12.0	52.0	2.4	54.5
						11				1.45	285.9	Alternately sunny and cloudy	12.9	50.4	2.8	55.8
18	Fall	13/06/'17	Lamar	2	Truck–trailer	16		263	323	1.48	280.0	Standard	14.2	67.6	8.2	57.4
						10				1.50	295.6	Alternately sunny and cloudy	13.9	65.9	7.6	56.8
19	Winter	13/10/'17	Sarel	2	Truck–trailer	19		461	380	1.32	337.0	Standard	15.9	70.9	10.5	60.0
						10				1.40	285.8	Cloudy, windy, sunny at arrival				
20	Winter	13/10/'17	Sarel	1	Truck	16	d	370	380	1.57	255.7	Diamond shaped in one surface				
						16				1.16	326.8	Cloudy, moist, sunny at arrival	16.5	67.3	10.3	60.9
21	Winter	17/10/'17	General Pico	2	Truck–trailer	11	f	632	700	1.16	351.0	Cloudy, rainy, moist	21.4	87.0	19.1	69.6
						23				1.28	332.0	Standard				
22	Winter	23/10/'17	Lamar	2	Truck–trailer	11	e	175	160	1.40		Alternately sunny and cloudy, rainy	22.4	47.5	10.5	68.0
						16				1.55		Standard	21.2	48.6	9.8	66.6
23	Winter	24/10/'17	Lamar	1	Tractor–trailer	26	a	200	180	1.48		Cloudy, rainy, sunny at arrival	27.2	40.7	12.7	74.0

Table A5. Required space per horse for road transport in Europe [7,43].

Type of Horse	Available Space
Adult horses	1.75 m^2 (0.7 × 2.5 m)
Young horses (6–24 months) (transport <48 hours)	1.2 m^2 (0.6 × 2.0 m)
Young horses (6–24 months) (transport >48 hours)	2.4 m^2 (1.2 × 2.0 m)
Ponies (<1.44 m)	1.0 m^2 (0.6 × 1.8 m)
Foals (0–6 months)	1.4 m^2 (1.0 × 1.4 m)

Table A6. Density per compartment of the transports with deviating density in one or more compartments. Densities in bold deviate from the norm. * = unknown distribution. T1 = front loading space; T2 = hind loading space; C1 = first compartment; C2 = second compartment; C3 = third compartment.

Transport Date	Slaughter Date	Slaughter-House	Number of Trailers	Compartment	Density (m^2/Horse)	Fighting
7/11/2016	8/11/2016	General Pico	1	T1	**1.72**	Average
15/11/2016	16/11/2016	Sarel	2	T1	**0.80**	Average
				T2	1.11	High
14/3/2017	15/3/2017	Lamar	2	T1*	1.54	Barely or not
				T2, C1	1.06	
				T2, C2	**2.48**	Average
				T2, C3	**2.23**	
20/3/2017	21/3/2017	General Pico	2	T1	**2.21**	Barely or not
				T2	**1.88**	Barely or not
6/6/2017	8/6/2017	General Pico	1	T1	**2.45**	
				T1, C1	**2.68**	High
				T1, C2	**2.25**	
23/10/2017	25/10/2017	Lamar	2	T1, C1	1.46	High
				T1, C2	1.33	
				T2, C1	**1.74**	
				T2, C2	1.49	High
				T2, C3	1.49	

References

1. BESTAT. Statistics Belgium. Available online: https://bestat.statbel.fgov.be/bestat/crosstable.xhtml?datasource=3488f928-86af-4067-92db-4fbe45c9a251 (accessed on 26 February 2020).
2. FAOSTAT. Food and Agricultural Organization of the United Nations. Available online: http://www.fao.org/faostat/en/#data/QL (accessed on 26 February 2020).
3. Vermeulen, L.; Van Beirendonck, S.; Van Thielen, J.; Driessen, B. A review: Today's practices about the fitness for travel on land of horses toward the slaughterhouse. *J. Vet. Behav.* **2019**, *29*, 102–107. [CrossRef]
4. Poznyakovskiy, V.M.; Gorlov, I.F.; Tikhonov, S.L.; Shelepov, V.G. About the quality of meat with PSE and DFD properties. *Foods Raw Mater.* **2015**, *3*, 104–110. [CrossRef]
5. Pawshe, M.D.; Badhe, S.R.; Khedkar, C.D.; Pawshe, R.D.; Pundkar, A.Y. Horse meat. In *Encyclopedia of Food and Health*; Elsevier: Amsterdam, The Netherlands, 2016; pp. 353–356. [CrossRef]
6. Cruz-Monterrosa, R.G.; Reséndiz-Cruz, V.; Rayas-Amor, A.A.; López, M.; La Lama, G.C.M. Bruises in beef cattle at slaughter in Mexico: Implications on quality, safety and shelf life of the meat. *Trop. Anim. Health Prod.* **2017**, *49*, 145–152. [CrossRef] [PubMed]
7. Consortium of the Animal Transport Guides Project. *Guide to Good Practices for the Transport of Horses Destined for Slaughter*; Publications Office of the European Union: Luxembourg, 2018; pp. 1–66. Available online: http://animaltransportguides.eu/wp-content/uploads/2016/05/EN-Guides-Horses-final.pdf (accessed on 3 March 2020).
8. Gorlov, I.F.; Pershina, E.I.; Tikhonov, S.L. Identification and prevention of the formation of meat with PSE and DFD properties and quality assurance for meat products from feedstocks exhibiting an anomalous autolysis behavior. *Foods Raw. Mater.* **2013**, *1*, 15–21. [CrossRef]

9.   Marlin, D.; Kettlewell, P.; Parkin, T.; Kennedy, M.; Broom, D.; Wood, J. Welfare and health of horses transported for slaughter within the European Union Part 1: Methodology and descriptive data: Transport to slaughter. *Equine Vet. J.* **2011**, *43*. [CrossRef]

10.   Leadon, D.; Waran, N.; Herholz, C.; Klay, M. Veterinary management of horse transport. *Vet. Ital.* **2008**, *44*, 16.

11.   Friend, T.H.; Martin, M.T.; Householder, D.D.; Bushong, D.M. Stress responses of horses during a long period of transport in a commercial truck. *J. Am. Vet. Med. Assoc.* **1998**, *15*, 838–844.

12.   Gibbs, A.E.; Friend, T.H. Horse preference for orientation during transport and the effect of orientation on balancing ability. *Appl. Anim. Behav. Sci.* **1999**, *63*, 1–9. [CrossRef]

13.   Grandin, T.; McGee, K.; Lanier, J. Survey of Trucking Practices and Injury to Slaughter Horses (unpublished). Available online: https://www.grandin.com/references/horse.transport.html (accessed on 3 March 2020).

14.   Friend, T.H. Dehydration, stress, and water consumption of horses during long-distance commercial transport. *J. Anim. Sci.* **2000**, *78*, 2568–2580. [CrossRef]

15.   Collins, M.N.; Friend, T.H.; Jousan, F.D.; Chen, S.C. Effects of density on displacement, falls, injuries, and orientation during horse transportation. *Appl. Anim. Behav. Sci.* **2000**, *67*, 169–179. [CrossRef]

16.   Gibbs, A.E.; Friend, T.H. Effect of animal density and trough placement on drinking behavior and dehydration in slaughter horses. *J. Equine Vet. Sci.* **2000**, *20*, 643–650. [CrossRef]

17.   Iacono, C.M.; Friend, T.H.; Johnson, R.D.; Krawczel, P.D.; Archer, G.S. A preliminary study on the utilization of an onboard watering system by horses during commercial transport. *Appl. Anim. Behav. Sci.* **2007**, *105*, 227–231. [CrossRef]

18.   Iacono, C.; Friend, T.; Keen, H.; Martin, T.; Krawczel, P. Effects of density and water availability on the behavior, physiology, and weight loss of slaughter horses during transport. *J. Equine Vet. Sci.* **2007**, *27*, 355–361. [CrossRef]

19   Werner, M.; Gallo, C. Effects of transport, lairage and stunning on the concentrations of some blood constituents in horses destined for slaughter. *Livest. Sci.* **2008**, *115*, 94–98. [CrossRef]

20.   Tateo, A.; Padalino, B.; Boccaccio, M.; Maggiolino, A.; Centoducati, P. Transport stress in horses: Effect of two different distances. *J. Vet. Behav.* **2012**, *7*, 33–42. [CrossRef]

21.   Padalino, B.; Hall, E.; Raidal, S.; Celi, P.; Knight, P.; Jeffcott, L.; Muscatello, G. Health problems and risk factors associated with long haul transport of horses in Australia. *Animals* **2015**, *5*, 1296–1310. [CrossRef]

22.   Abbott, A. Physiological Responses of Horses in Transit and the Effect on Welfare. Available online: http://essays.cve.edu.au/sites/default/files/vein_essays/content_2607/Abbott.pdf (accessed on 3 March 2020).

23.   Friend, T.H. A review of recent research on the transportation of horses. *J. Anim. Sci* **2001**, *79*, E32–E40. [CrossRef]

24.   Šímová, V.; Večerek, V.; Passantino, A.; Voslářová, E. Pre-transport factors affecting the welfare of cattle during road transport for slaughter—A review. *Acta Vet. Brno.* **2016**, *85*, 303–318. [CrossRef]

25.   Dai, F.; Dalla Costa, A.; Bonfanti, L.; Caucci, C.; Di Martino, G.; Lucarelli, R.; Padalino, B.; Minero, M. Positive reinforcement-based training for self-loading of meat horses reduces loading time and stress-related behavior. *Front. Vet. Sci.* **2019**, *6*, 350. [CrossRef]

26.   Luz, C.S.M.; Fonseca, W.J.L.; Vogado, G.M.S.; Fonseca, W.L.; Oliveira, M.R.A.; Sousa, G.G.T.; Farias, L.A.; Sousa Júnior, S.C. Adaptative thermal traits in farm animals. *J. Anim. Behav. Biometeorol.* **2015**, *4*, 6–11. [CrossRef]

27.   Morgan, K. Thermoneutral zone and critical temperatures of horses. *J. Therm. Biol.* **1998**, *23*, 59–61. [CrossRef]

28.   National Research Council. Farm animals and the environment. In *Effect of Environment on Nutrient Requirements of Domestic Animals*; National Academies Press: Washington, DC, USA, 1981; ISBN 978-0-309-03181-3.

29.   Stull, C.L.; Rodiek, A.V. Physiological responses of horses to 24 hours of transportation using a commercial van during summer conditions. *J. Anim. Sci.* **2000**, *78*, 1458–1466. [CrossRef] [PubMed]

30.   Schlatter, T.W. Temperature-humidity index. In *Climatology. Encyclopedia of Earth Science*; Springer: Boston, MA, USA, 1987. [CrossRef]

31.   Resolución N° 97/1999. Créase el Registro Nacional de Medios de Transporte de Animales. Características Técnicas. Habilitación. Condiciones Para el Embarque y Transporte. Lavado y Desinfección. Disposiciones Generales. Available online: https://www.argentina.gob.ar/normativa/nacional/resoluci%C3%B3n-97-1999-55716/texto (accessed on 3 March 2020).

32. Resolución N° 25/2013 SENASA. Available online: http://www.senasa.gob.ar/resolucion-252013 (accessed on 3 March 2020).

33. Resolución N° 581/2014. Creación del Registro Nacional Sanitario de Medios de Transporte de Animales Vivos. Available online: http://www.senasa.gob.ar/sites/default/files/normativas/archivos/res_581-2014.pdf (accessed on 3 March 2020).

34. Ley N° 18.471. Tenencia Responsable de Animales. Available online: https://legislativo.parlamento.gub.uy/temporales/leytemp9528174.htm (accessed on 3 March 2020).

35. Council Regulation (EC) No 1099/2009 of 24 September 2009 on the Protection of Animals at the Time of Killing. Available online: https://eur-lex.europa.eu/eli/reg/2009/1099/oj (accessed on 3 March 2020).

36. Horse Meat Federation. Manual for the Animal Welfare of Horses during Transport and Slaughtering. Available online: http://www.respectfullife.com/wp-content/uploads/2016/09/160202-Homefe-Guidelines-animal-welfare.pdf (accessed on 3 March 2020).

37. Kibler, H.H. Environmental physiology and shelter engineering. LXVII. Thermal effects of various temperature-humidity combinations on Holstein cattle as measured by eight physiological responses. *Res. Bull. Mo. Agric. Exp. Stn.* **1964**, *862*, 1–42.

38. Hartmann, E.; Hopkins, R.J.; Von Brömssen, C.; Dahlborn, K. 24-h sheltering behaviour of individually kept horses during Swedish summer weather. *Acta Vet. Scand.* **2015**, *57*, 45. [CrossRef]

39. Snoeks, M. Schuilgedrag van Paarden op de Weide. Ph.D. Thesis, KU Leuven, Leuven, Belgium, July 2017.

40. Du Preez, J.H.; Giesecke, W.H.; Hattingh, P.J. Heat stress in dairy cattle and other livestock under Southern African conditions. I. Temperature-humidity index mean values during the four main seasons. *Onderstepoort J. Vet. Res.* **1990**, *57*, 77–86.

41. Gantner, V.; Mijić, P.; Jovanovac, S.; Raguž, N.; Bobić, T.; Kuterovac, K. Influence of temperature-humidity index (THI) on daily production of dairy cows in Mediterranean region in Croatia. In *Animal Farming and Environmental Interactions in the Mediterranean Region*; Casasús, I., Rogošiç, J., Rosati, A., Štokoviç, I., Gabiña, D., Eds.; Wageningen Academic Publishers: Wageningen, The Netherlands, 2012; pp. 71–78. ISBN 978-90-8686-741-7.

42. Moran, J. Appendix 1: Temperature humidity index. In *Tropical Dairy Farming: Feeding Management for Small Holder Dairy Farmers in the Humid Tropics*; Landlinks Press: Oxford, UK, 2008; pp. 275–291. Available online: https://www.publish.csiro.au/ebook/chapter/SA0501275 (accessed on 3 March 2020).

43. Council Regulation (EC) No 1/2005 of 22 December 2004 on the Protection of Animals During Transport and Related Operations and Amending Directives 64/432/EEC and 93/119/EC and Regulation (EC) No 1255/97. Available online: https://eur-lex.europa.eu/eli/reg/2005/1/oj (accessed on 3 March 2020).

44. Grandin, T. Designing meat packing plant handling facilities for cattle and hogs. *Trans. ASAE* **1979**, *22*, 912–917. [CrossRef]

45. Grandin, T. Design of loading facilities and holding pens. *Appl. Anim. Behav. Sci.* **1990**, *28*, 187–201. [CrossRef]

46. Grandin, T. Behavioural principles of livestock handling. *Prof. Anim. Sci.* **1989**, *5*, 1–11. [CrossRef]

47. Maria, G.A.; Villarroel, M.; Chacon, G.; Gebresenbet, G. Scoring system for evaluating the stress to cattle of commercial loading and unloading. *Vet. Rec.* **2004**, *154*, 818–821. [CrossRef]

48. Draaisma, R. Tension shimmers through calming signals. In *Language Signs and Calming Signals of Horses*; CRC Press: Boca Raton, FL, USA, 2018; pp. 73–121.

49. Andronie, I.C.; Pârvu, M.; Andronie, V.; Ciurea, A. Risk assessment of welfare depreciation in horses during transport. In Proceedings of the XVth International Congress of the International Society for Animal Hygiene, Animal Hygiene and Sustainable Livestock Production, Vienna, Austria, 3–7 July 2011.

50. World Organisation for Animal Health. Terrestrial Animal Health Code (2019). Available online: https://www.oie.int/en/standard-setting/terrestrial-code/access-online/ (accessed on 12 March 2020).

51. Grandin, T. Safe handling of large animals. *Occup. Med.* **1999**, *14*, 195–212.

52. Grandin, T. Livestock—handling quality assurance. *J. Anim. Sci.* **2001**, *79*, E239–E248. [CrossRef]

53. Correa, J.A.; Torrey, S.; Devillers, N.; Laforest, J.P.; Gonyou, H.W.; Faucitano, L. Effects of different moving devices at loading on stress response and meat quality in pigs. *J. Anim. Sci.* **2010**, *88*, 4086–4093. [CrossRef] [PubMed]

54. Huertas, S.; Gil, A.; Piaggio, J.; Van Eerdenburg, F. Transportation of beef cattle to slaughterhouses and how this relates to animal welfare and carcase bruising in an extensive production system. *Anim. Welf.* **2010**, 281–285.
55. Gaughan, J.B.; Mader, T.L.; Holt, S.M.; Lisle, A. A new heat load index for feedlot cattle. *J. Anim. Sci.* **2008**, *86*, 226–234. [CrossRef]
56. McKeever, K.H.; Eaton, T.L.; Geiser, S.; Kearns, C.F.; Lehnhard, R.A. Age related decreases in thermoregulation and cardiovascular function in horses: Ageing and thermoregulation. *Equine Vet. J.* **2010**, *42*, 220–227. [CrossRef] [PubMed]
57. Stull, C.L. Responses of horses to trailer design, duration, and floor area during commercial transportation to slaughter. *J. Anim. Sci.* **1999**, *77*, 2925. [CrossRef]
58. Mars, L.A.; Kiesling, H.E.; Ross, T.T.; Armstrong, J.B.; Murray, L. Water acceptance and intake in horses under shipping stress. *J. Equine Vet. Sci.* **1992**, *12*, 17–20. [CrossRef]
59. Nielsen, B.L.; Dybkjær, L.; Herskin, M.S. Road transport of farm animals: Effects of journey duration on animal welfare. *Animal* **2011**, *5*, 415–427. [CrossRef]
60. World Hore Welfare; FVE; EEVA; Animal Transportation Association; Animals' Angels; The Donkey Sanctuary. Practical Guidelines on the Watering of Equine Animals Transported by Road. 2014. Available online: http://www.animaltransportationassociation.org/Resources/Documents/Best%20Practices/Watering-guidelines-equines%2026June15.pdf (accessed on 3 March 2020).

16

Effect of Linseed (*Linum usitatissimum*) Groats-Based Mixed Feed Supplements on Diet Nutrient Digestibility and Blood Parameters of Horses

Markku Saastamoinen * and Susanna Särkijärvi

Production Systems, Natural Resources Institute Finland (Luke), FI-31600 Jokioinen, Finland;
susanna.sarkijarvi@luke.fi
* Correspondence: markku.saastamoinen@luke.fi

Simple Summary: In this study, the effect of linseed groat-based fibrous feed supplements on diet digestibility was studied. In addition, possible detrimental health effects due to continuous feeding of such supplemental feeds containing linseed were examined by evaluating blood parameters. The supplemented diets had statistically significantly higher digestibility of crude protein compared to the control diet. In addition, the digestibility of fat (ether extract) was higher in the supplemented diets than in the basal feeding. There were no statistically significant differences or trends in the blood parameters between the treatments. It is concluded that linseed by-products (linseed groats 0.8 g/kg BW/d) combined with other fibre sources can be safely used, for example, in feeding strategies replacing grains in the horses' rations in order to reduce the intake of starch.

Abstract: Linseed (*Linum usitatissimum*) and its by-products are common supplements used in equine diets and are claimed to have beneficial health effects. In this study, the effect of linseed groat-based fibrous feed supplements on diet digestibility was studied. Also, possible detrimental health effects due to continuous feeding of supplemental feeds containing linseed were examined by evaluating blood parameters. The experimental design was arranged as two balanced 3 × 3 Latin Squares. The horses were individually fed at the maintenance energy level, the forage-to-concentrate ratio being 70:30, with three diets: (A) Control diet consisting of dried hay and whole oats; (B) Control diet + Feed 1; and (C) Control diet + Feed 2. Feed 1 contained 70% of linseed groats, 15% dried carrot, 10% dried garlic and 5% molasses. Feed 2 contained 65% linseed groats, 15% molassed sugar-beet pulp, 10% dried garlic, 5% dried carrot and 5% molasses. Digestibility data were obtained by using chromium mordanted straw as an indigestible external marker for the estimation of apparent digestibility. Blood samples were collected from the jugular vein at the end of each feeding period to evaluate the possible effects of the supplemented diets B and C on the health of the horses. Diets B and C had a higher digestibility of crude protein compared to the control diet A ($p < 0.05$). In addition, the digestibility of ether extract was higher in the supplemented diets than in the basal feeding ($p < 0.01$). There were no statistically significant differences or trends ($p > 0.05$) in the blood parameters between the treatments. It is concluded that linseed groat-based supplements (offering approximately 6.3%–6.7% linseed groats in the diet's dry matter (DM), or 0.8 g/kg BW/d), and feed containing soluble fibre sources (sugar-beet pulp, dehydrated carrot), improved the crude protein and fat digestibility of hay-oats diets of horses, and can be used, for example, in feeding strategies replacing grains in the horse rations in order to reduce the intake of starch without any adverse effects on the blood parameters and health of the horses.

Keywords: feeding; haematology; flax seed; fibre

1. Introduction

Linseed (*Linum usitatissimum*) or by-products (groats, cakes, meals) of linseed oil pressing have been used in human and animal nutrition for decades because they are believed to have numerous beneficial effects, many times without any scientific evidence. The "basic" horse nutrition literature in different countries [1–3] has recommended the feeding of linseed in various amounts for a long time as a supplemental feed to promote gut and skin health as well as coat quality. Thus, linseed products are commonly used in equine diets [4,5]. However, there is a knowledge gap on the nutritional and health effects of feeding linseed products to horses, because scientific research about this is scarce. For example, proper and safe supplementation levels are not given or known.

Linseed meal is high in crude fibre, acid detergent fibre (ADF) and neutral detergent fibre (NDF) [6]. Pectins and other dietary fibres of linseed have been proved to promote the health of the gastrointestinal tract in humans and dogs (e.g., [7,8]). The hull fraction (outer seed coat) contains 2%–7% polysaccharide mucilage [6,9]. Mucilage is readily water dispersible and forms a viscous slime, which is believed to have positive effects on the stomach and gut [10]. Further, in our preliminary study [11], linseed-based feed enhanced sand removal from the digestive tract of the horses.

Linseed oil is a good source of valuable fatty acids (omega-3) [12]. Groats and meal from cold-pressing may have an oil content of up to 20% [13]. Thus, linseed can be viewed as a cost-effective and economical way to boost omega-3 fatty acids in the feed [14]. In one study [15], a significant improvement in a skin test response to *Culicoides* spp. was reported due to linseed supplementation. Improved hair coat and skin condition scores have also been obtained in dogs after one month of linseed supplementation [16]. Horses have low fatty acid elongation activity, which is important for the inflammatory response, and there is speculation that linseed as a source of omega-3 PUFA may decrease signs of laminitis by inhibiting inflammatory mediators [17]. In a quite recent study [18], increased concentrations of red blood cells, haemoglobin and haematocrit, as well as improved n-3 fatty acid profiles, were reported as a result of linseed oil supplementation. Vineyard et al. [19] found that supplementing horses with milled linseed resulted in pronounced early inflammatory responses to phytohaemagglutinin injections. Both studies also showed increased fatty acid contents of red blood cell membranes. In addition, fats are an important source of energy for horses [20] and can be applied to reduce the starch content of the diet. However, the ether extract digestibility of linseed observed for horses was lower compared to oats and bran [21].

Linseed by-products are rich in protein [6], but a comparison of linseed meal to blended milk products showed that the growth and feed/gain were much better for milk products, the main reason being their better lysine content [22]. Linseed by-products have not been successfully used as protein sources in chicks either [23]. Instead, conflicting results on the effects on growth and health have been reported for pigs [24,25].

Consequently, there are several reasons for the interest in including linseed meal or oily linseed by-products in horse diets by horse owners [5]. However, linseed is known to contain compounds that may be toxic or have anti-nutritive properties [6,26], when the enzyme linase releases cyanide from the glycoside and diclucosides of the seeds [27]. Cyanide levels in linseed are below the level hazardous to humans [6], but there is some concern about the possibility of cyanide poisoning in horses, which are fed linseed [3]. However, intoxications or studies on this matter and where the daily intakes are given have not been reported in horses.

Williams and Lamprecht [28] reviewed studies where linseed oil has been fed to horses, but the feeding of linseed by-products has rarely been the subject of controlled studies with horses, or any other species for that matter. In addition, data regarding the effects (beneficial or detrimental) on diet digestibility and/or animal health when large amounts are fed is scarce. Instead of this, rather small amounts (only 50–120 g/d) are recommended to horse diets [2]. Science-based levels are not given, but recently Lindinger [29] concluded in his article in a veterinary science journal, based on a trade blog [4]), that the highest recommended amount for horses is 454 grams/d (1 pound). Neither research data exists regarding feeds in which linseed is combined with botanically diverse fibre sources.

Therefore, the objectives of this study were to: (1) investigate the effect of two linseed groat-based mixed feed supplements containing other fibrous ingredients on diet digestibility; (2) evaluate the possible detrimental health effects due to continuous feeding of linseed groats supplements in terms of blood parameters. The hypothesis was that there will be no detrimental effects due to the linseed supplementation on the diet digestibility and the haematological values of the horses. The results can be applied in the practical feeding of horses, or by the feed industry utilising linseed groats as a feed ingredient.

2. Materials and Methods

2.1. Horses and their Management

The influence of two linseed-based feed supplements on diet digestibility was examined with six Finnhorse mares (5–14 years; mean initial BW 636 kg, s.d. 37.8 kg), owned by MTT Agrifood Research Finland (currently Natural Resources Institute Luke). The experiment was conducted in the facilities of Luke. The experimental horses were individually housed in stalls (3 × 3 m) with peat as bedding. The horses had free access to water and a salt block and they were de-wormed before the experiment, and dental care and vaccinations had also been carried out regularly prior to the experiment. During the experiment, they were freely exercised daily in outdoor paddocks for four hours, and one hour by riding at a slow walk, to fulfil their needs of exercise and ensure their wellbeing.

The experimental design was arranged as two balanced 3 × 3 Latin Squares. Each experimental period consisted of 21 days: 16 days of adaptation to the new diet followed by a five-day period of collecting representative spot faecal samples. The BW (electronic animal scale Lahden Vaaka/Lahti Precision Ltd., Lahti, Finland) of the horses was monitored after each collection period to control possible changes and to adjust the individual energy intakes if necessary.

In animal handling and sample collection, the European Union recommendation directives (1999/275/EU) and national animal welfare and ethical legislation set by the Ministry of Agriculture and Forestry of Finland were followed carefully. The experimental procedures were evaluated and approved by The Animal Care Committee of MTT (Permit 9/2001) before the study was started.

2.2. Experimental Feeds and Feeding

The horses were randomly allotted to three dietary treatments: (A) Control diet consisting of dried hay dominated by timothy grass and whole oats; (B) Control diet + Feed 1; and (C) Control diet + Feed 2. Feed 1 contained 70% of linseed groats, 15% dried carrot, 10% dried garlic and 5% molasses. Feed 2 contained 65% linseed groats, 15% molassed sugar-beet pulp, 10% dried garlic, 5% dried carrot and 5% molasses. The hay and oats for this experiment were produced by Luke. Feed 1 and Feed 2 were manufactured in a single batch for this experiment by a Finnish medical and food factory (Neomed Ltd., Somero, Finland) and were in granulated form (granulated in 70–80 °C heat for 5 to 6 min). The linseed groats in the experimental feeds were by-products of cold-pressing of linseed oil with an average fat (oil) content of 20%. The other raw materials were included in the experimental feeds in order to improve the palatability as well as owing to their technological properties [30]. They are also common supplemental feeds included in horse diets. The average chemical composition of the feeds is presented in Table 1.

The horses were individually fed at the maintenance energy level according to the Finnish Feed Tables and Feeding Recommendations [13], the forage-to-concentrate ratio being 70:30. Each experimental feed ration was formulated and adjusted to be as isocaloric and isonitrogenous as possible. The average daily allowances of hay, oat and experimental feeds in each dietary treatment are presented in Table 2. About 8% of the oats was substituted with the experimental feeds in the treatments B and C.

Table 1. Average chemical composition of the experimental feeds (g/kg dry matter).

Composition	Hay	Oats	Feed 1	Feed 2
Dry matter g kg^{-1}	870.7	883.1	905.2	885.4
Organic matter	934.7	971.4	930.6	924.3
Crude protein	95.0	112.8	211.5	209.8
Ether extract	15.6	61.0	185.7	172.7
NDF	687.7	263.3	183.3	228.0
Crude fibre	339.0	89.0	96.0	105.3
Ash	63.5	28.6	69.4	75.7
NFE	485.1	708.6	437.5	436.5
ME MJ/kg DM	9.10	12.6	14.3	13.9

NDF = neutral detergent fibre; NFE = nitrogen free extract; ME = metabolisable energy.

Table 2. Daily allowances of hay, oats and experimental feeds (Feed 1 or Feed 2) fed in each dietary treatment (DM kg/day) (with ranges).

Feed	Diet A (Control)	Diet B (Feed 1)	Diet C (Feed 2)
Hay	5.89 (5.66–6.36)	5.44 (5.22–5.92)	5.43 (5.22–5.75)
Oats	1.85 (1.80–2.12)	1.74 (1.68–1.89)	1.74 (1.68–1.84)
Experimental feed	—	0.757 (0.710–0.780)	0.745 (0.730–0.750)

The change in rations between periods was made gradually during the first five days of the adaptation period. Feeds were offered in equal meal sizes three times a day at 06:30, 12:30 and 17:30. The grain ration was given about 30 min after the hay ration. The experimental feeds (Feeds 1 and 2) were aimed to be fed at a level of approximately 10% of the total dry matter (DM) intake, the average daily portion being 765 g DM/horse divided into three equal portions that were fed separately after the intake of oats. They were soaked in warm water (45–50 °C) before feeding to ensure their palatability. Mineral intakes were balanced with a commercial vitamin–mineral mixture (Suomen Rehu Ltd., Seinäjoki, Finland).

2.3. Feed and Faeces Sampling

Digestibility data were obtained by using chromium mordanted straw (68 mg Cr/g DM) with a daily dose of 1.6 g/kg feed DM as an indigestible external marker for the estimation of apparent digestibility. The chromium mordanted straw was prepared according to Udèn et al. [31]. The chromium dosage was calculated separately for every feed portion and served on the top of the concentrate three times a day, as described in detail by Särkijärvi and Saastamoinen [32]. Samples of hay and oats were collected for analysis over the last seven days of each period, and stored until the end of the five-day collection period. Faecal grab samples (500 g) were taken from each horse twice a day after the morning and mid-day feeding, during the five-day collection period. Samples were collected from the floor of the pen from a freshly produced pile. Daily faecal samples were stored at –24 °C until mixed, sub-sampled and dried (at 100 °C for 1 h + at 60 °C for 72 h) for laboratory analysis [32].

Feed and faeces samples were analysed in the feed laboratory of Luke (Luke Laboratories, Jokioinen, Finland) for dry matter (DM), organic matter (OM), neutral detergent fibre (NDF), crude fibre (CF), crude protein (CP), ether extract (EE) and ash with standard wet chemical methods as described by Särkijärvi et al. [33]. The nitrogen-free extract (NFE) was calculated: (100–CP–CF–EE–ash). The digestible CP (DCP) was calculated: DCP (g/kg DM) = CP (g/kg DM) × CP digestibility (g/kg CP)/1000, where the CP digestibility was taken from the Finnish Feed Tables and Feeding Recommendations [11]. The metabolisable energy value (ME) was calculated according to the British energy evaluation system [34].

2.4. Blood Sampling

Blood samples were collected at the end of each period to evaluate the possible effects of the diets on the health of the horses. The samples (2 × 10 mL) were collected 90 min after the morning meal from the jugular vein to heparinised blood collection tubes, and centrifuged. The blood analysis consisted of the contents of red blood cells (RBC), white blood cells (WBC), haemoglobin (Hb), haematocrit (HcT), and fibrinogen, as well as liver enzymes alanine aminotransferase (ALT) and γ-glutamyltransferase (GT), to indicate possible detrimental effects of linseed (cyanogenic glucosides) on the liver. All samples were analysed in the clinical laboratory of Luke.

2.5. Statistical Analysis

The data were analysed with a linear mixed model using the MIXED procedure of the SAS system using the REML estimation method. The following statistical model was applied:

$$Y_{ijk} = \mu + a_i + t_j + p(sq)_k + \varepsilon_{ijk} \tag{1}$$

where Y_{ijk} is the observation, μ is the overall mean, a_i is the random effect of ith animal (i = 1, ... 6), t_j is the fixed effect of jth dietary treatment (j = 1, ... 3), $p(sq)_k$ is the fixed effect of kth period within the square (k = 1, ... 3) and ε_{ijk} is the normally distributed error with a mean of 0 and the variance δ^2. Residuals were tested for normality. The differences were tested with Tukey's test, and the level of significance was set at the 5% level.

3. Results

3.1. Feed and Nutrient Intakes

The palatability of all the diets was good and there were no feed refusals. There were only very minor changes (± 0.8–2.2% between the measurements) in the body weights of the horses during the study. Diet intake was isocaloric and isonitrogenous between study periods (all diets combined; Table 3), but the intakes differed between the diets (Table 4). The average intakes of fat (EE) and CP of the supplemented horses (Diets B and C) were 58.0% and 14.1% higher than in the control group. Concerning ME, CF and NDF intakes, the differences were much smaller with +2.0%, –4.0% and –4.5%, respectively.

Table 3. Average daily ME, CP and DM intakes (± s.d.) for each period.

Intake	Period I	Period II	Period III
ME MJ	80.3 ± 4.8	81.8 ± 3.9	80.1 ± 3.2
CP g	832.4 ± 68.9	850.4 ± 73.1	835.2 ± 50.6
DM kg	7.86 ± 0.37	7.98 ± 0.29	7.86 ± 0.31

ME = metabolisable energy; CP = crude protein; DM = dry matter.

The proportions of linseed groats (on a DM basis) in diets B and C were 6.7% and 6.3%, respectively. On the BW basis, the intakes of linseed groats were approximately 0.8 g DM/kg BW/d.

3.2. Diet Digestibility

The supplementation of the experimental feeds improved only the digestibilities of total diet fat (ether extract, EE) and crude protein (CP) ($p = 0.0012$ and 0.0182, respectively) compared to the control diet (Table 5). In addition, the digestibility of ash (minerals) seemed to be numerically (but not statistically, $p = 0.2093$) somewhat higher in the supplemented diets than in the control diet. None of the digestibility values differed between the supplemental diets (Diet B versus Diet C) (p-values = 0.47–0.80).

Table 4. Average daily dry matter, metabolisable energy and nutrient intakes (±s.d.) in each dietary treatment.

Intake	Diet A (Control)	Diet B (Feed 1)	Diet C (Feed 2)
DM kg	7.77 ± 0.39	7.94 ± 0.28	7.91 ± 0.27
ME MJ	77.3 ± 3.9	82.9 ± 3.0	81.5 ± 2.6
EE g	206.4 ± 10.8	332.8 ± 9.7	318.9 ± 6.5
CP g	763.9 ± 40.2	877.5 ± 29.6	867.3 ± 26.5
DCP g	518.5 ± 26.5	618.2 ± 18.2	612.0 ± 23.2
NDF g	4556.3 ± 209.1	4338.6 ± 156.8	4358.9 ± 166.6
CF	2164.8 ± 108.0	2072.0 ± 74.4	2073.9 ± 83.0

DM = dry matter; ME = metabolisable energy; EE = ether extract (fat); DCP = digestible crude protein; CP = crude protein; NDF = neural detergent fibre; CF = crude fibre.

Table 5. Average apparent digestibility coefficients (%) and standard deviations (s.d.) for the total diet nutrients in the control and experimental diets.

Composition	Diet A (Control)	Diet B (Feed 1)	Diet C (Feed 2)	p Value (B and C vs. Control Diet)
Dry matter	54.8 (5.29)	55.3 (2.54)	57.0 (3.96)	0.5065
Organic matter	56.9 (5.05)	57.2 (3.24)	58.9 (3.92)	0.5710
Crude protein	61.4 (6.20)	64.0 (4.46)	65.7 (5.83)	0.0182
Ether extract	56.2 (7.99)	68.0 (5.02)	68.8 (5.46)	0.0012
NDF	47.2 (6.01)	47.8 (4.17)	47.9 (4.16)	0.8139
Crude fibre	44.6 (5.27)	43.9(4.45)	46.3 (4.87)	0.8484
Ash	19.7 (20.08)	25.1 (8.74)	27.4 (10.68)	0.2093
NFE	62.8 (5.09)	62.0 (2.83)	63.5 (3.32)	0.9601

NDF = neutral detergent fibre; NFE = nitrogen free extract.

3.3. Blood Parameters

There were no statistically significant differences or trends ($p > 0.05$) in the blood parameters between the diets (Table 6). The variation in the blood parameters (except ALT) was largest when the diet C was fed to the horses. γ-glutamyltransferase (GT) activity was numerically (but not statistically) somewhat higher, and the ALT activity lower, in the linseed supplemented diets. The average number of WBC and concentration of fibrinogen were also numerically higher in linseed supplemented horses but were within the reference values as well. Compared to the other horses, one horse had an exceptionally high GT activity (23–37 U/l), and another individual a low ALT activity (2.0 U/l) during the study period. The Hb and RBC values in all horses were low.

Table 6. Average values of blood parameters and standard deviations (s.d.) for the horses in the control and experimental diets.

Blood Parameters	Diet A (Control)	Diet B (Feed 1)	Diet C (Feed 2)	Reference Values [35,36]
Glutamyltransferase, U/l	17.0 (4.46)	19.7 (6.66)	18.0 (7.10)	10–70
Alanine aminotransferase, U/l	6.33 (2.02)	5.67 (2.57)	5.83 (1.95)	5–45
Haemoglobin, g/l	127.7 (7.27)	124.2 (5.64)	126.0 (11.03)	120–155
Haematocrit, %	34.7 (2.15)	34.3 (1.88)	34.4 (3.51)	34–43
Red blood cells, $\times 10^{12}$/l	6.99 (0.55)	6.86 (0.55)	6.92 (0.89)	7.0–9.0
White blood cells, $\times 10^{9}$/l	7.35 (1.02)	7.25 (0.69)	7.23 (1.72)	4.6–9.5
Fibrinogen, g/l	2.67 (0.28)	2.78 (0.25)	2.86 (0.34)	1.2–4.0

None of the differences between the groups were statistically significant (all $p > 0.05$; p-values 0.6–1.0).

4. Discussion

4.1. Feed and Nutrient Intakes

Maintaining BW showed that the feeding level used equalled the energy needs [11] of the horses. By period, intakes were isocaloric and isonitrogenous. However, the average fat (EE) intake of the supplemental diets (B and C) was more than 1.5-fold greater compared to the control diet, and the CP was 14.1% higher. In contrast, the differences were minor for energy and fibre components.

The palatability of the studied supplemental feeds was good. In our earlier unpublished study, we observed that the palatability of plain linseed groats was not good when fed in large (more than 6% in DM) portions. Delobel et al. [37] found no effects of flaxseed oil supplementation on diet palatability in horses. The other ingredients (sugar-beet pulp, carrot and molasses) of the linseed supplements fed in the present study likely improved the palatability of the supplemental feeds. Because of the 10% content of dried garlic in the supplemented diets, the horses had to ingest approximately 120 mg/kg BW dried garlic, which has been reported to be within the safe limits of garlic intake given by The National Academies [38].

4.2. Diet Digestibility

The improved digestibilities of dietary CP and EE when the supplemental feeds were fed is most likely due to the high concentrations and intakes of those nutrients in the feeds, and is supported by previous studies for CP [32,39,40] and EE [21,41]. In addition, Reitnour and Salsbury [42] found that the caecal administration of linseed meal increased the digestibility of total diet protein. There is, however, no evidence that fat and protein of linseed groats are more digestible than those of the control diet. The improvements may also be partly attributed to the dilution of endogenous faecal nitrogen and fat at higher intakes, which enhances their apparent digestibility [41,43–46].

Supplementing the diets with the experimental feeds caused a minor decrease in the NDF and CP intakes. The supplemental feeds contained properly digestible fibre sources, carrot and molassed sugar-beet pulp. Both of those ingredients are rich in dietary fibre of a soluble form, and have approximately the same amount of CF, but the NDF content of sugar-beet pulp is approximately two times that of carrot [30,47]. Sugar-beet pulp containing a lot of soluble and highly fermentable fibre [48] has been reported to be well utilised by horses [49–51]. No data on the digestibility of carrot is available. In addition, Snel et al. [52] (in their review of studies with different animal species) and Dongowski et al. [53] (in rats) have reported that the dietary fibre in sugar-beet pulp may have a prebiotic effect on intestinal flora improving the microbial activity and, thus, the digestibility. Murray et al. [54] reported in horses that sugar-beet pulp enhanced total diet digestibility fed together with forages. In addition, Lindberg and Palmgren Karlsson [44] explained their results with a positive effect of soluble and fermentable fibre in horses when sugar-beet pulp and fat was added to horse diets. As a potentially contributing factor, Clauss et al. [55] reported that nutrient supply to gut bacteria is the major digestive constraint in horses.

Because the experimental feeds contained many ingredients (being combinations of botanically diverse fibrous feeds), comparisons with previous studies including different diet compositions, and where linseed or variety of its by-products was used, are difficult. It is likely that the other components have their influences too, and that there are confounding and synergistic effects of the diet ingredients. Reitnour and Salsbury [42] found that the caecal administration of linseed meal decreased diet DM digestibility. In our unpublished study, we found that the supplementation of plain linseed groats from 0% to 10% (in the diet DM) gradually decreased the digestibility of the diet nutrients. This has also been reported in dogs [8], and may be due to the poor digestibility of linseed husks and the mucilage content of the hull. Linseed meal is also high in lignin [48], and most of the dietary fibre in linseed meal is in an insoluble form [8]. These findings are supported by Takagi et al. [21] (intakes were not given) who reported very low (21.8) digestibility for the crude fibre of linseed in horses. In one study, the inclusion of extruded linseed (20% in DM) in the diets of horses decreased the digestibility of nutrients compared to hay-only and hay/wheat bran diets [56]. In agreements with the results of the present study, Smolders et al. [57] found increased digestibility values of the diet nutrients when horses were fed compound feed containing (16%) linseed expeller plus more digestible ingredients (cereals). In the present study, the intake of linseed groats was 6.3%–6.7% in DM, and when combined with digestible fibre sources, also improved the diet's digestibility. Inconsistency of the results between studies is likely due to the different methods and processing of adding linseed or by-products, and different compositions of the diets.

Concerning other animal species, low levels (8%–10% in the feed) of linseed meal in pig feed may improve digestibility and growth rate, but 12% inclusion caused adverse effects [58]. Sled dogs can utilise up to 4.2% linseed cake as a source of fibre without severe reductions in nutrient digestibility or feed consumption [8]. In dairy cows, linseed supplementation improved total tract nutrient utilisation without any adverse effects on ruminal fermentation [59].

Regarding the method applied in determining digestibility, Palmgren Karlsson [60] suggested that chromium mordanted fibre could be an alternative to the administration of chromium, but may result in underestimated digestibility values. In Särkijärvi et al. [33], however, chromium mordanted silage gave quite precise digestibility values in horses (of a similar breed, gender and age as in this study).

4.3. Haematology Parameters

The serum concentrations of the liver enzymes alanine aminotransferase (ALT) and γ-glutamyltransferase (GT) were within the reference values of Finnhorses, but close to their lower limit [61]. Elevated ALT and GT values might have indicated possible detrimental effects of linseed (cyanogenic glucosides) on the liver. Additionally, the average number of WBC and concentration of fibrinogen were within the reference values. It can be concluded that there were no inflammatory reactions in the bodies of the horses due to the diet. The large variation in the haematology parameters when the Feed 2 diet was fed to the horses was likely due to the small number of horses used in this study.

The Hb, HcT and RBC values in all horses were low and close to the lower limit of the reference values of Finnhorses [61,62] regardless of the feeding group. The generally low Hb might be a result of the low feeding and exercise intensity of the horses [63].

Based on the blood analyses, we concluded that no adverse health effects in horses were caused from the supplementation of the diet with linseed groat-based feeds (offering approximately 6.3%–6.7% linseed groats in the diet DM or 0.8 g/kg BW/d) during the nine-week experimental period. This is supported by Vineyard et al. [19] who fed milled flaxseed for 70 days without any health problems.

No negative effects on health or performance of sport horses were found when linseed cake (990 g/d) was fed for 60 days [62], which agreed with O'Neil et al. [15] who observed no negative side effects when milled flaxseed was fed (1g/kg BW). The latter researchers concluded that stomach acid can inactivate enzymes within the seeds, which are required to interact with glycosides to form cyanide. Most glycosidases have a pH-optimum of around 7, so in herbivores with acid digestion, e.g., horses, they are usually inactivated [26]. In pigs, Batterham et al. [24] report lighter kidneys, pancreas and spleens for those animals given linseed, but no effects on the weight of livers were observed. They concluded that this may be a result of the anti-nutritional factors of linseed. Mazza and Oomah [26] concluded that most herbivores excrete the unhydrolysed cyanogenic compounds without harm to the animal.

The content of the anti-nutritive compounds in seeds depends on the cultivar, location and year of production, with the cultivar having the most significant effect [6,26]. The current and new *L. usitatissimum* varieties for human nutrition are rather low in toxic and detrimental compounds [26], as were the cultivars used in the present study (Neomed Ltd., Somero, Finland). According to Abraham et al. [63], in case of missing or inactivated glucosidase, the hazard potential (to humans) is low. Boiling is usually recommended to remove the potentially toxic cyanide components, and heat processing of linseed reduces its content of cyanogenic glycosides. Thus, in the present study, both the manufacturing process in the temperatures of 75–80 °C, and soaking into 45–50 °C water before feeding, might also have decreased the content of possible harmful compounds of the linseed groats [6,26]. In addition, HCN content is reduced when linseed is mixed with several ingredients and when the product is pelleted [64] as in the present study.

5. Conclusions

Linseed groat-based supplements (offering approximately 6.3%–6.7% linseed groats in the diet DM or 0.8 g/kg BW/d), and containing soluble fibre sources (sugar-beet pulp, dehydrated carrot), improved the crude protein and fat digestibility of hay/oats diets of horses, and had no effects on fibre digestibility. No adverse or anti-nutritional effects were observed on the availability of any component of the diet or the haematologic parameters and health of the horses. Linseed by-products combined with other fibre sources can be used, for example, in feeding strategies replacing grains in the horse rations in order to reduce the intake of starch. There is a need to investigate the synergetic and confounding effects of diet ingredients of different sources, especially botanically diverse fibrous feeds.

Author Contributions: M.S. and S.S. contributed methodology, investigation and data curation; S.S. contributed formal analysis and data analyzing; M.S. contributed resource, writing—original draft preparation, supervision and project administration. All authors have read and agreed to the published version of the manuscript.

References

1.   Cunha, T.J. *Horse Feeding and Nutrition*, 2nd ed.; Academic Press Inc.: San Diego, CA, USA, 1991.

2.   Meyer, H. *Pferdefütterung*, 3rd ed.; Blackwell Wissenschafts-Verlag: Berlin, Germany; Wien, Austria, 1995.

3.   Frape, D. *Equine Nutrition and Feeding*, 3rd ed.; John Wiley & Sons: Bodmin, UK, 2008.

4.   Janickie, K.M. 5 Facts of About Flax. The Horse.com. Available online: https://thehorse.com/149473/5-facts-about-flax (accessed on 15 July 2019).

5.   Christopherson, K. Tha Facts on Flax – the Value of Flaxsees in Your Horse's Diet. Equine Wellness Magazine.com. Available online: https://equinewellnessmagazine.com (accessed on 15 July 2019).

6.   Wanasundra, P.K.J.P.D.; Shadidi, S. Process-induced compositiona cganges of flaxseed. In *Process-Induced Chemical Changes in Food*; Plenum Press: New York, NY, USA, 1998; pp. 307–325.

7.   Reinhart, G.A.; Moxley, R.A.; Clemens, E.T. Source of dietary fiber and its effects on colonic microstructure, function and histopathology of beagle dogs. *J. Nutr.* **1994**, *124*, 2701S–2703S. [CrossRef] [PubMed]

8.   Kempe, R.; Saastamoinen, M. Effect of linseed cake supplementation on digestibility and faecal and haematological parameters in dogs. *J. Anim. Physiol. Nutr.* **2007**, *91*, 319–325. [CrossRef] [PubMed]

9.   Bhatty, R.S.; Cherdkiatgumchai, P. Compositional analysis of laboratory-prepared and commercial samples of linseed meal and hulls isolated from flax. *J. Am. Oil Chem. Soc.* **1990**, *67*, 79–84. [CrossRef]

10. Bhatty, R.S. Physiochemical properties of roller-milled barley bran and flour. *Cereal Chem.* **1993**, *70*, 397–402.

11. Särkijärvi, S.; Hyyppä, S.; Saastamoinen, M. Effect of linseed based feed supplementation on sand excretion in horses. In *The Impact of Nutrition on the Health and Welfare of Horses*; Wageningen Academic Publishers: Wageningen, The Netherlands, 2010; pp. 266–268.

12. Cunnane, S.C.; Ganguli, S.; Menard, C.; Liede, A.C.; Hamadeh, M.J.; Chen, Z.-Y.; Wolever, T.M.S.; Jenkins, D.A. High-α-linolenic acid flaxseed (*Linum ussitatissimum* L.); some nutritional properties in humans. *Br. J. Nutr.* **1993**, *69*, 443–453. [CrossRef]

13. Luke. *Finnish Feed Tables and Feeding Recommendations*; Natural Resources Institute Finland: Helsinki, Finland, 2015. Available online: http://urn.fi/URN:ISBN:978-952-326-054-2 (accessed on 18 December 2019).

14. Elghandour, M.M.Y.; Reddy, P.R.; Salem, A.Z.M.; Reddy, P.P.R.; Hyder, I.; Barbabosa-Pliego, A.; Yasawini, D. Plant bioactives and extracts as feed additives in Horse nutrition. *J. Equine Vet. Sci.* **2018**, *69*, 66–77. [CrossRef]

15. O'Neil, W.; McKee, S.; Clarke, A.F. Flaxseed (*Linum usitatissimum*) supplementation associated with reduced skin test lesional area in horses with *Culicoides* hypersensitivity. *Can. J. Vet. Res.* **2002**, *66*, 272–277.

16. Rees, C.; Bauer, J.; Burkholder, W.; Kennis, R.; Dunbar, B.; Bigley, K. Effects of dietary flaxseed and sunflower seed supplementation on normal canine serum polyunsaturated fatty acids and skin and hair coat condition scores. *Vet. Dermat.* **2001**, *12*, 111–117. [CrossRef]

17. Tinworth, K.D.; Harris, P.A.; Sillence, M.N.; Noble, G.K. Potential treatments for insulin resistance in the horse: A comparative multi-species review. *Vet. J.* **2010**, *186*, 282–291. [CrossRef]

18. Patoux, S.; Istasse, L. Incorporation of sunflower oil or linseed oil in equine compound feedstuff: 1 Effects on haematology and fatty acid profiles in the blood cells membranes. *J. Anim. Physiol. Anim. Nutr.* **2016**, *100*, 828–835. [CrossRef]

19. Vineyard, K.R.; Warren, L.K.; Kivipelto, J. Effect of dietary omega-3 fatty acid source on plasma and red blood cell membrane composition and immune function in yearling horses. *J. Anim. Sci.* **2010**, *88*, 248–257. [CrossRef]

20. Harking, J.D.; Morris, G.S.; Tulley, R.T.; Nelson, A.G.; Kamerling, S.G. Effect of added dietary fat on racing performance in thoroughbred horses. *J. Equine Vet. Sci.* **1992**, *12*, 123–129. [CrossRef]

21. Takagi, H.; Hashimoto, Y.; Yonemochi, C.; Ishibashi, T.; Asai, Y.; Watanabe, R. Digestibility of cereals, oil meals, brans and hays in thoroughbreds. *J. Equine Sci.* **2003**, *14*, 119–124. [CrossRef]

22. Hintz, H.F.; Schryver, H.F.; Lowe, J.E. Comparison of a blend of milk products and linseed meal as protein supplements for young growing horses. *J. Anim. Sci.* **1971**, *33*, 1274–1277. [CrossRef]

23. Treviño, J.; Rodríguez, M.L.; Ortiz, L.T.; Rebolé, A.; Alzueta, C. Protein quality of linseed for growing broiler chicks. *Anim. Feed Sci. Technol.* **2000**, *84*, 155–166. [CrossRef]

24. Batterham, E.S.; Andersen, L.M.; Baigent, D.R.; Green, A.G. Evaluation of meals from Linola™ Loe-linolenic acid linseed and conventional linseed as protein sources for growing pigs. *Anim. Feed Sci. Technol.* **1991**, *35*, 181–190. [CrossRef]

25. Romans, J.R.; Johnson, R.C.; Wulf, D.M.; Libal, G.W.; Costello, W.J. Effects of ground flaxseed in swine diets on pig performance and on physical and sensory characteristics and omega-3 fatty acid content of pork: I. Dietary level of flaxseed. *J. Anim. Sci.* **1995**, *73*, 1982–1986. [CrossRef]

26. Mazza, G.; Oomah, D.B. Flaxseed dietary fibre and cyanogens. In *Flaxseed in Human Nutrition*; American Oil Chemists' Society: Champaign, IL, USA, 1995; pp. 56–81.

27. Oomah, B.D.; Mazza, G.; Kenaschuk, E. Cyanogenic compounds in flaxseeds. *J. Agric. Food Chem.* **1992**, *40*, 1346–1348. [CrossRef]

28. Williams, C.A.; Lamprecht, E.D. Some commonly fed herbs and other functional foods in equine nutrition: A review. *Vet. J.* **2008**, *178*, 21–31.

29. Lindinger, M. Ground flaxseed—How safe is it for companion animals and for us? *Vet. Sci. Res.* **2019**, *1*, 35–40. [CrossRef]

30. Sharma, K.D.; Karki, S.; Thakur, N.S.; Attri, S. Chemical composition, functional properties and processing of carrot – a review. *J. Food Sci. Technol.* **2012**, *49*, 22–32. [CrossRef]

31. Uden, P.; Colucci, P.E.; Van Soest, P.J. Investigation of chromium, cerium and cobalt as digesta flow markers in rate of passage studies. *J. Food Agric. Sci.* **1980**, *31*, 625–632. [CrossRef]

32. Särkijärvi, S.; Saastamoinen, M. Feeding value of various processed oat grains in equine diets. *Livest. Sci.* **2006**, *100*, 3–9. [CrossRef]

33. Särkijärvi, S.; Sormunen-Cristian, R.; Heikkilä, T.; Rinne, M.; Saastamoinen, M. Effect of grass species and cutting time on *in vivo* digestibility of silage by horses and sheep. *Livest. Sci.* **2012**, *144*, 230–239. [CrossRef]

34. MAFF. *Energy Allowances and Feeding Systems for Ruminants*; Reference Book 433; Ministry of Agriculture, Fisheries and Food (MAFF), Her Majesty's Stationary Office: London, UK, 1984.

35. Movet. 2018. Available online: www.movet.fi. (accessed on 15 November 2019).

36. Pösö, A.R.; Soveri, T.; Oksanen, H.E. The effect of exercise on blood parameters in Standardbred and Finnish-bred horses. *Acta Vet. Scand.* **1983**, *24*, 170–184.

37. Delobel, A.; Fabry, C.; Schoonheere, N.; Istasse, L.; Hornick, J.L. Linseed oil supplementation in diet for horses: Effects on palatability and digestibility. *Livest. Sci.* **2008**, *116*, 15–21. [CrossRef]

38. The National Academies. *Safety of Dietary Supplements for Horses, Dogs and Cats*; The National Academies Press: Washington, DC, USA, 2008.

39. Gibbs, P.G.; Potter, G.D.; Schelling, G.T.; Kreider, J.L.; Boyd, C.L. Digestion of hay protein in different segments of the equine digestive tract. *J. Anim. Sci.* **1988**, *66*, 400–406. [CrossRef]

40. Ragnarsson, S.; Lindberg, J.-E. Nutritional value of timothy haylage in Icelandic horses. *Livest. Sci.* **2008**, *113*, 2020–2028. [CrossRef]

41. Lindberg, J.E.; Essen-Gustavsson, B.; Dahlborn, K.; Gottlieb-Vedi, M.; Jansson, A. Exercise response, metabolism at rest and digestibility in athlete horses fed high-fat oats. *Equine Vet. J. Suppl.* **2006**, *36*, 626–630. [CrossRef]

42. Reitnour, C.M.; Salsbury, R.L. Digestion and utilization of cecally infused protein by the equine. *J. Anim. Sci.* **1972**, *35*, 1190–1193. [CrossRef]

43. Julen, T.R.; Potter, G.D.; Greene, L.W.; Scott, G.G. Adaptation to a fat-supplemented diet by cutting horses. *J. Vet. Sci.* **1995**, *15*, 436–441. [CrossRef]

44. Lindberg, J.E.; Palmgren Karlsson, C. Effect of partial replacement of oats with sugar beet pulp and maize oil on nutrient utilisation in horses. *Equine Vet. J.* **2001**, *33*, 585–590. [CrossRef] [PubMed]

45. McCann, J.S.; Meacham, T.N.; Fontenot, J.P. Energy utilization and blood traits of ponies fed fat-supplemented diets. *J. Anim. Sci.* **1987**, *65*, 1019–1026. [CrossRef] [PubMed]

46. Farley, E.B.; Potter, G.D.; Gibbs, P.G.; Schumacher, J.; Murray-Gerzik, M. Digestion of soybean meal protein in the equine small and large intestine at various levels of intake. *J. Equine Vet. Sci.* **1995**, *15*, 391–397. [CrossRef]

47. Fadel, J.G.; DePeters, E.J.; Arosemena, A. Composition and digestibility of beet pulp with and without molasses and dried using three methods. *Anim. Feed Sci. Technol.* **2000**, *85*, 121–129. [CrossRef]

48. Bach Knudsen, K.E. Carbohydrate and lignin contents of plant materials used in animal feeding. *Anim. Feed Sci. Technol.* **1997**, *67*, 319–338. [CrossRef]

49. Lindberg, J.E.; Jacobsson, K.G. Effects of barley and sugar-beet pulp on digestibility, purine excretion and blood parameters in horses. *Pferdeheilkunde Sonderausgabe* **1992**, *12*, 116–118.

50. Hyslop, J.J.; Jessop, N.S.; Stefansdottir, G.J.; Cudderford, D. Comparative degradation *in situ* of four concentrate feeds in the caecum of ponies and the rumen of deters. In Proceedings of the 15th Equine Nutrition and Physiology Symposium, Ft. Worth, TX, USA, 28–31 May 1997; pp. 116–117.

51. Palmgren Karlsson, C.; Jansson, A.; Essén-Gustavsson, B.; Lindberg, J.-E. Effect of molassed sugar beet pulp on nutrient utilisation and metabolic parameters during exercise. *Equine Vet. J. Suppl.* **2002**, *34*, 44–49. [CrossRef]

52. Snel, J.; Harmsen, H.J.M.; van der Wielen, P.W.J.J.; Williams, B.A. Dietary strategies to influence the gastrointestinal microflora of young animals, and its potential to improve intestinal health. In *Nutrition and Health of the Gastrointestinal Tract*; Wageningen Academic Publisher: Wageningen, The Netherlands, 2002; pp. 37–69.

53. Dongowski, G.; Plass, R.; Bleyl, D. Biochemical parameters of rats fed dietary fibre preparation from sugar-beet pulp. *Zeits. Lebensm. Unters. Forsc.* **1998**, *206*, 393–398. [CrossRef]

54. Murray, J.-A.M.D.; Longland, A.; Hastie, P.M.; Moore-Colyer, M.; Dunnett, C. The nutritive value of sugar beet pulp-substituted Lucerne for equids. *Anim. Feed Sci. Technol.* **2008**, *140*, 110–124. [CrossRef]

55. Clauss, M.; Schiele, K.; Ortmann, S.; Fritz, J.; Codron, D.; Hummel, J.; Kienzle, E. The effect of very low food intake on digestive physiology and forage digestibility in horses. *J. Anim. Physiol. Anim. Nutr.* **2014**, *98*, 107–118. [CrossRef]

56. De Marco, M.; Miraglia, N.; Peiretti, P.G.; Bergero, D. Apparent digestibility of wheat bran and extruded flax in horses determined from the total collection of feces and acid-insoluble as ash an internal marker. *Animal* **2011**, *6*, 227–231. [CrossRef]

57. Smolders, E.A.A.; Steg, A.; Hindle, V.A. Organic matter digestibility in horses and its prediction. *Netherl. J. Agric. Sci.* **1990**, *38*, 435–447.

58. Bell, J.M.; Keith, M.O. Nutritional evaluation of linseed meals from flax with yellow or brown hulls, using mice and pigs. *Anim. Feed Sci. Technol.* **1993**, *43*, 1–18. [CrossRef]

59. Gonthier, C.; Mustafa, A.F.; Berthiaume, R.; Petit, H.V.; Martineau, R.; Quellet, D.R. Effects of feeding micronized and extruded flaxseed on ruminal fermentation and nutrient utilization by dairy cows. *J. Dairy Sci.* **2004**, *87*, 1854–1863. [CrossRef]

60. Palmgren Karlsson, C. Nutrient utilization in horses—Effect of oat replacement on ration digestibility and metabolic parameters. Doctoral Thesis, Swedish University of Agricultural Sciences, Uppsala, Sweden, 2001.

61. *Laboratory Diagnosis for Sport Horses*; Lindner, A. (Ed.) Wageningen Pers: Wageningen, The Netherlands, 1998.

62. Świstowska, A.; Kuleta, Z.; Stopyra, A.; Minakowski, D.; Tomczyński, R. The use of linseed cake in sport horse nutrition. *Annales Universitatis Marie-Curie-Skłodowska, Sectio DD, Medicina Veterinaria* **2006**, *61*, 103–114.

63. Abraham, K.; Buhrke, T.; Lampen, A. Bioavailability of cyanide after consumption of a single meal foods containing high levels of cyanogenic glycosides: A crossover study in humans. *Arch. Toxicol.* **2015**, *90*, 559–574. [CrossRef]

64. Feng, D.; Shen, Y.; Chavez, E.D. Effectiveness of different processing methods in reducing hydrogen cyanide content of flaxseed. *J. Sci. Food Agric.* **2003**, *83*, 836–841. [CrossRef]

PERMISSIONS

LIST OF CONTRIBUTORS

Jenny Yngvesson, Juan Carlos Rey Torres, Jasmine Lindholm, Annika Pättiniemi and Hanna Sassner
Department of Animal Environment & Health, Swedish University of Agricultural Sciences SE-53223 Skara, Sweden

Petra Andersson
Department of Philosophy, Linguistics and Theory of Science, University of Gothenburg, SE-40530 Gothenburg, Sweden

Miriam Baumgartner, Margit H. Zeitler-Feicht and Theresa Boisson
Ethology, Animal Husbandries and Animal Welfare Research Group, Chair of Organic Agriculture and Agronomy, TUM School of Life Sciences Weihenstephan, Technical University of Munich; Liesel Beckmann-Str. 2, 85354 Freising, Germany

Michael H. Erhard
Chair of Animal Welfare, Ethology, Animal Hygiene and Animal Husbandry, Department of Veterinary Sciences, Faculty of Veterinary Medicine, Ludwig-Maximilians-University Munich, Veterinärstr. 13, 80539 Munich, Germany

Seppo Hyyppä
Ypäjä Equine College, Opistontie 9, 32100 Ypäjä, Finland

Wanda Górniak, Martyna Wieliczko and Mariusz Korczyński
Department of Environment Hygiene and Animal Welfare, Wroclaw University of Environmental and Life Sciences, Chelmonskiego 38C, 51-630 Wroclaw, Poland

Maria Soroko
Institute of Animal Breeding, Wroclaw University of Environmental and Life Sciences, Chelmonskiego 38C, 51-630 Wroclaw, Poland

Sara Ringmark and Anna Jansson
Department of Anatomy, Physiology and Biochemistry, Swedish University of Agricultural Sciences, SE-75007 Uppsala, Sweden

Anna Skarin
Department of Animal Nutrition and Management, Swedish University of Agricultural Sciences, SE-75007 Uppsala, Sweden

Chloe Ready, Leanne Farkas and Abigail Hodder
Department of Animal Biosciences, University of Guelph, Guelph, ON N1G 2W1, Canada

Katrina Merkies
Department of Animal Biosciences, University of Guelph, Guelph, ON N1G 2W1, Canada
Campbell Centre for the Study of Animal Welfare, University of Guelph, Guelph, ON N1G 2W1, Canada

Agata Rzekęć
Research Unit MOISA (Marchés, Organisations, Instituts et Stratégies d'acteurs)-French National Research Institute for Agriculture, Food and Environment (INRAE), CIHEAM-IAMM, CIRAD, Montpellier Supagro, Univ Montpellier, 34060 Montpellier, France

Céline Vial
Research Unit MOISA (Marchés, Organisations, Instituts et Stratégies d'acteurs)-French National Research Institute for Agriculture, Food and Environment (INRAE), CIHEAM-IAMM, CIRAD, Montpellier Supagro, Univ Montpellier, 34060 Montpellier, France
Pôle Développement, Innovation, Recherche-French Institute for Horse and Horse Riding (Ifce), 61310 Exmes, France

Geneviève Bigot
Université Clermont Auvergne, AgroParisTech, French National Research Institute for Agriculture, Food and Environment (INRAE), VetAgro Sup, Research Unit Territoires, 63000 Clermont-Ferrand, France

Tayler L. Hansen, Elisabeth L. Chizek, Olivia K. Zugay, Jessica M. Miller, Jill M. Bobel, Jessie W. Chouinard, Angie M. Adkin, Leigh Ann Skurupey and Lori K. Warren
Department of Animal Sciences, University of Florida, Gainesville, FL 32611, USA

Joaquín Bull and Fernando Bas
Departamento de Ciencias Animales, Facultad de Agronomía e Ingeniería Forestal, Pontificia Universidad Católica de Chile, Avda. Vicuña Mackenna 4860, Santiago 7820436, Chile

Macarena Silva-Guzmán
Private statistical consultant, Guardia Vieja 441, Santiago 7510318, Chile

Hope Helen Wentzel
Escuela de Graduados, Facultad de Ciencias Agrarias, Universidad Austral de Chile, Valdivia 5110566, Chile

Juan Pablo Keim
Instituto de Producción Animal, Facultad de Ciencias Agrarias, Universidad Austral de Chile, Independencia 641, Valdivia 5110566, Chile

Mónica Gandarillas
Departamento de Ciencias Animales, Facultad de Agronomía e Ingeniería Forestal, Pontificia Universidad Católica de Chile, Avda. Vicuña Mackenna 4860, Santiago 7820436, Chile
Instituto de Producción Animal, Facultad de Ciencias Agrarias, Universidad Austral de Chile, Independencia 641, Valdivia 5110566, Chile

Katrin M. Lindroth, Johan Dicksved, Jan Erik Lindberg and Cecilia E. Müller
Department of Animal Nutrition and Management, Swedish University of Agricultural Sciences, 750 07 Uppsala, Sweden

Astrid Johansen
NIBIO, Norwegian Institute of Bioeconomy Research, 1431 Ås, Norway

Viveca Båverud
National Veterinary Institute, 751 89 Uppsala, Sweden

Nicoletta Miraglia and Elisabetta Salimei
Dipartimento Agricoltura, Ambiente e Alimenti, Università degli Studi del Molise, Campobasso 86100, Italy

Francesco Fantuz
Scuola di Bioscienze e Medicina Veterinaria, Università degli Studi di Camerino, Camerino MC 62032, Italy

Elisa Valtonen
Department of Animal Science, University of Helsinki, FI-00790 Helsinki, Finland

Dominique-Marie Votion and Marie-Catherine Bouquieaux
Equine Pole, Fundamental and Applied Research for Animals & Health (FARAH), Faculty of Veterinary Medicine, University of Lieège, 4000 Liège 1 (Sart Tilman), Belgium

Anne-Christine François, Benoit Renaud and Pascal Gustin
Department of Functional Sciences, Faculty of Veterinary Medicine, Pharmacology and Toxicology, Fundamental and Applied Research for Animals & Health (FARAH), University of Liège, 4000 Liège 1 (Sart Tilman), Belgium

Caroline Kruse
Department of Functional Sciences, Faculty of Veterinary Medicine, Physiology and Sport Medicine, Fundamental and Applied Research for Animals & Health (FARAH), University of Liège, 4000 Liège 1 (Sart Tilman), Belgium

Arnaud Farinelle
Fourrages Mieux asbl, 6900 Marloie, Belgium

Christel Marcillaud-Pitel
Réseau d'Epidémio-Surveillance en Pathologie Équine (RESPE), 14280 Saint-Contest, France

Malin Connysson
Wången National Center for Education in Trotting, Vången 110, S-835 93 Alsen, Sweden

Marie Rhodin
Department of Anatomy, Physiology and Biochemistry, Swedish University of Agricultural Sciences, SE-75007 Uppsala, Sweden

Béke Nivelle and Bert Driessen
Laboratory of Livestock Physiology, Department of Biosystems, KU Leuven, 3001 Heverlee, Belgium
Dier&Welzijn vzw, 3583 Paal, Belgium

Liesbeth Vermeulen
Westvlees NV, 8840 Westrozebeke, Belgium

Sanne Van Beirendonck
Bioengineering Technology TC, KU Leuven, 2440 Geel, Belgium

Jos Van Thielen
Bioengineering Technology TC, KU Leuven, 2440 Geel, Belgium
Thomas More, 2440 Geel, Belgium

Markku Saastamoinen and Susanna Särkijärvi
Production Systems, Natural Resources Institute Finland (Luke), FI-31600 Jokioinen, Finland
Natural Resources Institute Finland (Luke), Production Systems, Tietotie 2, 31600 Jokioinen, Finland
Production Systems, Natural Resources Institute Finland (Luke), FI-31600 Jokioinen, Finland

Index

Printed in the USA
CPSIA information can be obtained
at www.ICGtesting.com
JSHW062237071123
51533JS00031B/96